TOOLS FOR LAND USE ANALYSIS ON DIFFERENT SCALES

System Approaches for Sustainable Agriculture Development

VOLUME 8

Aims and Scope
The book series *System Approaches for Sustainable Agriculturel Development* is intended for readers ranging from advanced students and research leaders to research scientists in developed and developing countries. It will contribute to the development of sustainable and productive systems in the tropics, subtropics and temperate regions, consistent with changes in population, environment, technology and economic structure.
The series will bring together and integrate disciplines related to systems approaches for sustainable agricultural development, in particular from the technical and the socio-economic sciences, and presents new developments in these areas.
Furthermore, the series will generalize the integrated views, results and experiences to new geographical areas and will present alternative options for sustained agricultural development for specific situations.
The volumes to be published in the series will be, generally, multi-authored and result from multi-disciplinary projects, symposiums, or workshops, or are invited. All books will meet the highest possible scientific quality standards and will be up-to-date. The series aims to publish approximately three books per year, with a maximum of 500 pages each.

The titles published in this series are listed at the end of this volume.

Tools for Land Use Analysis on Different Scales

With Case Studies for Costa Rica

Edited by

BAS A.M. BOUMAN

International Rice Research Institute,
Los Baños, The Philippines

HANS G.P. JANSEN

Agricultural Economics Research Institute,
The Hague, The Netherlands

ROBERT A. SCHIPPER

Department of Economics and Management,
Wageningen University, Wageningen, The Netherlands

HUIB HENGSDIJK

Department of Crop Science,
Wageningen University, Wageningen, The Netherlands

ANDRÉ NIEUWENHUYSE

ZONISIG Project,
La Paz, Bolivia

SPRINGER SCIENCE+BUSINESS MEDIA, B.V.

Library of Congress Cataloging-in-Publication Data is available.

Additional material to this book can be downloaded from http://extras.springer.com.

ISBN 978-0-7923-6479-5 ISBN 978-94-011-4074-4 (eBook)
DOI 10.1007/978-94-011-4074-4

Printed on acid-free paper

Contents

Preface

This book reflects the results of more than ten years of cooperative research involving Wageningen Agricultural University (WAU) in the Netherlands, the Tropical Agricultural Research and Higher Education Center (CATIE; *Centro Agronómico Tropical de Investigación y Enseñanza*) in Costa Rica and the Costa Rican Ministry of Agriculture and Livestock (MAG; *Ministerio de Agricultura y Ganadería*) as part of the Research Program on Sustainability in Agriculture (REPOSA) in the Central American country. The type of cooperation was unusual as it focused on both research and the education of students undertaking either M.Sc. thesis projects or a program of practical training in the various aspects of studying land use. Since funding was provided by WAU, a high degree of scientific autonomy was created that has clearly benefited the independent, scientific rigor of the work. Over the ten-year period, the program has changed from being a patchwork of various insulated specialist projects, into a truly interdisciplinary effort, leading to the development of innovative tools for analyzing land use on a number of geographical scales. These tools are presented in this book. Besides CATIE and MAG, cooperation with other Costa Rican partner institutions has been essential from the beginning, and this process of interaction has also evolved considerably over time. Courses were occasionally given by REPOSA staff, and Costa Rican students actively participated in our work, but the main thrust fell on the discussion of concepts and on the development and application of approaches that would do justice to the particular context, questions and challenges involved in the main study region (*i.e.*, the Atlantic Zone in Costa Rica). Facilities provided and experiences that our collaborators shared with us were crucial contributors to the progress made in the highly complex field of land use analysis. As time moved on, interaction improved, and we now feel confident that the results of our joint work will continue to be used and further developed. Even though we could have continued the collaboration for many years to come in order to help implement and fine-tune the methodologies developed, we believe that sufficient seeds have been sown to ensure continuity in the local context. The very fact that several years are needed to develop meaningful interaction among international research partners and students demonstrates the limitations of many short-term projects which often end before they can realistically be expected to make an impact. We hope and trust that the research results presented in this book will find acceptance among the international scientific community, as well as be increasingly appreciated by policy makers. Last but not least, we are grateful to all our collaborators and cherish the experience of having worked with them for so many years.

Prof. Dr. Kees Karssen, Rector Magnificus, WAU
Dr. Ruben Guevara, Director-General, CATIE
Dr. Esteban Brenes, Minister of Agriculture and Livestock, Government of Costa Rica

1 Introduction

JOHAN BOUMA, HANS G.P. JANSEN, ARIE KUYVENHOVEN,
MARTIN K. VAN ITTERSUM, and BAS A.M. BOUMAN

1.1 Scope

This book offers an overview of the methodologies of studying actual and future land use on different scales that were developed over a twelve year period (1986-1998) in Costa Rica by an interdisciplinary team of Wageningen Agricultural University (WAU) of the Netherlands. The work was carried out in association with the Tropical Agricultural Research and Higher Education Center (CATIE; *Centro Agronómico Tropical de Investigación y Enseñanza*) and the Costa Rican Ministry of Agriculture and Livestock (MAG; *Ministerio de Agricultura y Ganadería*). While complementary in nature, the methodologies are carefully defined in terms of their specific objectives, terminology and use of quantitative, well developed methods and techniques, some of which are strongly process-oriented. In addition, the methodologies are applicable to the study of land use systems on the full spectrum of geographical scales: field, farm, sub-region, region and nation. In addition to their respective roles in supporting agricultural policy in the Atlantic Zone of Costa Rica, this methodological flexibility ensures universal applicability to all location-specific agro-ecological and socio-economic conditions. Applicability beyond specific case studies is an important aspect of the methodologies presented here and explains why considerable effort has been spent on clearly defining the different components of each one. Rather than considered in isolation, the various methodologies discussed are linked through a string of well-defined objectives that are logically inter-related.

The work in Costa Rica did not constitute a research project in the narrow sense since it also involved a significant education component. Some 250 graduate students were essential participants in the project's activities throughout its existence, including students from Wageningen Agricultural University, other universities and higher education institutions in the Netherlands, as well as from similar educational institutions in Costa Rica and other European countries. However, the focus in this book will be on the content and application of the research methodologies developed, rather than on knowledge transfer and student training. Still, Information and Communication Technology plays a central role in both the development and dissemination of the research. The decision to include a CDROM in this book to allow interactive work with the data is meant to offer students, research specialists and decision makers opportunities to familiarize themselves with the research methods in creative ways.

1.2 Choosing between methodologies

The set of methodologies for analyzing land use presented and discussed in this book is the product of twelve years of intensive research. During this period a significant shift occurred from an essentially fragmented multidisciplinary to a truly interdisciplinary approach, while interaction with stakeholders also increased as time progressed. Questions and practical problems associated with land use were the starting point of methodology development. Such questions and problems vary according to the (bio-physical and socio-economic) conditions of the area being considered, the particular stakeholders involved, and the time-scale as well as the geographical scale of analysis. In addition, the problems perceived by the various stakeholders may differ considerably: they include the short-term problems with which farmers and plantation owners are confronted; the medium-term questions on which policy-makers tend to focus; and the long-term problems that draw the attention of environmental protection and nature conservation agencies.

The main study area was the Atlantic Zone of Costa Rica which, until a few decades ago, was covered by tropical forests. Extensive agricultural development took place as the forests were progressively cut. Extensive meadows are intermingled with banana plantations and small areas of crop land and, increasingly over time, with farms specializing in high-value products mainly for export, such as flowers and ornamental plants. Even though, until recently, the Atlantic Zone was not a prime focus of attention for policy makers in the capital city of San José, political interest in the area has grown as a result of the increasingly conflicting policy objectives concerning agricultural production, environmental quality, the establishment and maintenance of nature reserves, and the effective settling of landless farmers on the subdivided large farms that were bought by the state.

In this book, we discuss methodologies that are capable of quantifying, analyzing and, hopefully, resolving conflicts of the kind broadly described above. Again, such problems are widespread throughout the tropics and the methodologies presented are potentially applicable beyond the national boundaries of Costa Rica. Before briefly introducing the various methodologies, some attention should be paid to the process of methodology selection in general. Given the tendency of many researchers to stick to the methodologies with which they are most familiar, often relatively little attention is given to the procedure of choosing a methodological orientation. Therefore, a seven-point sequence is advocated for use when initiating any land use analysis project:

(I) problem definition in interaction with stakeholders, including definition of the geographical unit of analysis;

(II) selection of a research methodology (*i.e.*, explanatory/projective, exploratory, predictive/policy-oriented, or focused on prototyping/decision support) and identification of participating disciplines;

(III) development of models and methods, explicitly taking account of scale hierarchies (in time as well as in space);

(IV) establishment of data requirements to be satisfied with existing data and/or with newly collected data;

(V) model application;

(VI) assessment of results in terms of quality, accuracy, sensitivity and risk; and

(VII) presentation of results with due attention to the possibilities created by Information and Communication Technology.

1.3 Methodologies for land use analysis

Most of the analytical methodologies presented in this book are meant to operate on the regional level, the level on which planners, policy makers and regulatory agencies tend to focus their attention. However, regional considerations are affected by the possible alternatives and likely decisions involved in farm-level operations. On the other hand, regional development can not be addressed in isolation without taking higher geographical levels into account, *e.g.*, the whole nation. Over the years, five broad types of methodologies for land use analysis have been developed and applied. Even though each of these will be presented in detail in the chapters that follow, it is useful to briefly consider them individually in terms of their nature, methods used, and key outputs delivered (see Table 1.1):

Table 1.1. Methodologies for land use analysis

Nature	*Name*	*Tools and methods*	*Key outputs*
A. Projective	CLUE	Statistical regression + GIS	Possible future developments in land use
B. Exploratory	SOLUS	Linear programming + GIS	Technological options + trade-off analysis + aggregate policy effects
C. Predictive	UNA-DLV	Linear programming + GIS + econometrics	Technological options + farmers' reactions + policy effectiveness
	SEM	Non-linear (quadratic) programming + econometrics	Quantification of trade flows + policy effectiveness
D. Generation of land use systems	Technical Coefficient Generator (LUCTOR, PASTOR)	Process-based and expert knowledge + literature + field experiments	Quantification of input-output relationships
E. Prototyping	BanMan	GIS + field experiments	Precision farm management

A. Starting with current land use, future developments can be *projected* by extrapolating current trends in land use as defined by the CLUE (Conversion of Land Use and its Effects) methodology (see Chapter 3). The underlying question is: "What will be the likely changes in land use if current trends are extrapolated into the mid-term future

(10-15 years ahead)?" In the CLUE methodology, statistical methods are employed in a spatially explicit manner in which the "drivers" (*i.e.*, explanatory variables) of land use operating on the different geographical levels are correlated with major kinds of land use. The key output is a description of possible future developments in land use. However, there is no explicit relationship with underlying bio-physical and/or socio-economic processes, nor are current land use trends and conditions explained.

B. A more process-oriented methodology for analyzing land use considers bio-economic *explorations* of alternative and/or current land use systems using scenario analysis. Land use options are defined in a systematic way and serve as input for an optimization model that considers both biophysical and socio-economic constraints. This approach is embodied in the SOLUS (Sustainable Options for Land Use) (see Chapters 6 and 7) methodology, which can be used to explore the options for land use with a time frame ranging from relatively short term (1-5 years) to relatively long term (15-25 years). The underlying question in this approach is: "What are the options for land use when optimizing for potential income, employment, environmental quality and risk, and what are the trade-offs involved in attaining these goals?" The outputs produced by the SOLUS methodology consist of technological options for land use, aggregate effects of alternative policies on a particular region (including the possibilities of realizing multiple objectives) and quantification of the trade-offs involved in goal attainment. Both currently practiced land use systems as well as so-called alternative land use systems are taken into account. The latter include theoretical constructs as well as land use systems that, even though validated in the field, thus far have been adopted only to a limited extent if at all. Rather than on actually prevailing conditions and practices, the focus in this methodology is on what might occur if a series of (policy-defined) changes are brought about. Some of these modifications may be far from realistic, but they can help to stimulate the imagination of land users and policy makers. At the same time, by varying parameters, the limits of what may be technically possible in the given region can be found. In Chapter 6 an application of this type of exploratory land use analysis (using a linear programming model called REALM [Regional Economic and Agricultural Land use Model]) is described, one that explores both the bio-physical and socio-economic constraints affecting land use. In Chapter 7, an application is presented (using a linear programming model called GOAL-AZ [General Optimal ALlocation for the Atlantic Zone]) that focuses more on the exploration of bio-physical restrictions on land use.

C. Exploration of land use options is logically followed by policy-oriented land use *predictions* using the so-called UNA-DLV (*Universidad Nacional Autónoma — Duurzaam Landgebruik en Voedselvoorziening*; Autonomous National University of Costa Rica - Sustainable Land Use and Food Security program of Wageningen Agricultural University) methodology (see Chapter 8). This methodology predicts short term (1-5 years) effects of policy measures intended to change the ways in which farmers make land use decisions by modeling farmer behavior. The underlying question is: "Which effective policy instruments induce the changes in land use to achieve certain farm and regional objectives?" The preferences and behavior of farmers, and the socio-economic context underlying them are highly relevant factors that need to be explicitly modeled. This requires a methodology that is capable of taking into account resource endowments and farm household objectives, as well as the prevailing market conditions and institutions. The UNA-DLV policy-oriented land use analysis methodology combines linear programming techniques for assessing the performance of production options with econometric farm household models for farmer behavior. Aggregation methods have been developed for regional equilibrium analysis and to model and assess interactions among different types of representative farm households. The latter's reactions to policy measures constitute the main output of this methodology.

The Spatial Equilibrium Modeling (SEM) approach discussed in Chapter 4 constitutes another type of policy-oriented methodology for the analysis of land use, but it is different from the other methodologies discussed in this book in at least two respects. First, it is highly economically oriented, with little involvement of other disciplines. Second, it is the only application in this book that addresses the national level (although CLUE was originally developed to undertake nation-wide analysis as well, using Costa Rica as a test case). The main question that SEM can help to answer is: "Which effective trade and infrastructural policies improve welfare in the agricultural sector, reduce regional imbalances and increase product profitability?" The time frame of SEM is short term (1-5 years). Key outputs include quantified evaluations of (consumer and producer) welfare, land use and trade flows (domestic as well as international) under alternative policy scenarios.

D. While methodologies A, B and C are intended to assess land use options and to analyze the effects of land use policies, methodology D develops well-defined models for the systematic *generation* of a large number of land use systems (including actual ones and alternative ones that may not (yet) be in use). These so-called Technical Coefficient Generators for crop and livestock activities, respectively called LUCTOR (Land Use Crop Technical coefficient generatOR) and PASTOR (Pasture and Animal System Technical coefficient generatOR), (see Chapter 5), provide the building blocks for methodologies B and C. Building upon familiar FAO terminology, the concept of land use system is defined as a combination of a land unit and a land utilization type (see the section at the end of the book entitled Concepts and definitions in land use analysis for a full explanation). *Actual* systems represent land use systems as currently practiced by farmers, whereas *alternative* land use systems incorporate technological progress. Alternative systems are generated using the target-oriented approach: target production levels are predefined and the combination of inputs required to realize these target levels is subsequently quantified. Both process-based and expert knowledge play a crucial role in determining which alternative land use systems are technically feasible and sustainable from a biophysical point of view, *i.e.*, can be repeated over time without changing input requirements.

E. Finally, methodology E involves the *designing* and implementation of sustainable farm-level production systems without losing sight of their repercussions on higher levels. Both expert and empirical knowledge are crucial factors when designing such systems by using a so-called *prototyping* approach. A decision support system with a considerable Information and Communication Technology component presented in Chapter 9 illustrates how prototyping may be employed in designing new types of land use systems. The developed prototype applies the principles of precision agriculture to banana plantations, and is called BanMan (Banana Management). In fact, when exploring alternative land use systems from a regional perspective (via methodology B), local knowledge is also necessary to make sure that realistic regional scenarios are developed. This methodology defines the operational aspects of the real farming systems that satisfy both production and environmental requirements in a given socio-economic context.

1.4 Outline of the book

Each of the methodologies presented in this book is illustrated by case studies on the northern part of the Atlantic Zone (AZ) of Costa Rica (except for Chapter 4 which presents the Spatial Equilibrium Model (SEM) and which uses the whole of Costa Rica as its case study). Chapter 2 therefore presents a detailed bio-physical and socio-economic description of the northern part of the AZ, complete with basic background data and GIS-produced maps. The projective CLUE methodology is presented in Chapter 3, demonstrating trends in land use dynamics from different geopolitical points of view. Chapter 4 deals with an application of the SEM methodology to the entire country of Costa Rica and as such is able to provide useful information on the effect of alternative trade and other policies on welfare, land use and commerce in agricultural products. Chapter 5 discusses the concepts behind the generation of technical coefficients (i.e., the inputs and outputs of production systems) and their implementation in LUCTOR and PASTOR. Both crop and livestock activities are given consideration. The technical coefficients of land use systems generated by LUCTOR and PASTOR are employed in the three subsequent chapters (Chapter 6, 7 and 8). LUCTOR as well as PASTOR constitute an integral part of the explorative SOLUS methodology, which is explained in Chapter 6. After an introduction to the principles of integrating biophysics and socio-economics in SOLUS, Chapter 6 continues with an application of SOLUS that focuses on the combined biophysical and socio-economic constraints on land use. Chapter 7, on the other hand, provides an application of SOLUS that determines the maximal level of production in a given set of biophysical conditions. Chapter 8 explains the predictive UNA-DLV methodology developed to analyze a region's land-use policy. The case study presented uses optimal single-farm models to determine how a region may optimize its overall production. Taken together, Chapters 5 through 8 analyze the trade-offs between biophysical and economic sustainability considered on levels varying from field and farm to sub-region and region. Chapter 9 presents an example of farm-level prototyping involving the decision support system in large-scale banana plantations. The developed prototype is based on the principles of precision agriculture which is meant to improve both economic results and the state of the environment by reducing losses of agro-chemicals into the environment. Finally, Chapter 10 critically discusses the presented methodologies and highlights their complementarity in land use studies. Scientific issues common to each of the methodologies are summarized and reviewed, including scope, terminology, transgression of levels, and aggregation.

This book contains a list of Concepts and definitions in land use analysis, an explanation of Abbreviations used[1], and an Introduction to the CDROM. The list of Concepts and definitions in land use analysis encapsulates the common language used in all methodologies throughout the book. A good set of definitions is a prerequisite for fruitful cooperation among disciplines, especially between biophysicists and socio-economists, and we hope that the list provided in this book will contribute to improved mutual understanding. The complementary CDROM that accompanies this book contains the models, data bases and users' guides relevant to the methodologies presented here. The Introduction to the CDROM lists these models, the associated computer and software requirements, and explains how to access the ÇDROM. The CDROM itself contains a self-instruction module and guides the user through the installation of the necessary software. Even though all applications presented are for Costa Rica, we encourage readers to apply the methodologies in other areas with different agro-ecological and socio-economic environments.

[1] Even thoug it was deliberately tried to avoid abbreviations as much as possible, the use of some abbreviations and acronyms proved unavoidable.

2 The physical and socio-economic setting: the northern Atlantic Zone of Costa Rica

ANDRÉ NIEUWENHUYSE, BAS A.M. BOUMAN, HANS G.P. JANSEN, ROBERT A. SCHIPPER, and RODRIGO ALFARO

Abstract

The biophysical and socio-economic setting of the northern part of the Atlantic Zone of Costa Rica, insofar as relevant for the other chapters of this book, is described. The flat topography, the perhumid tropical climate, and the soils that vary considerably in fertility and drainage conditions, are important biophysical factors that influence land use. From a socio-economic point of view, the area is characterized by its colonization history, an expanding but still limited road infrastructure outside the main regional centres, and agricultural practices dominated by large plantations and cattle holdings, along with a large number of smallholders both inside and outside state-organized settlements. Abolishment of subsidies for basic grains in the 1980s and the increasing exposure to world markets led to important changes in land use. Salient development issues in the area concerning questions of land use include: unequal land distribution between small and medium farm holdings on the one hand, and large cattle farms and banana plantations on the other; conflicts about forest use and protection; intensive use of agro-chemicals; greenhouse gas emissions; and low, sometimes declining, productivity of agriculture in certain parts of the region as a result of improper resource management.

2.1 Introduction

The northern part of the Atlantic Zone of Costa Rica (AZ) covers the northern half of the province of Limón, roughly between 10°00' and 11°00' latitude and 83°00' and 84°00' longitude (Figure 2.1). It includes the northern Caribbean lowlands and the bordering areas of the Central and Talamanca *Cordilleras* (*i.e.*, mountain range) (Figure 2.2), and encompasses 447 000 ha (Table 2.1), of which about 22 % is protected for nature conservation. The infrastructure is relatively well developed in the centre of the area, which has many paved roads and all weather gravel roads. The southern mountainous part is poorly accessible, while infrastructural development in the northern part is hampered by poor drainage. The main destinations for the zone's agricultural products are the densely populated Central Valley located at about 60 km south-west of the region and the harbor of Limón in the south-east of the area. Administratively, the region is subdivided into five counties, which in turn are divided into 20 districts. The current size of the region's population is about 259 000 and is concentrated in the central part of the AZ (DGEC, 1997a; Table 2.1). Between 1984 and 1996 the region's population grew at an annual rate of 4.4 %, considerably more than the national growth rate of 2.9 % per year. Nearly half of the active population works in agriculture, about 37% in services and commerce, while 14% is engaged in manufacturing and construction (DGEC, 1987b).

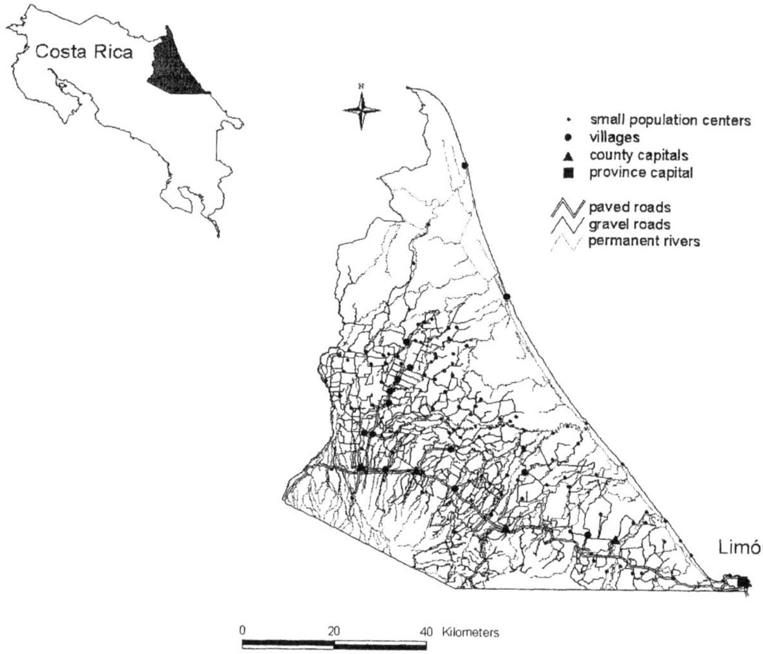

Figure 2.1. Location of the study area, the northern part of the Atlantic Zone of Costa Rica, and its main rivers, roads, and villages.

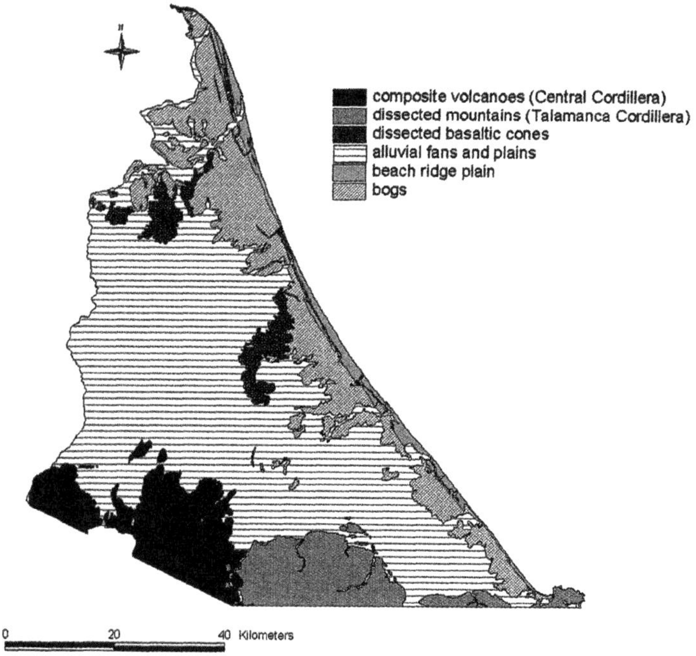

Figure 2.2. Main geomorphological units of the northern part of the Atlantic Zone of Costa Rica.

Table 2.1. Area and demographic data for the northern part of the AZ, each of its five constituent counties, the province of Limón, and Costa Rica

	Area (km^2)	*Population 1973*[1]	*Population 1984*[1]	*Population 1996 (excluding migration)*[2]	*Population 1996 (including migration)*[3] *(%)*	*Annual population growth 1973-1984 (%)*	*Annual population growth 1984-1996*	*Active population in agriculture 1996*	*Population density 1996 (persons km^{-2})*
Northern part of AZ	4468	91097	143386	215244	258667	4.2	5.0	41949	57.9
Pococí	2409	20688	44187	69611	80993	7.1	5.2	14392	33.6
Guácimo	579	11572	16472	24915	33157	3.3	6.0	6967	57.3
Siquirres	867	18133	29079	43609	60530	4.4	6.3	10770	69.8
Matina	529[4]	10489	14723	23293	26278	3.1	4.9	5117	49.7
Limón	84[4]	30215	38925	53816	57709	2.3	3.3	4267	68.7
Limón province	9189	115138	168262	258369	281262	3.5	4.4	46075	30.6
Costa Rica	51100	1871780	2416809	3398008	3398008	2.4	2.9	247871[5]	66.5

[1] Source: DGEC (1987b).
[2] Source: DGEC (1997a).
[3] Source: extrapolation of 1996 population data (based on birth and death statistics since 1984; DGEC, 1997a), adjusted for internal migration, estimated by applying 1979-1984 migration rates (DGEC, 1987b).
[4] Data refer to the respective parts of the country and the province included in the northern part of the AZ. The total area of Matina county is 773 km^2, and that of Limón province 1776 km^2.
[5] Source: DGEC (1997b).

2.2 Geology and geomorphology

Three major geographic units can be distinguished in the northern AZ: *i*) the Caribbean lowlands bordering the Caribbean in the east[1], *ii*) the northern slopes of the active Irazú and Turrialba volcanoes belonging to the Central *Cordillera* in the south, and *iii*) the northern part of the Talamanca *Cordillera* (Figure 2.2). Elevation in the AZ varies from sea level to about 2500 m on the slopes of the Central *Cordillera*.

The Caribbean lowlands form the southeastern part of the Nicaragua depression and constitute the largest part of the study area. This depression is a back-arc basin formed by crustal thinning as a consequence of the subduction of the Cocos plate beneath the Caribbean plate (Weyl, 1980; Seyfried *et al.*, 1991). Since the late Cretaceous, marine sediments have predominantly filled in the basin, while the Quaternary deposits are composed mainly of fluvio-volcanic sediments derived from the Central *Cordillera*.

Alluvial fan and plain deposits make up the major part of the Caribbean lowlands. While fan deposits vary in age between Pleistocene and Holocene, most of the alluvial plain deposits are young and probably have been deposited during the last 6000 years. Inundations occur regularly, especially in the lower parts of the Caribbean lowlands (less than 20 m above sea level). The almost straight coastline is bordered by an up to 3 km wide sandy beach ridge plain, while further landwards, extensive Holocene peat swamps overlay marine sand deposits (Nieuwenhuyse and Kroonenberg, 1994).

Scattered throughout the Caribbean lowlands, small hills with flat tops rise from 5 to 30 m above the actual river floodplains. They are the remains of an older Pleistocene terrace level. Peat deposits frequently occur in poorly drained depressions between these hills (Nieuwenhuyse, 1996). In the northeastern part of the lowlands, dissected remains of small early Pleistocene basaltic cones rise up to maximally 300 m above the plain. They are the remains of local eruptions through fissures and are composed of olivine basalt (Sprechmann, 1984).

The Central *Cordillera* is composed of irregularly formed strato volcano complexes that are still active, such as the Turrialba and the Irazú. Historic records indicate that both volcanoes remain active with mainly pyroclastic eruptions (Alvarado, 1989). Although northeastern winds deposited ash mainly south of the region (Melson *et al.*, 1985), thick ash deposits are also found on the northern flanks of both volcanoes at altitudes above 600 m. At lower altitudes, mud- and large lava flows without a recognizable ash cover dominate the volcano flanks (Reagan, 1987; Nieuwenhuyse, 1996). The transition between the Central *Cordillera* and the lowland is gradual, and extensive alluvial fans have been formed (Kesel and Lowe, 1987).

The Talamanca *Cordillera* is the largest and highest mountain range of Costa Rica, and stretches southeast of the Central *Cordillera* into Panama. Its geological structure is rather complex, being built up of folded Tertiary sedimentary rocks, with intercalated volcanic and Middle Miocene plutonic rocks (Seyfried *et al.*, 1991). The morphology reflects the resistance of the various formations against erosion: soft silt and clay stones form low relief areas with smooth slopes in which frequent landslides occur, while sandstones and conglomerates form escarpments with steeper slopes. Major rivers, such as the Pacuare and Chirripó, are deeply incised. The transition between the Talamanca *Cordillera* and the lowland is abrupt, with relatively small alluvial fan development.

---

[1] The Caribbean is part of the Atlantic Ocean, hence the name The Atlantic Zone.

2.3 Climate

Due to the close proximity to oceans at any point, the weather in Costa Rica is determined mainly by atmospheric perturbations that originate over either the Caribbean or the Pacific Ocean. Three large-scale systems are important (Portig, 1976; Herrera, 1985):

1. North - south movements of the Intertropical Convergence Zone (ITCZ, *i.e.*, the zone in which northern and southern trade winds meet), generating large cloud systems and, as a consequence, frequent rainfall. Between December and April, the ITCZ is located south of the Equator and does not influence the weather in Costa Rica. Between May and November, the ITCZ is located close to Costa Rica and brings rain to the Pacific side of the country, while moderating trade winds on the Caribbean side.
2. Strong, relatively cool northeastern trade winds, which affect Central America during winter in the northern hemisphere. Between December and March, these winds occasionally transport cold (polar) air masses towards Central America. The mountain systems which separate the Caribbean from the Pacific side of the country retain these air masses, causing prolonged rainfall on the Caribbean side, while the Pacific side of the country remains dry. When the trade winds do not transport cold cloudy air masses, they generate dry weather in the northern and Caribbean lowlands. Between May and November northeastern trade winds are warmer and weak, and only affect the Caribbean side of the country, where they generate rainfall.
3. Unstable low-pressure belts, which travel from the eastern part of the Caribbean Sea in a westerly direction between June and October. Especially when retained by mountains, they cause heavy rainfall mainly in the Caribbean part of Costa Rica.

Besides these large-scale systems, small-scale air movements such as marine or mountain winds influence the local weather of the AZ.

As a result of these processes, the climate of the AZ is hot and humid throughout the year. Although between December and April dry spells of up to several weeks may occur and may cause some water shortages for crops and pastures, invasions of cold air during this period produce moderate to heavy rainfall, thus preventing serious drought.

Most of the region receives a mean annual rainfall of about 3500-4000 mm, although variation is large and may vary as much as 100% between different years. Mean annual rainfall diminishes along the coast, from 5700 mm in the extreme northeast to 3500 mm in the city of Limón. From the coastal area toward the west, there is another decreasing rainfall gradient that increases again to about 7000 mm on the middle slopes of the *Cordilleras*. In areas above 1500 m, rainfall diminishes to about 2800 mm at 3300 m altitude.

Mean annual temperature in the lowlands is about 26 °C, with a difference of about 2 °C between the warmest (April - June) and coolest months (December-January) (Figure 2.3). Temperature decreases with altitude at a rate of 0.52 °C per 100 m (Herrera, 1985).

Relative air humidity is high, with average daily values (based on hourly observations) of about 85% throughout the year. Even during dry spells, noon-time air humidity never falls below 60%. Wind speed is low, and usually does not exceed 3-6 km h^{-1} in the interior, and 7-10 km h^{-1} along the Caribbean coast (Zarate, 1978). Occasionally, during thunderstorms, local gusts of wind may damage crops, especially on banana plantations.

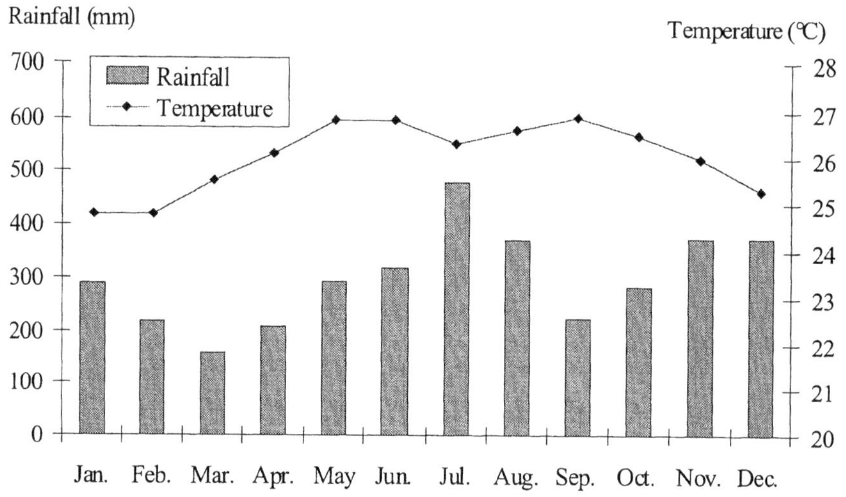

Figure 2.3. Mean monthly rainfall (bars) and mean monthly daily temperature (line) at El Carmen weather station which is 15 m a.s.l.

Average potential Penman evapotranspiration is about 3.5 to 4.5 mm d^{-1} in the lowland and decreases with altitude to about 2.5 to 3.5 mm d^{-1} on the footslopes of the *Cordilleras* (Castro, 1985). Mean monthly water balances for various weather stations in the AZ indicate that no water shortages for crops occur (Castro, 1985). Day length in the area varies from approximately 11.5 hours on the 21st of December to 12.5 hours on the 21st of June.

2.4 Soils

The soil survey carried out by REPOSA originally identified 74 soil types in the northern part of the Atlantic Zone (Wielemaker and Vogel, 1993). Since, for land use analysis, this number of soil types is too large, they were aggregated into four major soil groups (Figure 2.4), based on fertility and drainage criteria: *i*) young, well drained soils of relatively high fertility, classified as Soil Fertile Well drained (SFW), *ii*) old, well drained soils of relatively low fertility, classified as Soil Infertile Well drained (SIW), iii) young, poorly drained, soils of relatively high fertility, classified as Soil Fertile Poorly drained (SFP), and *iv*) soils not suited for agriculture. In Table 2.2, the key characteristics of the three cultivable soil types are presented. A more detailed description of the four major soil groups, based on Nieuwenhuyse (1996), is provided in Appendix 2.1 of this chapter. Together with the land characteristics "slope" and "stoniness", these major soil classes were the basis for distinguishing six land units,[2] which could be used as variables in quantified land use models in the AZ (see also Chapter 5).

[2] See the definitions at the end of this book.

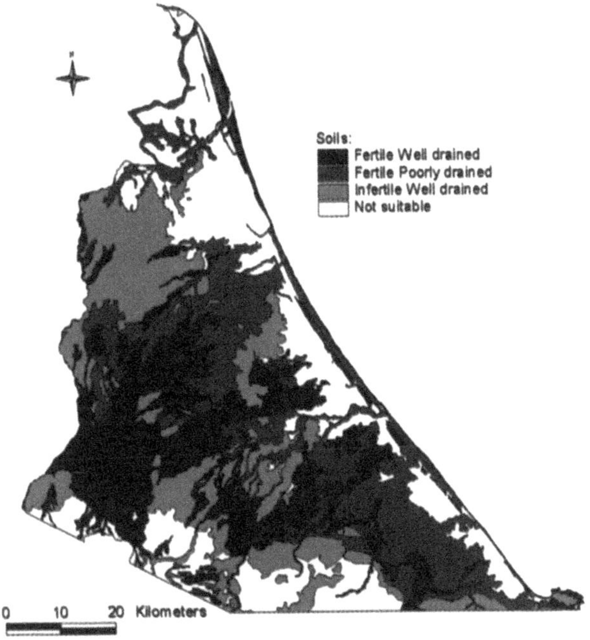

Figure 2.4. Major soil groups in the northern part of the AZ.

Table 2.2. Area and key characteristics of the three major soil groups[1] in the northern part of the AZ

Main soil group	*Fertile Well drained Soils (SFW)*	*Fertile Poorly drained Soils (SFP)*	*Infertile Well drained Soils (SIW)*	*Soils not suitable for agriculture*
Area (ha)	123 000	149 500	86 300	88 200
No. of profiles	74	34	36	n.r.
Clay (%)	16	20	42	n.r.
Silt (%)	32	45	31	n.r.
Sand (%)	52	35	27	n.r.
Bulk density (g cm^{-3})	0.75	0.88	0.93	n.r.
Organic C (%)	3.7	3.9	2.8	n.r.
PH-H_2O	5.9	6.0	5.2	n.r.
Exchangeable (Ca+Mg+K+Na) (meq $100g^{-1}$)	9	15	5	n.r.
P-retention (%)	84	67	80	n.r.
P-Olsen (ppm)	9	17	11	n.r.

[1] REPOSA data. Values are the weighted means of the upper 30 cm from 144 analyzed soil profiles (courtesy R. Plant).

2.5 Land use

Widespread evidence of pre-Columbian settlements is found on several well-drained sites throughout the region (Willey, 1971). However, when the first Spaniards arrived around 1500, they found a sparsely populated area. Colonization of the forest started about 300 years ago along the Matina and San Juan rivers. The process accelerated about 130 years ago, when railroad construction in the footslopes of the *Cordilleras* started (in 1871), connecting the Caribbean Coast with the Central Valley. Former railroad construction workers settled in the area and began to cultivate private fields or to work in newly established banana plantations. The foundation of the major towns in the AZ is closely related to the development of the railway infrastructure. For instance, the seaport of Limón was built for shipping mostly coffee and banana to Europe and the east coast of the United States.

A major influx of settlers occurred in the second half of this century, partly spontaneously, partly stimulated by government policies that induced the clearing of potential agricultural land. Since then, deforestation has accelerated (Veldkamp *et al.*, 1992). A last major impulse of settlement in the AZ was the construction of a highway in the late 1980s, providing a direct link between the capital of San José and the harbor of Limón.

Current land use distribution in the northern AZ, as estimated from a 1996 LANDSAT Thematic Mapper image, is given in Table 2.3, while some key characteristics of the main current land use systems are given in Table 2.4. The geographical distribution of forest and banana land use and protected areas is presented in Figure 2.5. The following sections briefly describe the main land use systems in the study area (see also Chapter 5).

Table 2.3. Land use in the northern part of the Atlantic Zone, plus about 105 000 ha of the neighboring Sarapiqui county (total 550 000 ha), based on an October 1996 LANDSAT Thematic Mapper image (Driese and Reiners, 1999).

Major kind of land use	*Area (%)*
Forest	28.7
Swamp forest	19.7
Tree plantation	1.0
Pasture	33.2
Banana	8.7
Bamboo	0.7
Palm heart	1.6
Ornamentals	0.3
Other agriculture	2.0
Pineapple	0.1
Water	2.6
Urban	1.4

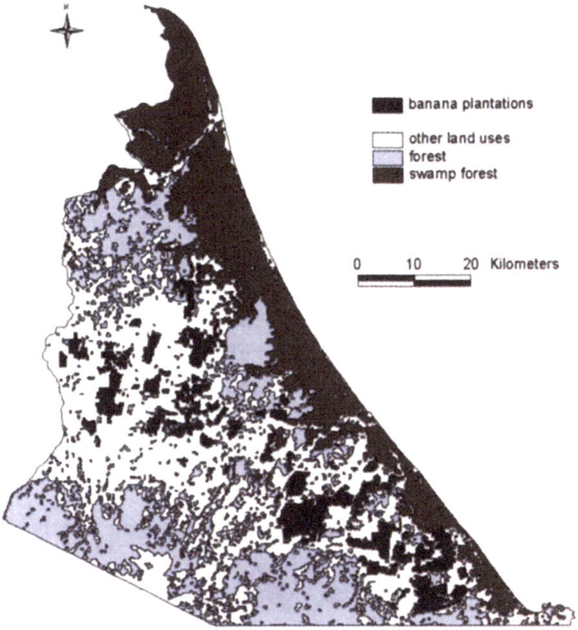

Figure 2.5. Forest and banana land use, and protected areas in the northern part of the AZ.

Table 2.4. Key characteristics of actual land use systems in the AZ (see also Chapter 5).

Crop	*Major land units*	*Yield*[1] ($t\ ha^{-1}\ y^{-1}$)	*Biocide use*[2] ($kg\ a.i.\ ha^{-1}\ y^{-1}$)	*N fertilizer use* ($kg\ ha^{-1}\ y^{-1}$)	*Labor use*[3] ($d\ ha^{-1}\ y^{-1}$)
Banana	SFW, SFP	35.2 - 43	57.9 - 61.2	500	140 - 170
Plantain	SFW, SFP	16.3 - 20	26.0 - 40.8	77	110 - 140
Palm heart	SFW, SIW	4.7 - 5.5	0 - 1.9	250	29 - 38
Cassava	SFW, SIW	2.3 - 6	0.2 -1.2	0	24 - 37
Maize grain	SFW	2.1 - 4.1	0.3 - 2.2	65	19 - 29
Maize cob	SFW	1.4 - 2.8	0.3 - 2.2	65	13 - 22
Bean	SFW	0.34 - 0.68	0.4 - 1.3	0	15 - 32
Local pineapple	SFW, SIW	14.4 - 24.2	4.3 - 5.6	110	38 - 95
Export pineapple	SFW, SIW	50.9 - 76.8	10.5 -11.6	465	100 - 190
Natural pasture	SFW, SFP, SIW	10 - 12	0.9 - 1.1	0	2
Grass-legume	SFW, SIW	15 - 20	0.1	0	3 - 4
Managed natural forest	SFW, SFP, SIW	1 - 1.3[4]	0	0	0.2

[1] Yields are first class fresh products; second class or rejected products are not included. Yields of maize and beans are based on one crop per year.

[2] Biocides comprise all kinds of agrochemicals used, including herbicides to combat weeds , and nematicides, fungicides and insecticides to fight pests and diseases ; a.i. = active ingredients.

[3] In 8-hours work days.

[4] Wood yield in m^3.

Source: based on calculations using LUCTOR (Hengsdijk *et al.*, 1998a) and PASTOR (Bouman *et al.*, 1998a), see also Chapter 5.

Besides the information on current land use derived from the 1996 LANDSAT Thematic Mapper, secondary statistics constitute another potential source of data for assessing land use. However, the last agricultural census was held in 1984 (DGEC, 1987a), while more recent information on areas, production and yields is difficult to obtain, as it is either irregularly collected, incomplete, or scattered over different institutions. The most recent information concerns 1995 and was assembled as input for a Spatial Equilibrium Model (Chapter 4) of the whole of Costa Rica. The information relates not just to the northern part of the AZ, but to the entire AZ as a governmental planning region, an area that includes the entire province of Limón and thus the part south of Limón that contains parts of the Matina and Limón counties (Table 2.1) as well as the county of Talamanca. This 1995 land use information is presented in Table 2.5. Areas used to cultivate various crops and pastures in 1984 are also included.

Table 2.5. Main agricultural products in the AZ: 1984 area, and 1995 area, yield, price and value of production, and share in 1995 national production

	1984	*1995*					
Product	*Area* ha	*Area*[1] ha	*Yield* t ha^{-1}	*Production* t	*Price* \$ t^{-1}	*Value of production*[2] \$ 10^3	*Value of production as % of national production*
Crops							
Banana	22 713	49 599	39.0	1 933 475	305	590 256	95.1
Plantain	4 684	5 879	21.0	123 734	205	25 332	90.4
Palm heart	1 050	2 100	5.0	10 501	252	2 643	50.0
Coffee	927	297	7.8	2 317	362	838	0.3
Maize	8 842	645	1.8	1 152	180	207	4.1
Pineapple	200	12	27.0	318	271	86	0.2
Bean	724	350	0.4	123	537	66	0.5
Cassava[3]	775	19	15.0	282	189	53	0.3
Rice	7 243	0	2.3	1	246	0	0.0
Cacao	12 755	Decreased strongly due to *Monilia* disease and low prices					
Coco	4 322	Similar as in 1984					
Macadamia	1 500	No data but decreased in area					
Ornamental	1 500	No data but increased in area and economic importance					
Sub-total[4]	67 448	60 837				620 931	49.5
Livestock							
Meat	n.d.	240 766	0.05	12 038	1565	18 843	24.5
Milk	n.d.	40 556	1.5	60 834	277	16 847	11.3
Sub-total	106 026	281 322				35 690	15.8
Total[4]	173 474	342 160				656 621	44.3

[1] In the case of meat and milk, the area refers to the area of pasture land.
[2] Valued at the average 1995 exchange rate of ¢ 179.14 = \$ 1.00.
[3] The 1995 estimated cassava area is likely to be an underestimation; furthermore, other root and tuber crops, in particular yam, are important in the AZ as well.
[4] Sub-totals are not equal to the sums of individual data due to the fact that insignificant crops are excluded; the same applies to the overall totals.
Sources: 1984 data are from Kruseman *et al.* (1994), and are based on DGEC (1987a) and MIDEPLAN (1991); 1995 data are mainly based on Roebeling *et al.* (1999b; see also Chapter 4).

The comparison of 1984 and 1995 area data shows increases in the areas of banana, plantain and palm heart cultivation, but decreases in the areas devoted to maize, beans and cacao. In general, the total area of crop land has not increased, while the total area of pasture land has more than doubled. The latter expansion comes mainly at the expense of the forest area, and occurred mostly between 1984 and 1990, when forest laws permitted a certain amount of forest to be converted to farm land. Currently, the conversion of forest to agricultural land is not permitted. Although forest areas are often directly converted into pasture, this process is often more complicated, *i.e.*, part of the converted forest area is used in crop production (*e.g.*, for banana) while abandoned crop area, including old banana plantations, is often converted to pastures.

In terms of national agricultural production, the AZ is in particular important for its banana, plantain and palm heart crops, as well as (though to a lesser extent) for meat production. Furthermore, due to the overwhelming importance of banana, the AZ contributes 44% to the total earnings from agricultural production in Costa Rica (see also Chapter 4). Below, the various land uses in the northern part of the AZ are described.

Natural forest

In 1996, forests in the AZ of Costa Rica are estimated to cover about 217 000 ha, including some 92 000 ha of palm-dominated swamp forests. It is especially the poorly drained areas in the lowlands and on the slopes of both *Cordilleras* that still maintain a considerable forest cover. About 98 000 ha of the forested areas are protected to some degree, but only 21 000 ha are contained in national parks (Driese and Reiners, 1999). Except for the national parks, most forests have suffered at least one intervention. Most of the forests outside the swamp and protected areas occurs in small patches of an average size of about 29 ha.

Tree plantations

About 1% of the study area is planted with trees for timber production, mainly melina (*Gmelina arborea*), teak (*Tectona grandis*), and local species among which laurel (*Cordia alliadora*) is the most important. Moreover, fast growing pioneer trees such as laurel and cedar (*Cedrela odorata*) that grow on pasture and agricultural land are usually left in the field and supply a considerable amount of wood.

Pasture

As observed earlier, pasture is often the first land use after deforestation. It is mainly used for beef cattle ranching, as 68% of the total herd animals are dedicated to this purpose (CNP, 1990). Double-purpose cattle, constituting 19% of the herd, represent the second most important form of livestock kept in the AZ, while 13% of all animals are held on dairy farms. In 1988, the total herd size was about 273 000 animals (CNP, 1990). Given the national decline in cattle ranching during the past 4 years (Montenegro and Abarca, 1998), it is likely that the current beef cattle herd size is decreasing. Furthermore, dairy farming has become less important due to the closure of the milk receiving plant in the AZ. Beef cattle are typically Zebú breeds raised on low productive native and naturalized grasses with stocking rates of between 0.6 and 1.5 animals units (AU) of 400 kg live weight per hectare (Van Loon, 1997). Current pasture production is estimated to vary between 8 t dry matter (DM) ha^{-1} y^{-1} from old, degraded pastures, and 18 t DM ha^{-1} y^{-1} from young pastures converted relatively recently from forest land (CATIE, 1989; Veldkamp, 1993; Villareal, 1992; Hernandez *et al.*, 1995). Although there have been many attempts to introduce improved grass species, most of these species disappeared after a short time, often due to poor management (overgrazing) and to the fact that, given the current prevailing prices of fertilizer and beef, the use of fertilizer in beef cattle ranching is not economically feasible (Bouman and Nieuwenhuyse, 1999). Lately, promising research results have been obtained with the introduction of grass-legume mixtures (*e.g.*, *Brachiaria brizantha* with *Arachis pintoi*), which, when properly managed, are able to fix up to 100 kg of N per ha per year (Ibrahim, 1994). This amount of N is sufficient to maintain the more demanding improved grass species for a considerable period of time. Apart from the higher biomass production of such improved pastures, fodder quality is also better, permitting higher beef production by the grazing cattle. Dairy farms often devote a part of their land to the cultivation of fertilized improved grass species, on which stocking rates of up to 3 AU ha^{-1} can be achieved. Most breeds are typical milk breeds such as Holstein and Jersey, with some Zebú influence to increase resistance against climatic conditions.

Banana

Banana (*Musa* cvs, AAA group) plantations have played an important role in the AZ since the end of the 19th century. Being a non-indigenous crop to the Americas, banana was first introduced in the region as a food crop for railroad construction workers. During the 19th century, it became clear that a market existed for banana in Europe and the United States. As a result, the first multinational fruit companies were formed, and they established banana plantations in Central and South America. Using the constructed railroads, the export (and consequently, the quantity) of banana grown in the AZ quickly grew from zero before 1870 to 10 million bunches in 1910. Initially, commercialization was carried out by one company, the United Fruit Company. Soil degradation, the economic world crisis in the 1930s and increasing problems with the "Panama disease" (caused by the fungus *Fusarium oxysporum* and affecting the dominant "Gros Michel" cultivar) caused commercial production to end in 1942. Production shifted to a large extent to the southern part of the Pacific coast of Costa Rica. By the end of the 1950s, the "Cavendish" cultivars, which are resistant to the Panama disease, were introduced, and the multinational companies returned to the AZ. Currently, three large multinational companies account for some 36% of total banana production in the AZ of Costa Rica, although their share in commercialization is about 67% (through the purchase of a large part of the production of so-called "independent" producers) (CORBANA, 1996). Modern plantations are typically 100-600 ha large and have a well-organized infrastructure: extensive drainage systems run throughout the plantations, and aerial cable systems are used to transport the fruit to a central packing plant. The use of external inputs is high: an average plantation applies about

500 kg N ha^{-1} y^{-1} and between 45 kg ha^{-1} y^{-1} (Castillo *et al.*, 1997) to 61 kg ha^{-1} y^{-1} (Hengsdijk *et al.*, 1998a) of active ingredients in biocides, depending on the number of applications and type of biocides used. As a result, average production is among the highest in the world, and consists of about 39 ton ha^{-1} y^{-1} of exportable fruit. At the end of the eighties, government incentives and favorable market prospects caused a strong expansion of the area devoted to banana cultivation. In the late 1990s, markets were less favorable, mainly due to import quotas imposed by the EU, and some plantations were closed. Great concern still exists over the black Sigatoka (*Mycosphaerella fijiensis*) fungus, whose control requires aerial spraying every 10-14 days and which shows more and more resistance against existing fungicides. Often, bamboo plantations (Table 2.3) occur next to banana plantations to supply the bamboo stalks needed to support fruit-bearing plants.*

Maize

Until 1989, maize (*Zea mays*) was one of the most important crops for many small farmers in the area because of its subsidized price and secure market. When the subsidy was removed, the crop rapidly lost importance, and currently it is grown mostly for own consumption, or sold in the form of fresh cobs. In the AZ, maize cultivation decreased from about 18 000 ha in 1987 to less than 1000 ha in 1994 (SEPSA, 1992; Abarca, 1995, pers. comm.).

Beans

Beans (*Phaseolus vulgaris*) constitute an important part of the daily Costa Rican diet and are grown throughout the country. However, the perhumid climate makes bean cultivation in the AZ risky: according to farmers, about one out of two crops gets severely damaged during flowering or harvest because of heavy rains. Mostly, beans are grown on small plots for own consumption, and the use of external inputs is either absent or low.

Palm heart

Palm heart (*Bactris gasipaes*) is considered one of the promising "non-traditional" crops for export. The crop is well adapted to the environmental conditions of the AZ and requires relatively few inputs. Stimulated by favorable market outlooks, the area of its cultivation increased nation-wide from about 1800 ha in 1988 to 3500 in 1992. Currently, it is estimated that, in the northern AZ, at least 2000 ha are under palm heart cultivation. However, the international market for this crop is still limited, and because other countries in Central and South America have increased their production as well, it is uncertain to what extent production can grow any further in the future.

Roots and tubers

Probably dating back to the first colonization, root and tuber crops have always been important food crops in the area. With improving infrastructure, their importance as

* More information on the cultivation of bananas is given in Chapter 9.

export crops has increased. For example, the export of cassava (*Manihot esculenta*) increased from 3200 tons in 1978 to 12 000 tons in 1988 (SEPSA, 1989). This increase in export volume was associated with an increase in area cultivated, for which, however, no supporting data are available. Other root and tuber crops, especially yam (*ñame; Dioscorea* spp.), show similar tendencies. The cassava variety used in the AZ is geared towards the production of a high quality root, but has a relatively low yield, with a relatively short growth cycle of nine to twelve months.

Plantain

Since colonization, plantain (*Musa* cvs, AAB group) has been a food crop, of which the commercialization has increased as the country's infrastructure has improved. The Costa Rican plantain production is concentrated in Limón province, but the most important areas are located south of the northern part of the AZ (Talamanca county). Unlike banana, plantain is usually grown in small plots of 0.5 to 5 ha on small to medium sized farms. The use of fertilizer and biocides is lower than on banana plantations. Most of the production is for the domestic market, though exports have increased significantly in the 1990s.

Pineapple

The most important pineapple (*Ananas comosus*) producing regions of Costa Rica are located in the southwestern part of the country and in the northern plains west of the AZ, and consisted in 1992 of a total area of 5500 ha (SEPSA, 1992). Until recently, there was no pineapple production for export in the AZ, although the "Montelirio" cultivar was grown for domestic consumption. In 1989, only 50 ha of pineapple were grown in Limón province (Den Daas, 1993). Recently, however, several hundreds of hectares have been planted with new cultivars for export and for the juice industry. As is the case with banana, inputs of fertilizer and biocides are high.

2.6 Farm structure

As the last agricultural census dates back to 1984 (DGEC, 1987a), more recent official information on farm structure in the AZ is not available. Table 2.6 shows the farm structure for Limón province in 1973 and 1984. Between these two census years, the number of farms increased by 67%, while total cultivated area increased by only 18%. Average farm size decreased from 46.2 ha to 30.7 ha, while farm land became more evenly distributed, as indicated by a decrease in the Gini ratio from 0.76 in 1973 to 0.71 in 1984 (Kruseman *et al.*, 1984). This was mainly caused by a land distribution program in the context of which the IDA created a large number of small settlements (see also Section 2.8), made up of farms between 4 and 20 ha in size.

Table 2.6. Farm structure in the Atlantic Zone

Farm size	*1973 Census*[1]				*1984 Census*[2]				*1987 base survey*[3]
	farms		*area*		*farms*		*area*		*farms*
	#	*%*	*ha*	*%*	*#*	*%*	*ha*	*%*	*%*
Without land	156	3.0			283	3.0			
0-4 ha	1065	20.1	1956	0.8	1749	18.8	3170	1.1	7
4-20 ha	2168	41.0	20 524	8.4	4448	47.7	43 084	15.1	48
20-50 ha	1223	23.2	33 661	16.2	1690	18.1	48 472	17.0	22
50-200 ha	501	9.5	42 982	16.0	893	9.6	77 007	27.0	15
> 200 ha	177	3.3	145 391	59.5	253	2.8	113 583	39.8	7
Total	5290	100.0	244 531	100.0	9316	100.0	285 316	100.0	100

[1] Applies to Limón province (DGEC, 1976).
[2] Applies to Limón province (DGEC, 1987a).
[3] Based on a REPOSA survey taken in January/February 1987 of the Neguev settlement (n=50), Río Jiménez district (n=46) and Cocori area (n=50), three areas considered to be representative for the northern part of the AZ (Schipper, 1993).

During January and February 1987 REPOSA conducted a survey of 149 farms in three research areas considered representative of the agricultural situation in the northern part of the AZ. The three survey areas consisted of the Neguev settlement, representative of many settlements created by IDA; the district of Río Jiménez, representative of "older" agricultural settlements (dating from the beginning of the 20th century); and Cocori, an area in the northeastern part of the AZ, representative of the recent "spontaneous" colonization on the forest frontier. The results of this survey show that distribution of farm land did not change significantly between 1984 and 1987.

According to the survey in 1987, average farm size in Pococí and Guácimo counties (in which the three research areas are located) is 38.5 ha (Schipper, 1993), which is statistically not different from the average farm size of 34.9 ha recorded in 1984 (DGEC, 1987a). Comparing the average farm size with a median farm size of 17.0 ha, points towards a relatively skewed distribution of farm land.

In 1992 a survey of 96 farms in the same three research areas was conducted to determine farm forest areas (Van Leeuwen and Hofstede, 1995). Average farm size increased in Neguev and Río Jiménez but decreased in Cocori. The assumption of identical numbers of farms in the three research areas leads to a weighted estimation of an overall average farm size of 36.5 ha, which is not statistically different from the average farm size in the 1987 survey.

Based on the 1984 census, Kruseman *et al.* (1994) developed a farm classification consisting of four categories. Of a total number of 9316 farms (with a total area of 285 316 ha), 6480 were classified as small farms between 0-20 ha (total area 46 254 ha), 1690 as medium farms between 20-50 ha (total area 48 472 ha), 1010 as extensive *haciendas* with more than 50 ha and more than 50 head of cattle (total area 169 526 ha) and 136 as banana plantations (total area 21 064 ha). The above farm structure is still more or less valid, even though the importance of banana plantations has increased (in 1995 there were 187 plantations with a total area of 49 599 ha). On the other hand, the number of small farms has probably decreased, particularly in settlements. This is supported by a 1996 survey of 39 farms in the Neguev settlement (Kuiper, 1996), all of which had been included in the 1987 survey. While the total number of farms in the Neguev decreased from 310 in 1987 to about 200 in 1996, average farm size increased from about 12 ha in 1987 to about 20 ha in 1996. With the exception of those farmers who concentrate on palm heart cultivation, many settlers have sold (part) of their land and are earning most of their living as (plantation) laborers, while a few others acquired land from their neighbors for cattle raising. Developments similar to those that occurred in the Neguev are a general phenomenon in many other settlements in the AZ, indicating a considerable decrease in the number of small farms (see also Section 2.8), although there might still be a large number of "farms" without land or farms smaller than 4 ha (*minifundios*), whose main function is to provide a house for their occupants and a homegarden. Finally, the area of pasture land and the number of cattle increased considerably between 1984 and 1995. However, the consequences of this development for the number and area of medium farms and *haciendas*, when considered in conjunction with the increased number and size of banana plantations and the decreased number and size of small farms, are not obvious.

2.7 Macro-economic and agricultural policy

Historically, the Costa Rican economy has performed quite well. Between 1957-1977, per capita GDP (Gross Domestic Product or national income) grew at an average rate of 2.7% per year, with corresponding substantial improvements in all social indicators, in particular those related to health and education (Céspedes, 1998). During this period, Costa Rica's economic policy was largely based on the so-called import substitution model supported by foreign aid agencies (OFIPLAN, 1982). Salient characteristics of this model include: strong direct government interference via a range of protection measures meant to shelter both the agricultural sector and manufacturing industries from foreign competition; high dependency on imported capital goods; and discouragement of export initiatives. As a result of the large and increasing role of the state, government expenditures grew from 26% of GDP in 1950 to 54% in 1980 (Céspedes, 1998).

Agricultural policy during this period was directed primarily towards the production of traditional export crops, such as banana and coffee, along with the achievement of self-sufficiency in the basic food crops such as maize, rice and beans (González, 1994). Policies applied to achieve food security included government regulation of prices through the National Production Council (CNP; *Consejo Nacional de Producción*) as well as investments in infrastructure. Furthermore, the Ministry of Agriculture and Livestock (MAG; *Ministerio de Agricultura y Ganadería*) provided technical assistance to farmers, and the IDA turned many landless laborers into land owners by establishing settlements of small and medium-sized farms. Credit (often subsidized) was mainly provided by the public banking system (Cartín and Piszk, 1980; Quirós *et al.*, 1997). The marketing of basic food crops was regulated by the CNP, which guaranteed fixed producer prices for any quantity supplied. Most of the produce was sold on the domestic market at below the guaranteed producer prices, while any surplus production was exported to other Central American and Caribbean countries. Import of basic food grains was only allowed in times of shortages (Guardia *et al.*, 1987). At the same time imports of inputs such as agrochemicals and agricultural machinery were taxed, thus providing an additional incentive for relatively input-extensive basic food crop production, as compared to the more intensive production of export crops.

At the end of the 1970s, the agriculture-led economic growth slowed as a result of a decrease in the growth of agricultural production, culminating in the 1980-1982 crisis during which the Costa Rican economy suffered a reduction in GDP and high inflation (Figure 2.6). It was increasingly realized that the size of the domestic market is too small to serve as a base for rapid and sustained growth in the agricultural sector. To overcome this crisis, a number of policy changes were implemented to achieve a better integration of the Costa Rican economy into the world economy. Structural reform, aimed primarily at lowering inflation rates and balancing fiscal (*i.e.*, government revenues and expenditures) and external (*i.e.*, imports and exports) accounts, consisted in a lowering of trade barriers (mainly tariff and non-tariff measures aimed at controlling imports) and reform of both the financial and state sector. Exports were stimulated by an exchange rate policy which aimed at maintaining the competitive position of Costa Rica *vis-a-vis* its main trading partners through a system of mini-devaluations.

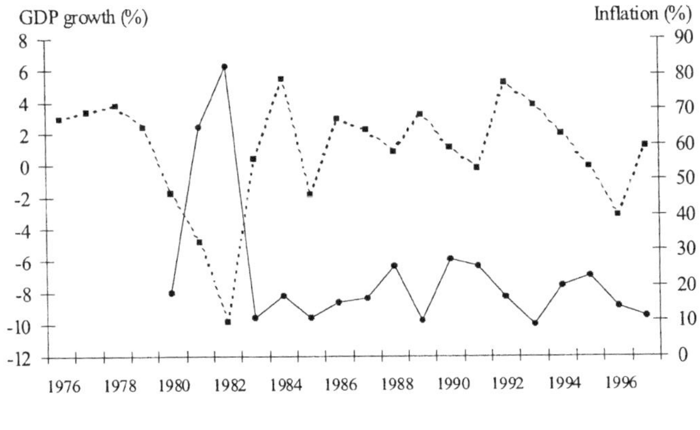

Figure 2.6. The annual growth rate in the Costa Rican Gross Domestic Product (dotted line) and the inflation rate (solid line) between 1977 and 1997.

The main consequence for the agricultural sector of the market liberalization policies was a much higher degree of integration into the world market (Pomareda, 1995; SEPSA, 1997), and, in particular, increased exports of a larger range of commodities. The system of guaranteed producer prices and consumer subsidies for basic grains was gradually phased out, while production of non-traditional export crops was promoted through a system of tax credits (essentially export subsidies), as well as credit on favorable terms for export activities (Mora *et al.*, 1994). These measures favored production of both traditional export crops (*e.g.*, banana) and non-traditional export crops (*e.g.*, palm heart, roots and tubers, ornamental plants), by comparison with basic food crops. Some of the small and medium-scale farmers who were not capable of making the transition to these favored crops were placed in difficulty and had to look for alternative income-generating activities (Mora *et al.*, 1994; SEPSA, 1997).

Besides these social costs, during the 1994-97 period it became increasingly clear that implementation of structural reform and market liberalization policies involved other costs (Mesalles, 1998). For example, interest rates surged (at least partially because of a government bail-out of the depositors of a major bank which collapsed in 1994), resulting in reduced investment which, combined with a contraction in consumption, caused a recession and a lower GDP in 1996 (Figure 2.6). As a result, the fiscal deficit did not improve between 1995 and 1996 (due to lower government income and higher debt service payments), remaining at around 5% of GDP.

In 1997, new policies adopted to combat this new crisis, such as an increased flexibility in monetary policy (Mesalles, 1998), resulted in an increase in GDP to just over 3%, a lower inflation of about 11% and lower real interest rates of about 7%. However, both the current account (exports and imports of goods and services) and balance of payments (which includes financial flows as well) developed deficits of some 4% of GDP, reflecting increased economic activity (higher imports of capital goods). Also the internal debt (amounting to 28% of GDP; Latin America Monitor, June 1998) remains problematic, with interest payments accounting for some 30% of total government expenditure (Vargas, 1998).

General consensus exists that implementation of the structural reforms since 1983 has been both incomplete and insufficiently consolidated (Céspedes, 1998; Hausmann, 1998; Mesalles, 1998; Vargas, 1998). Nevertheless, some notable successes have been achieved. For example, both economic growth and employment have increased since the 1980 crisis, the export base has been diversified, and foreign investment has increased. Moreover, no widespread bankruptcies occurred, the share of wages in GDP increased slightly, and income distribution has not deteriorated. Even though public expenditures on social programs have decreased, many social indicators, such as education and life expectancy, remain relatively high. However, inflation is not yet under control and the performance of the government sector (still accounting for 45% of total GDP in the late 1990s, although down from 54% in 1980) is still problematic. For example, figures in Vargas (1998) indicate that, whereas tax income between 1985 and 1997 fluctuated between 14 and 17% of GDP, government expenditure in this period fluctuated between 18 and 22% of GDP. Besides pushing up interest rates, interest payments on government debt (accounting for about 5% of GDP) put pressure on such other expenditures as education, infrastructure, health, *etc.* (Hausmann, 1998). As it is often claimed that government expenditures in education and health have a regressive character (*i.e.*, benefiting particularly the poorer segments of society; Céspedes, 1998), this imbalance is a serious matter. Current policy measures that hurt the poor are exemplified by the high tariff duties (on average 40%) on a large number of basic food products (see also Chapter 4). These tariffs are highly regressive, reducing the purchasing power of the relatively poor households in particular (Céspedes, 1998).

Lately, agricultural policy also emphasizes issues involved in food security and promoting efficient basic food production, without significantly compromising the attention placed on export-led agricultural growth (SEPSA, 1997). On the other hand, the system of subsidies for non-traditional agricultural exports is incompatible with international trade agreements and are to be phased out by 1999. Concern for increasing rural poverty and degradation of the natural resource base caused the concept of sustainable development to be adopted as official government policy (Quesada Mateo, 1990; SEPSA, 1995). This measure should facilitate programs that improve the competitiveness of small and medium-scale farms (SEPSA, 1997), that stimulate better organization to strengthen the ability of farmers to market their produce (Jansen and van Tilburg, 1996), that enhance farmer creditworthiness and access to formal loans, that introduce a quality certification system for agricultural exports, and that promote the use of modern agro-industrial technologies to increase the profitability of exports.

2.8 Social and institutional factors

In the 1960s a combination of factors caused a strong immigration of settlers into the Atlantic Zone. Difficult economic and ecological conditions in the Central and Pacific regions of the country forced a considerable number of farmers to sell their land and move elsewhere. The existence of large areas of virgin forest made it relatively easy to obtain land in the Atlantic Zone (UNA, 1986). Furthermore, the government through the IDA (earlier known as ICTO, *Instituto de Colonización de Tierras y Ordenamiento*, which was established in 1961) actively stimulated the influx of new settlers into the frontier areas that previously had been identified as promising for agriculture (Nuhn, 1962). In the AZ, IDA undertook a number of colonization projects (*e.g.*, Bataan, Cariari, El Indio and Neguev) and was active in land titling programs. Since then, various processes have influenced the development of the AZ (Kruseman *et al.*, 1994; De Vries, 1990).

1. While formerly many farmers had limited access to markets and produced mainly for own consumption, intrastructural improvements not only considerably reduced the distances to both export and national markets, but also made it possible for farmers to increase their income. All the same, marketing conditions for small and medium farmers remain a long way from being satisfactory (Jansen and van Tilburg, 1996).
2. The increase in the area of banana cultivation, especially between 1988 and 1992, increased economic activity in the region, but also caused the disappearance of many small and medium sized farms, which were bought out by plantations. Some of these farmers bought land in more remote areas, others moved to urban centers and quit farming altogether, while again others became laborers on banana plantations.
3. An important proportion of farmers shifted to less intensive production techniques (using less capital and labor) as a consequence of the policy of market liberalization and abolishment of subsidies for basic grain production (Pomareda, 1998). A number of these farmers have tried to produce export crops, some successfully, others without success. Transformation from basic grain production for own consumption and for a subsidized market to the production of non-traditional export crops is impeded by insecure markets, strong price fluctuations, and lack of knowledge about marketing

opportunities (Jansen and van Tilburg, 1996). Technical constraints include pests, diseases and insufficient agronomic knowledge. Farmers often lack sufficient capital and knowledge to grow these new crops successfully in the ecological and socio-economic setting of the northern part of the AZ. Programs designed to transform smallholder agriculture in the AZ by involving institutions like MAG, IDA and CNP have promoted agricultural diversification by stimulating crops which have a high demand for capital, technology and labor. Although some small and medium sized farmers managed to adapt themselves successfully to the new market conditions, many of them did not succeed to change from producers of basic grains into producers of non-traditional export crops (Alfaro, 1993; Mora *et al.*, 1994). The adjustment crisis of many small and medium sized farmers coincided with a strong expansion of plantations growing banana, ornamental plants and palm heart (see Section 2.6). As a result many of the younger farmers choose to become laborers on these plantations instead of full-time farmers.

4. Since the late 1980s, the number of squatters has decreased, and land transactions in the more remote (recently colonized) areas appear to have increased. This may be interpreted as a consequence of a reduced interest in farming by younger people, who now opt to look for work on plantations and in urban centers. On the other hand, it can be considered a sign that the frontier region no longer exists, and that property rights have become more or less established.

2.9 Issues affecting sustainability

2.9.1 Biophysical and socio-economic sustainability

The conversion of forest to agricultural land in the AZ has raised a number of issues concerning both the biophysical and socio-economic sustainability of farm practices. Negative environmental effects of converting tropical forests into farm land include the loss of bio-diversity, increased land degradation, and higher emissions of trace and greenhouse gasses such as CO_2, N_2O, and NO (Detwiler and Hall, 1988; Houghton, 1991; Keller *et al.*, 1993). Detwiler and Hall (1988) estimated that the annual net emission of carbon in the form of the greenhouse gas CO_2 caused by tropical deforestation may be only second to the global carbon release from the burning of fossil fuels. For the AZ of Costa Rica, Keller *et al.* (1993) reported that soils from recently established pastures emitted one order of magnitude more N_2O than forest soils. A similar trend was found for nitric oxide (NO), which is a precursor to the formation of tropospheric ozone, yet another greenhouse gas.

Land degradation resulting from the agricultural use of land after forest clearing can have various dimensions. In the AZ, nearly two-thirds of the deforested area is currently estimated to be under pasture (Driese *et al.*, 1999). Pastures are dominated by relatively unproductive naturalized and native grass species, and management is characterized by zero fertilizer use and low levels of other external inputs (Hernández *et al.*, 1995). Immediately after forest clearing, soils are relatively rich in nutrients such as N (Veldkamp, 1993), leading to relatively high grass yields. Invading weeds are resisted by a combination of manual weeding and herbicides. With continued pasture use, economic and environmental sustainability may change in time. When stocking rates are adapted to the carrying capacity of the environment - as determined by climate, soil properties and natural nutrient inputs - pasture production levels can be sustained for long periods of time. However, with higher stocking rates, the removal of nutrients (especially N) in cattle products and by leaching and gaseous losses may be higher than nutrient inputs, resulting in soil mining (Haynes and Williams, 1993). Values of annual soil N depletion rates in humid and sub-humid tropical pastures have been reported to be as high as 65-94 kg ha^{-1} (Cadisch *et al.*, 1994; Thomas *et al.*, 1992). Bouman *et al.* (1999b) estimated values of 50-65 kg ha^{-1} for the AZ. In the long run soil mining leads to pasture degradation, as evidenced by productivity decline and weed invasion (Myers and Robbins, 1991; Williams and Chartres, 1991). 't Mannetje and Ibrahim (as quoted in Jansen *et al.*, 1997b) estimated that over 70% of the pastures in the AZ are in an advanced stage of degradation, with overgrazing and lack of sufficient N input identified as the principal causes (Hernández *et al.*, 1995). With decreasing pasture yields, farmers' income decreases (Bouman *et al.*, 1999b) and degraded pastures may eventually be abandoned because of non-profitability (Uhl *et al.*, 1988; Haynes and Williams, 1993). Relatively low beef prices during 1995 and 1996 have aggravated the situation for cattle farmers in Costa Rica, and a steadily declining cattle stock and abandonment of pasture land have been reported by Montenegro and Abarga (1998). Even though degrading pastures, even at an advanced stage of degradation, may be economically optimal from an individual farmer's point of view (Bulte *et al.*, 1999b), this may not be the case from a more general social perspective, particularly not in areas where the agricultural frontier has been reached and where the possibility of shifting cultivation (which would allow regeneration of degraded lands over time) no longer exists. In such cases, there usually are significant external effects associated with land degradation (*e.g.*, regional impoverishment, unemployment, and enhanced emissions of the greenhouse gas CO_2, all of which are not taken into account by private decision makers), justifying government intervention. Finally, it should be noted that degradation of pastures also can have a beneficial effect on sustainability indicators. For example, Keller *et al.* (1993) reported that, after a decade following forest clearing, N_2O and NO emissions from pastures in the AZ dropped below original forest levels.

In cropped land, like in pastures, soil nutrient mining is one of the problems affecting biophysical sustainability. Soil nutrient balances for most currently prevailing cropping systems are negative (Jansen *et al.*, 1995; Hengsdijk *et al.*, 1998a). With negative nutrient balances, the production potential of the soil declines over time, leading to lower yields, or higher costs of fertilization to sustain the original yield levels. In the AZ, the use of biocides is common for most crops and pastures. In banana plantations and the cultivation of ornamental plants, the use of biocides is even imperative and consequently very high, despite an average estimated use for Costa Rica as a whole of 6 kg a.i. ha^{-1} (von Düszeln, 1990). Biocides contain toxic substances that threaten the health of ecosystems (Castillo *et al.*, 1997), depending on the quantity used, toxicity, persistence in the environment, and mode of transportation. Regarding the latter, in banana plantations, large quantities of fungicides are sprayed about every 10-14 days from airplanes to combat the Black Sigatoka fungal disease. Drift losses from airplane spraying may be as high as 50% and can affect ecosystems many kilometers away from the region of application (Van der Werf, 1996). However, to our knowledge, no initiatives have been undertaken thus far to measure drift losses or effects on neighboring ecosystems in the AZ. A second mode of transportation of biocides involves ground and surface water. This type of water pollution potentially threatens downstream areas, including nature reserves and remote rural settlements that use ground water as a source of drinking water. Due to high amounts of rainfall and permeable soils in the AZ, average concentrations of biocides (and nutrients) in ground and surface water can be expected to be relatively low (Jansen *et al.*, 1995). However, concentrations in rivers and ground water may be much higher in certain localities immediately after biocide application. In addition to the effects on ecosystems, biocides may pose a human health problem (Wesseling, 1997). Even though the economic costs of short-term occupational intoxications related to the application of biocides in banana plantations in the AZ are relatively minor, long-term negative health effects and other social costs may be substantial (Jansen *et al.*, 1998). In addition, knowledge about the incidence of occupational intoxications on smallholder farms is very limited, even though it can be expected to be well above that on plantations.

Changes in biophysical sustainability parameters may negatively affect socio-economic sustainability as well. For instance, even though soil nutrient mining increases profitability in the short to medium term, it may negatively affect economic sustainability on account of declining yields and increased costs for weeding (in pastures) and for fertilizer. Furthermore, since maintaining or increasing rural employment is an important government objective, the presence of the banana companies in the AZ is important. With about 36 000 ha of plantations in the northern part of the AZ (CORBANA, 1996) and with some 24 000 workers, as well as associated post-harvest processing and transportation facilities, the banana industry is the largest employer of agricultural labor and is by far the largest generator of regional economic surplus and foreign exchange (see also Section 2.5).

2.9.2 Sustainable land use options

A number of efforts are being undertaken to develop production technologies that are more sustainable than the ones currently employed in the AZ. These initiatives vary from basic research at Costa Rican universities and public and private research institutes, to trial-and-error like experimentation by (groups of) farmers, non-governmental organizations and plantation owners. Environmental awareness on banana plantations is on the rise (Faber, 1997; Jansen *et al.*, 1998), while some significant changes in the crop and husbandry practices of smallholder agriculture have also taken place over the past few years. The Costa Rican government has adopted the conservationist agriculture (*agricultura conservacionista*) as an official policy, and the extension services assist smallholder farmers in implementing technologies that conserve natural resources (MAG-FAO, 1997).

In agricultural research, substantial work is being undertaken to halt the degradation of tropical pastures (accounting for some 75% of agricultural land use in Costa Rica), mostly focusing on the use of grass-legume mixtures and fertilized improved grass species (Ibrahim, 1994; Hernández *et al.*, 1995; Miller and Stockwell, 1991; Teitzel *et al.*, 1991). These new technologies are meant to halt soil N depletion by supplying extra N, either as a results of the fixation by micro-organisms in symbiosis with legumes, or directly in fertilizers. These alternative technologies, however, have other implications for sustainability as well, each of which might be perceived as positive or negative. First of all, since economic viability is a necessary requirement for farmers (Jansen *et al.*, 1997b), such technologies need to be profitable on an individual basis. Second, those new technologies affect a field's emission/sequestration of trace and greenhouse gasses, and thus the environmental sustainability becomes an issue. Improved grass species have been reported to produce a relatively high amount of deep root biomass with a low turnover time, thus acting as a sink for CO_2 (Veldkamp, 1993; Fisher *et al.*, 1994; Van Dam *et al.*, 1998). On the other hand, measurements on fertilized pastures in the AZ indicated high emissions of N_2O and NO relative to unfertilized naturalized pastures (Veldkamp *et al.*, 1998). Using a modeling approach, Bouman *et al.* (1999c) computed that the use of grass-legume mixtures and fertilized improved grass species would lead to CO_2 sequestration but enhanced NO and N_2O emission rates in comparison with the current pastures in the AZ. Environmental sustainability is also affected by the ways in which fertilizer may contaminate ground and surface water as a result of nutrient leaching.

In banana cultivation (the second major land use in Costa Rica), the private banana sector, together with public and semi-public agricultural research institutes, is studying the potential of controlling pests and diseases biologically (González *et al.*, 1997; Ruiz-Silvera *et al.*, 1997). Another on-going effort is the ECO-OK/Banana project, which certifies farms in the tropics that meet a comprehensive set of criteria related to the conservation of wildlife habitat, environmentally responsible cultivation practices, and the well-being of farm workers and local communities. In 1998, nearly 20% of banana production in Costa Rica was awarded such certification, including both independent and multinational producers (source: www.rainforest-alliance.org, October 1998). A third example in the quest for more sustainable production techniques in banana production is the recent experimentation with precision agriculture (see also Chapter 9). In precision agriculture, the application of external inputs such as agro-chemicals is optimized in time and space so as to meet crop requirements with minimum losses to the environment.

The examples given above regarding efforts to develop more sustainable production technologies in the AZ are far from exhaustive. However, they do illustrate the full range of current practices. In pasture-based beef production, research has resulted in a set of well-defined, promising alternatives that can be quantified and subsequently evaluated in various land use models. The quantification of grass-legume and fertilized pasture production systems is illustrated in Chapter 5 of this book, and their use in regional land use exploration and policy support models in Chapters 6, 7 and 8 (see also Bouman *et al*, 1999b). On the other hand, research on more sustainable production technologies in the cultivation of banana was begun only relatively recently, and has not yet advanced to a stage where promising alternatives may be reliably quantified in terms of input-output relationships. The example of precision agriculture is elaborated in Chapter 9, and may ultimately lead to quantified input-output relationships on the field or farm level that can subsequently be investigated in a regional context, just as has been done with the pasture alternatives. Both pasture land and the banana cultivation, however, illustrate various issues concerning socio-economic and biophysical sustainability, and highlight the need for an adequate quantification procedure to arrive at positive, rather than negative, trade-offs between economic and biophysical sustainability.

Finally, Costa Rica is one of the countries that increasingly regard agriculture as being inherently multi-functional in nature. The concept of multi-functional agriculture recognizes that the primary agricultural production process consisting of efficiently producing foodstuffs, fodder, fiber, and hides and skins, generates a number of joint products which can be considered as externalities. For both research and policy purposes, such externalities can be either negative or positive in nature. While a number of negative externalities are explicitly addressed in this book (*e.g.*, soil nutrient depletion, environmental contamination through biocide use), much less attention is being given to the positive externalities that are byproducts of primary agricultural production. Besides regulatory functions related to soil and water conservation, positive externalities may include provision and maintenance of rural welfare, landscapes, biodiversity and carbon sequestration. Regarding the latter, part of the research carried out in the context of REPOSA has addressed a number of carbon-related issues (see, *e.g.*, Veldkamp 1993; Keller *et al.*,1993; Bulte *et al.*, 1999a), and Costa Rica is developing innovative mechanisms in order to benefit fully from this positive externality.

Appendix 2.1 Major soils groups in the northern part of the Atlantic Zone of Costa Rica

SIW: Infertile, well-drained soils (Haploperox[3] and Humitropept)

SIW soils cover 86 300 ha, and are found on older landforms such as the remnants of Pleistocene terraces, and older deposits in both *Cordilleras* (Figure 2.4). Slopes vary between flat and steep (up to 60%). Especially in the *Cordilleras*, steep slopes and stoniness limit possibilities for mechanized agriculture (see also Chapter 5). Due to their high position in the landscape, these soils are well drained.

Usually such soils have a thin (<15 cm) brown A horizon which is underlain by a thick homogeneous reddish brown B horizon. Texture of the profiles is usually clayey. Below a depth of 1 to 2 m the B horizon gradually changes into a CB horizon which is several meters thick.

Soils of this group generally have rather low fertility (Table 2.2). This is partially due to the low pH and consequently high exchangeable acidity, and also to their advanced stage of weathering which has left little or no mineral reserves.

Under natural vegetation or certain crops (*e.g.*, palm heart), these soils have a high permeability and favorable structure for rooting. Structure varies usually from strong subangular blocky in the topsoil to weak blocky or massive in the subsoil. Due to

[3] All classifications are according to the Soil Survey Staff (1992).

the high clay content and perhumid climate, the topsoil is susceptible for compaction, *e.g.*, under grazed pasture.

SFW: Fertile, well-drained soils (Udands, Tropopsamments, Dystropepts)

The SFW soils cover 123 000 ha and have formed mainly on sandy deposits, and are consequently found mainly on the alluvial fans and on mixed alluvial and mudflow deposits at altitudes of 50 to 300 m, as well as on choked river channels and adjacent overbank deposits throughout the alluvial plains (Nieuwenhuyse, 1996). Furthermore, these soils are found on the sandy beach ridge deposits along the Caribbean coast. Parent material of alluvial soils south of the Reventazón river originates from the Talamanca *Cordillera*. Sediments differ in their mineralogical composition from sediments which originate in the Central *Cordillera*, and contain some free $CaCO_3$ in the silt and sand fraction (Van Seeters, 1993). Slopes vary from almost flat in the alluvial plains to maximally 10% in the upper alluvial fan areas. Many of the soils on the alluvial fans contain up to 50% stones throughout the profile, a factor which limits the possibilities for mechanized agriculture. Drainage varies from moderately well to excessively drained.

Soils have usually a thick (30 to 100 cm), brownish black A horizon with a loamy texture. The yellowish brown B horizon has a loamy to loamy-sandy texture, and is 30 to 70 cm thick. The youngest soils are AC profiles.

Due to andic properties (Soil Survey Staff, 1992), P availability for crops grown on sand derived from the Central *Cordillera* is low, and P recovery from fertilizer is often low as well.

The soil structure is dominantly fine (sub)angular blocky and crumby. This is reflected in high porosity (often 65 to 75%), high hydraulic conductivity, and a high content of plant-available water. The high porosity is stimulated by the dominant presence of short-range order clay minerals, high year-round organic matter production and fauna activity. Land use induced topsoil compaction may change the favorable structure and consequently some of the soil physical properties (Spaans *et al.*, 1989).

SFP: Fertile, poorly-drained soils (Tropaquepts, Dystropepts and Eutropepts)

Moderately to poorly drained soils which developed on fine textured sediments cover 149 500 ha and dominate in the lower parts of the alluvial plains, with poorest drainage in areas at elevations below 20 m (Figure 2.4). Slopes vary from flat to almost flat (*i.e.*, less than 2%), and stones are absent.

The somewhat better drained soils have a 5 to 15 cm thick, brown A horizon, followed by a yellowish brown (mottled) cambic B horizon that grades into a mottled CB horizon (or buried soil). Poorly drained soils have a weaker horizon differentiation and are mostly gray AC profiles. Texture varies from loamy to clayey, and the profiles often show textural differences between or within soil horizons as a result of sedimentary layering.

Fertility of the SFP soils is usually higher than that of the other two major soil groups (Table 2.2), and the loamy texture favors slower leaching and higher retention of fertilizer. As is the case for the well-drained soils, parent material of alluvial soils south of the Reventazón river differs from sediments that originate in the Central *Cordillera*, and soils have slightly different chemical characteristics.

Structural development is related to drainage: under conditions of moderate to imperfect drainage, soil structure is angular or subangular blocky, while the permanently waterlogged soils and deeper horizons have weak structures or are even structureless. A study by Mantel (1993) indicates that in many cases drainage is not limited by the hydraulic conductivity of the soil, but rather by a lack of natural drainage canals in the flat landscape. This is evidenced by the fact that as soon as areas with these soils are drained by open channels, the groundwater level lowers rapidly and the soils become suited for a variety of agricultural uses, such as banana plantations.

Soils not suited for agriculture

Soils not suited for agriculture cover 88 200 ha and contain a variety of soils grouped into the following three categories:

A. Clay soils developed on sedimentary rock in the Talamanca *Cordillera* (Tropaquepts, Humitropepts). Soils developed on these rocks have a dense, impermeable structure, causing severe drainage problems. Soils are often very acid (pH - H_2O < 4). Together with their position on unstable slopes, these factors make them unsuitable for agriculture.

B. Soils developed on ash deposits in the Central *Cordillera* (Hydrudands). Under the prevailing extremely humid and relatively cool conditions, Hydrudands have formed on ash deposits on the higher slopes of the Turrialba and Irazú volcanoes (Buurman *et al.*, 1997). Upon deforestation these soil are vulnerable to structure degradation. Factors that further hamper the use of these soils are their position on steep slopes, and to a lesser degree their chemical properties.

C. Peat soils (Tropohemists). Peat grows in permanently waterlogged sites where sediment supply is low or absent. For most peat soils, their position on almost sea level and their highly unconsolidated nature makes it difficult and costly to reclaim them for agriculture.

3 Using the CLUE framework to model changes in land use on multiple scales

KASPER KOK, and TOM (A.) VELDKAMP

Abstract

The analysis of changes in land use and land cover has recently received ample attention in the scientific literature. So far, dynamic and integrated modeling approaches which are essential for the modeling of complex systems are relatively few. CLUE (Conversion of Land Use and its Effects), an empirically based framework for modeling changes in land use on various scales was developed to fill part of this gap. CLUE combines a statistical description of land use with scenarios of changes in the demand for regional commodities to model the possible pathways of future land use development. In the Atlantic Zone of Costa Rica CLUE was applied for the first time on sub-national scale. Multiple regression analyses produced equations with coefficients of determination between 0.58 and 0.91 for the major kinds of land use in the area (forest, pasture and banana). The statistical analysis demonstrated the importance of considering both biophysical and socio-economic variables as the driving forces of land use changes. Predicted changes in geographic patterns between 1984 and 2005 under different scenarios could be related to processes that are already known to take place in the Atlantic Zone. Forest was predicted to be largely replaced by pasture and to become limited to areas unfavorable for agriculture. Model validation yielded highly significant results with correlation coefficients ranging from 0.87 to 0.95 for the major kinds of land use. The study demonstrates that dynamic, multi-scale empirical modeling is a suitable tool to model land use changes on sub national levels.

3.1 Introduction

3.1.1 General introduction

Changes in land use and land cover have been regarded as important topics of concern for the past 20 years (Turner *et al.*, 1995). Besides local and direct effects, like loss of biodiversity through deforestation or soil degradation through unsustainable use, increasing importance is given to the global impact of the more indirect effects, like greenhouse gas emissions and carbon fixation. Boosted by Information and Communication Technology, a rapidly increasing number of land use models have been developed, addressing a huge spectrum of different topics on different scales and with different aims (comprehensive overviews are given by Sklar and Constanza [1991] and Lambin [1994]). However, the majority of these models address a small range of topics or scale levels (Dumanski *et al.*, 1998; Veldkamp and Fresco, 1997). When examining complex systems, like land use and its processes of change, often only limited but tractable topics are investigated (Rosswall *et al.*, 1988). The problem then becomes how to transfer the results of detailed studies to higher levels, as relationships established on one hierarchical level typically can not be used on higher or lower ones (Easterling, 1997). Moreover, bottom-up models are often incomplete because of our lack of knowledge about interactions and feedbacks (Jarvis, 1993). It is especially socio-economic systems that operate on multiple spatial-temporal scales and, in the context of land use analysis, can not be exhaustively described on any single scale or by any single discipline. Finding ways to translate information up and down is one of the fundamental challenges faced by researchers who analyze land use (Levin, 1993). In several disciplines awareness of the effects of scale is growing. For example, in ecology (Holling, 1992; O'Neill, 1988), soil science (Bouma, 1997), and the field of greenhouse gas emissions (Sellers *et al.*, 1997) understanding about how to deal with scale effects is increasing. Although many underlying processes act at the lowest possible, visible or workable scale (Reynolds *et al.*, 1993), proxies for these processes become apparent at higher aggregation levels (Fresco and Kroonenberg, 1992; Veldkamp and Fresco, 1996b). Given that changes in properties of the system occur when scaling up, there arises a need for empirical modeling (besides process based modeling) to correctly address the relationships between land use change and (the proxies for) its drivers.

The CLUE (Conversion of Land Use and its Effects; Veldkamp and Fresco, 1996a; Schoorl *et al.*, 1997) modeling framework was designed to deal with several of the issues mentioned above. CLUE attempts to describe land use patterns as viewed on several different geographic scales by using a set of socio-economic and biophysical data. Using this description, the dynamics of the system are examined by analyzing the spatial effects of various possible future pathways of land use change. The framework is based on a set of statistical relationships and a predominantly top down modeling approach is followed (Figure 3.1): changes in demand for agricultural commodities on the highest level of aggregation determine the change in land use. Local effects are

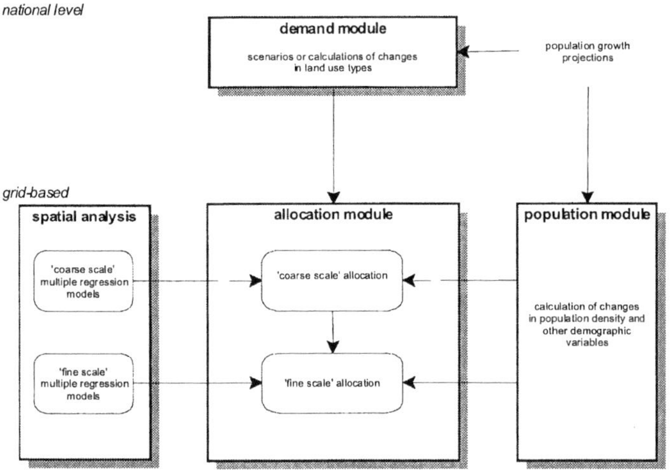

Figure 3.1. General structure of the CLUE modeling framework (source: Verburg *et al.*, 1999b).

incorporated by bottom-up feedback mechanisms. CLUE was tested on the nation-wide level in Costa Rica (Veldkamp and Fresco,1996b), China (Verburg and Veldkamp, 1997), Equador (De Koning *et al.*, 1998a; De Koning *et al.*, 1998b), and on the island of Java, Indonesia (Verburg *et al.*, 1999a). Maximal spatial resolution varied between 7.5 x 7.5 km for Costa Rica to 32 km^2 for China. The present study is the first attempt to apply CLUE to smaller geographic units.

3.1.2 CLUE Modeling framework

The CLUE modeling framework consists of three major modules: *i)* a multi-scale statistical description of present land use; *ii)* a general analysis of demand, and the definition of possible future demand pathways; and *iii)* a multi-scale allocation module, in which total demand in the study area, as defined in the demand scenarios, is spatially distributed over the study area using the statistical description.

i) A multi-scale statistical analysis is made of major kinds of land use[1] in terms of area in two different years. The analysis for the first year is used as model input, and the analysis of the second year is used to check the stability of the derived statistical relationships. The analysis consists of a multiple regression between major kinds of land use as dependent variables and a number of possible driving forces as independent variables. Driving forces range from biophysical characteristics, such as mean annual temperature and altitude, to socio-economic variables, such as rural population density and illiteracy rate. The analysis is conducted on various geographic scales, which are constructed by aggregating data from the basic (grid) unit to higher levels. For each level of aggregation multiple regression equations are constructed.

ii) Relationships are developed for the study area between commodity supply and demand. These relationships are general in the sense that they apply to the whole study area and have no geographic reference. Supply is assumed to equal demand and the analysis thus concentrates on demand. Furthermore, we assume that all commodities produced in the area are freely distributed across the entire zone, thus effectively ignoring transportation costs. The basic role of the demand module is to define a number of future demand pathways on a yearly basis. As it is difficult to predict future changes in some of the key variables needed to calculate demand (commodity prices, income per capita), scenarios are defined in such a way that a wide range of possible future pathways is covered. It is not our intention to predict what is going to happen, rather the aim is to explore a variety of possibilities. The demand module provides a broad estimate of the yearly demand — and thus required supply — in the study area for every commodity from the first year of analysis onwards (which is usually somewhere in the 1980s or 1990s) until a year somewhere in the future, typically between 2005 and 2015.

iii) The multi-scale allocation module combines the results of the statistical analysis and the demand module. A three-step top-down procedure with bottom-up feedbacks is followed, where aggregated total demand (of the whole study area) is allocated to smaller geographic units on two different scale levels by using the results of the statistical analysis. Selection of the second scale level is based

---

[1] Here, broad classes of land use are meant, such as forest, pasture, annuals *ect.* See also the definition in the Appendix at the back of the book.

on the coefficients of determination and explanatory variables in the different regression equations. A complete description of the allocation procedure is given by Verburg *et al.* (1999c). Regional aggregated demand is first distributed on the intermediate scale level, and then, in a second step, on the lowest spatial level. Because the modeling procedure uses information on a sub-grid level, implying that more than one major kind of land use can exist in every grid cell, competition is an inherent part of the allocation module. The competitiveness of every major kind of land use in every grid cell is defined by combining the statistical analysis, with an account of the relative changes in total demand, with expert knowledge. Bottom-up feedbacks are actuated when allocation fails because of lack of space or competitive strength.

3.1.3 Objectives

The overall objective of this study was to analyze the present land use pattern and to project possible future patterns of change in the use of land in the northern Atlantic Zone (AZ) of Costa Rica, by using an empirical multi-scale approach. Besides this overall objective, three secondary objectives were defined:

1. Analyze the pattern of land use and its driving forces in the AZ at two points in time, by using multiple regression techniques.
2. Analyze regional demand for the major agricultural commodities produced in the AZ and define a limited number of possible future changes in land use until the year 2005.
3. Analyze in a spatially explicit way the effects of those pathways on land use dynamics.

3.2 Methods and materials

In accordance with the objectives mentioned above, this section is divided into five parts: the statistical analysis, supply/demand analysis, scenario development, adaptation of the CLUE allocation module to the specific requirements of the AZ, and model validation.

3.2.1 Statistical analysis

The data used for the statistical analysis originated from two sources: the atlas of the northern AZ compiled by Stoorvogel and Eppink (1995), and the Costa Rican population and agricultural censuses conducted in 1984 (DGEC, 1987a, b). The latter were only used as additional sources when the maps from the atlas did not contain sufficient information.

The statistical analysis was done in two steps. First, a set of variables was selected using a stepwise regression procedure. Subsequently, the selected variables were used to construct multiple regression equations. The adjusted coefficient of determination ($\overline{R}^2$) was employed as a measure of the amount of variation explained. The standardized betas (regression coefficients) were taken to indicate the relative importance of individual variables in a given equation.

Data format

A grid-based approach with a pixel size of 2 km^2 was used as the basic unit of analysis. Instead of using uniform grid cells with one dominant major kind of land use and one type of soil, sub-grid information was present for most of the driving forces and each major kind of land use. For example, in every cell, the percentage of fertile soils present and of the cover of each major kind of land use was known. This meant that although local specificity was affected, we did not lose any information when either scaling down or up the data.

Dependent variable: major kind of land use

Two land use maps of the AZ were available in the atlas: a 1984 map of the entire area (scale 1:80 000), and a 1992 map covering 60% of the total area (scale 1:60 000), both based on the interpretation of aerial photographs. Since 1984 is the year of the most recent agricultural and population censuses, it was decided to use that year as the base year for statistical analysis. The spatial distribution of the following eight major kinds of land use was analyzed[2]: beans, maize, rice, cassava, annuals (total of the previous four), banana, forest, and pasture. The spatial distribution of maize, rice, beans, and cassava was derived from a combination of census data and the "annuals" class from the 1984 land use map; the spatial distribution of the other major kinds of land use was taken directly from the 1984 land use map. A ninth major kind of land use, palm heart, was taken into account in the demand and allocation part, but could not be statistically analyzed due to its absence in 1984.

Independent variables

A total of 40 potential driving forces were identified and grouped into four categories to facilitate interpretation. Numbers between brackets indicate the number of variables. A complete list of variables is given in Appendix 3.1:

1. Eighteen soil variables: base saturation (1), pH (2), cation-exchange capacity (2), acidity class (2), soil development stage (2), soil depth (2), soil fertility (2), soil drainage (2), and soil texture (3). Each variable was subdivided into 2 to 4 classes (*e.g.*, high, medium and low) and the percentage of each class was used as input. Soil data are described in detail by Wielemaker and Vogel (1993) and Stoorvogel and Eppink (1995).
2. Ten other bio-geophysical variables: mean altitude (1), average yearly rainfall (1), slope steepness (2), stoniness (2), flooding risk (2), and distance to nearest river (2).
3. Six demographic variables: rural population density (1), urban population density (1), agricultural labor force (1), distance to nearest road (2), and distance to nearest urban centre (1). Densities were based on district-level census data, a town distribution map and national yearly population growth rates derived from the FAOSTAT database (FAO, 1998), and other international data bases (US Census Bureau, 1998; USDA, 1998). Yearly growth rates were used to calculate annual population increases, which were spatially distributed according to census data and the location of towns.

[2] In this case areas of individual crops are considered as major kinds of land use since biological, socio-economic as well as technological conditions are not further specified.

4. Six policy related variables: presence and age of settlements (4) created by the Institute for Agricultural Development (IDA; *Instituto de Desarrollo Agropecuario*) and the presence of national parks or nature reserves (2). IDA is a government institute that bought considerable tracts of land mainly between 1960 and 1990 in order to stimulate smallholder development (see Chapter 2). About 50% of all IDA settlements in Costa Rica are located in the AZ (CCE, 1990).

Spatial and temporal dynamics

Three aggregation levels were created in order to examine the effects of scale. The basic level was set at a pixel size of 2.0 x 2.0 km, while aggregations were made to 3.7 x 3.7 km and 7.5 x 7.5 km, the latter being the basic grid cell size used for the analysis of the entire country by Veldkamp and Fresco (1996b).

The statistical relationships are important factors in the allocation module. Although mostly in an adapted form, relations derived from the 1984 situation are used until the arbitrarily chosen year 2005. It is thus important to analyze whether spatial relationships between components of land use change over time. If they do, those changes have to be accounted for in the allocation procedure. To check if the relationships describing land use changed between 1984 and 1992, an additional subset, covering the same area as the 1992 map, was created from the 1984 data. Only part of the area was characterized, but a direct comparison between 1984 and 1992 was enabled.

Interpretation pitfalls

One of the pitfalls of the interpretation of a multiple regression analysis is caused by the multicollinearity between variables. Although a stepwise regression procedure was followed to minimize multicollinearity effects, great care should be taken with interpretation. The ordinary least squares (OLS) estimation procedure requires independent explanatory variables for unbiased coefficient estimation. However, such independence can not be expected in our data set due to the complexity of land use and its drivers. Various obvious and less obvious interdependencies exist between explanatory variables. On higher aggregation levels, correlations between almost all potential drivers increase, thus increasing multicollinearity effects. Every analysis that was made and every statistical model that was eventually used in the allocation procedure, was carefully scrutinized for multicollinearity problems. The effects of multicollinearity were minimized by excluding highly correlated variables from the initial set of possible drivers. The decision to exclude certain variables was based on an analysis of (two by two) correlation matrices. Despite these efforts, individual statistical models remain hard to interpret directly. Multicollinearity effects as well as the lack of explanatory power of multiple regression analysis weakened the validity of statements about a single regression equation. It was therefore decided to group the results of several statistical models in order to strengthen interpretation. Specific cases will be used to illustrate changes between years and different levels of aggregation. An important question to be answered is whether the different categories of variables contributed equally to the regression models on different scales and in different years. To test the hypothesis that an integrated approach is necessary (*i.e.,* different types of variables are important drivers), the share of different categories of variables in the total set of input variables was compared with the share in the different models. As an additional test, comparisons were made between the input percentages and the most important variables in the correlation matrices.

3.2.2 Development of the demand module

The basic assumption underlying our estimates of the yearly supply in the AZ was that, after accounting for imports and exports, supply equals demand. Total demand for a specific commodity is assumed to consist of eight elements: food, feed, seed, stock changes, waste, processing, exports, and imports (FAO, 1998). We concentrated on modeling food demand, including its influence on imports and exports. Changes in the remaining categories were analyzed using national data, and certain assumptions regarding future changes were made. In this way processes, like the increasing of feed demand for maize and the decreasing of waste losses for banana, were recognized and accounted for. Feed, seed, waste, processing and stock changes accounted for at most 30% of total demand (supply).

To estimate future demand for a food commodity one needs to know present population size, demand per capita, income and price elasticities, along with the expected relative population, income and price changes. Since no unambiguous price change relationships could be established, it was decided to assume constant prices. Using the foregoing data, the following relationships were used to estimate future demand, where the suffix c stands for the type of commodity, and y for year:

$$\text{Food\_demand}_{cy} = \text{Population}_{y} \times \text{Demand\_per\_capita}_{cy} \quad (1)$$

$$\text{Population}_{y} = \text{Population}_{y-1} \times (1 + \text{Popgrow}_{y}) \quad (2)$$

$$\text{Demand\_per\_capita}_{cy} = (1 + [\text{Relative\_income\_change}_{y} \times \text{Income\_elasticity}_{c} + \text{Relative\_price\_change}_{cy} \times \text{Price\_elasticity}_{c}]) \times \text{Demand\_per\_capita}_{cy-1} \quad (3)$$

A division was made between export commodities (banana and palm heart) and commodities that are consumed primarily within Costa Rica (annuals, milk and beef)[3]. Changes in income, price, and population in the European Union (EU) and the United States of America (US) influence the demand for export crops, while changes within Costa Rica are more important for the demand for commodities of the second group.

Population

Population data for 1984 were obtained from the population census of that year (DGEC, 1987b). Low, medium, and high population growth scenarios exist (Heilig, 1996; UN, 1997; US Census Bureau, 1998; USDA, 1998) in which population growth for

[3] The major kind of land use "pasture" was subdivided in the commodities "beef" and "cow milk" for demand purposes.

Central America between 2000 and 2025 ranges between 1.5% per year and 2.1% per year. Annual population growth rates of 1.8% for Costa Rica and 0.4% for the EU and US were assumed, based on yearly data (US Census Bureau, 1998) and in accordance with the medium variant of the UN (UN, 1997).

Income

Geurts *et al.* (1997) provided estimates for income elasticities of demand by income quartile for all the major agricultural commodities in Costa Rica except banana. For the latter, the relevant income elasticity estimate was based on Hallam (1995). Estimates of income changes were based on future Gross Domestic Production (GDP) projections. The effect of changes in income on demand was derived by assuming that GDP growth can be directly translated into income growth and by adopting an average income elasticity. Two growth scenarios were distinguished, a medium GDP growth of 3% per year and a high GDP growth of 5% per year. Income growth per capita per year is derived by correcting for population growth. Per capita income growth was subsequently multiplied by the income elasticity to obtain the effect of income on demand per capita.

Imports and exports

In this study population growth as well as income changes are assumed to influence demand. Import and export quantities determine whether supply will come from the AZ or from elsewhere, or, in the case of export-oriented crops, whether demand for products from the AZ will be altered by external factors. Recent changes in the banana import policy of the EU, for instance, significantly influenced supply from the AZ (Hallam, 1995). A second well understood phenomenon is the virtual disappearance of annuals from the AZ as a result of the various structural adjustment programs implemented in Costa Rica since 1987 (see Chapter 2). Estimates of the importance of imports and exports are based on the response of commodity supply to similar events in the recent past, or on statements provided by USDA (USDA, 1998). Estimates of the percentage changes in imports and exports, population growth, and income are given in Table 3.1.

3.2.3 Scenario formulation

Two types of scenarios were formulated: scenarios that influence the total area to be allocated to production based on developments in demand (demand controlled scenarios), and a scenario that restricts allocation possibilities (controlled allocation scenarios). Scenarios were calculated for the period 1997 until 2005. Palm heart was not included in the analysis, as quantitative information was lacking. Instead, possible future changes were either based on the response of the area of banana cultivation, or on expert guesses about likely future developments in demand.

Table 3.1. Effects of changes in population, GDP, and import/export on the demand for products from the Atlantic Zone (percentages per year).

	Population growth	*GDP growth*		*Import/export*	
		Medium	*High*	*Decrease*	*Increase*
Annuals	1.8[1]	0.3	0.7	-35	-35
Banana	0.4[1]	0.7	1.6	-2.5	4
Milk	1.8	0.5	1.1	0	0
Beef	1.8	0[2]	0[2]	0	-2
Palmheart[3]	0.4	1.0	2.0	0	4

[1] Demand for commodities produced for the national market is influenced by Costa Rican population growth (1.8% y^{-1}); demand of export oriented commodities is governed by the average population growth of the EU and US (0.4% y^{-1}).

[2] Even though meat demand has been estimated as quite income-elastic (Geurts *et al.*, 1997), the replacement of beef by chicken meat is not incorporated in that study. Total meat demand increase in recent years was accounted for entirely by increased chicken meat consumption. It is not likely that similar income effects (1% and 2.1% for medium and high GDP growth, respectively) will be realized for beef.

[3] Quantitative information on demand for palm heart is lacking. It was assumed that its responses resemble the demand for banana.

Demand scenario formulation

Based on studies dealing with past changes in Costa Rican land use (*e.g.*: Easterly *et al.*, 1997; May and Bonilla, 1997; Fields, 1988), three macroeconomic policy scenarios were defined, each of which had different effects on GDP and on imports and exports, and all three were used as input scenarios in the CLUE allocation module:

1. **Base scenario.** Medium GDP growth and import/export growth similar to the recent past. The main driving force is the increasing demand of a growing population.
2. **Market protection scenario.** The EU will continue to favor ACP countries, a practice that slows banana exports from Costa Rica, while imports from the US are impeded as well. It is postulated that the decrease in the supply of annuals from the AZ continues. GDP growth remains at the medium level.
3. **Market liberalization scenario.** The EU will completely open its market, boosting banana (as well as palm heart) exports, while staple crops and beef will increasingly be imported from the US. Income growth will be higher than in the previous two scenarios.

The developments in demand for commodities from the Atlantic Zone according to the three scenarios is summarized in Table 3.2. Numbers are derived by adding up the separate growth rates given in Table 3.1. The effect of export/import changes is noticeable in banana and palm heart. Demand for milk is constant, while beef demand will decrease if markets are opened further. Annuals will disappear both when markets are more protected and when markets are more open.

Table 3.2. Developments in demand for commodities from the Atlantic Zone according to three scenarios. Percentage change per year between 1997 and 2005.

Scenario:	Annuals	Banana	Milk	Beef	Palm heart
Base	2.1	1.1	2.3	1.8	1.1
Protection	-32.9	-1.4	2.3	1.8	1.1
Liberalization	-32.9	6	2.9	-0.2	6

The effects of the three scenarios on the areas dominated by the major kinds of land use are presented in Table 3.3. Prior to 1996 supply in the AZ is known and is translated directly into an area estimate. Absolute changes are largest for pasture and forest. The area of banana cultivation grows very fast in the liberalization scenario. Despite these changes, pasture will continue to be the dominant land cover and forest the second largest land cover.

Table 3.3. Quantitative effects of three scenarios on the area containing the major kinds of land use in the Atlantic Zone (in ha). Differences indicate total growth between 1996 and 2005 in ha.

	Annuals	Banana	Pasture	Palm heart	Forest	Rest[1]
1996 situation	484	51 330	243 298	2000[2]	123 369	65 496
2005 base	891	54 962	262 701	4700	92 727	69 996
2005 protection	31	44 985	262 701	2700	110 064	65 496
2005 liberalization	401	80 261	223 766	6500	109 953	65 496
Difference Base	407	3632	19 403	2700	-30 642	4500
Difference Protection	-453	-6345	19 403	700	-13 305	0
Difference Liberalization	-83	28 931	-19 532	4500	-13 416	0

[1] The rest group is composed of other plantations, urban areas, lakes, and a large area of secondary vegetation.
[2] Based on an extrapolation of the 1992 situation.

Allocation controlled scenario

There is one important way in which allocation of land for agricultural use in the near future could be restricted. A large part of the AZ is protected land and, although only 5% of the AZ is considered to be national park, a further 25% has some kind of protected status (Stoorvogel and Eppink, 1995). Maintaining 30% of the area under complete protection in the future will have large consequences for land use. Full protection of any area would be difficult to realize, considering the land involved and the means available. This point is confirmed by the fact that, despite their status and despite all efforts undertaken, every year a part of the area within the national parks is deforested (Ramirez and Madonado, 1988). Nevertheless, international aid and/or pressure might change this situation. Thus, the effect of the total protection of national parks and the partial protection of other areas was examined in an adapted version of the base scenario.

3.2.4 CLUE allocation module

A detailed description of the allocation module is given by Verburg *et al.* (1999c) and will not be repeated here. The existing allocation module was adapted and parameterized for the AZ, and run for the four scenarios described above. For the first period (1984 - 1996), existing data on regional supply from the AZ were used, while for the second period (1997 - 2005) data as calculated above were incorporated.

3.2.5 Validation

A statistical analysis was conducted for a subset of the 1984 land use map that only included the part for which 1992 data were available. Area changes between 1984 and 1992 for every major kind of land use were calculated by subtracting the two land cover maps. The small time gap of eight years was not considered a limitation as, in this period, changes were extremely rapid. Next, the parameterized CLUE was run, starting in 1984, which resulted in a prediction for the land cover situation in 1992. Correlation between the actual and predicted map was analyzed by aggregating major kinds of land use to administrative unit averages. The performance of these aggregated results was examined as we were interested in capturing patterns of change and not detailed grid to grid changes.

3.3 Results

3.3.1 Statistical analysis

Model performance

The performance of the multiple regression models was acceptable to very good (Figure 3.2). Adjusted coefficients of determination ($\overline{R}^2$) within one major kind of land use invariably increased with aggregation level. For the most important major kinds of land use, $\overline{R}^2$ ranged between 0.25 for annuals at level 1 (2 km grid) to 0.92 for forest at level 3 (7.5 km grid). The spatial distribution of forest and pasture, together accounting for more than 80% of the total cover, was especially well described. The models for individual annuals were less satisfying, but the cover of all annuals amounts to no more than 8% of the total area.

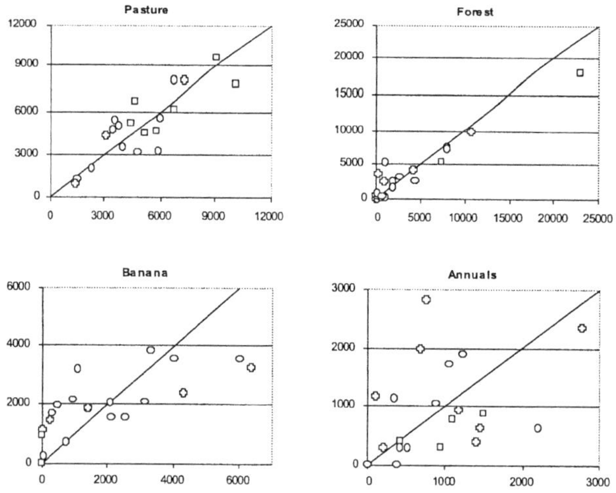

Figure 3.2. Coefficients of determination for three aggregation levels and major kinds of land use.

Variable importance

Complete regression results for level 1 are given in Appendix 3.2 and for level 3 in Appendix 3.3. The relative contribution of the various groups of driving forces in the multiple regression analyses and correlation analyses is summarized in Tables 3.4 and 3.5. The correlation analyses between major kinds of land use and driving forces were summarized by adding up the five highest correlating variables (always > 30%) for every analysis of any given year, aggregation level, or major kind of land use. The proportion of every variable group in the sum was subsequently divided by the proportion of the variable group in the input data. Multiple regression models were summarized in a similar manner, except that all variables significantly contributing ($P < 0.05$) were included in the sum of variables. In Table 3.4 differences between years and aggregation levels are analyzed; in Table 3.5, differences among major kinds of land use are examined. The plus/minus sign indicates whether a group of variables contributed more or less than expected based on the input proportion. For instance, when counting contributing variables of all multiple regression equations in 1984, 25% were policy variables, although only 15% (6 out of 40) of the input variables was policy oriented. A plus sign is thus placed in the appropriate section.

The results from the correlation analysis indicate that at least a number of variables from all groups correlate strongly with the major kinds of land use. This suggests that all groups have a potential importance, a potential that is confirmed by the summarized results from the multiple regression analysis. It can therefore be generally concluded that the distribution of land use can not be analyzed successfully without including various types of potential driving forces. Considering the major kinds of land use separately (Table 3.5), the population variables hardly contributed to explain the distribution of pasture, banana and annuals. Biophysical variables (stoniness, flooding), soil variables (fertility) or policy variables (parks, IDA settlements) had a greater influence.

Table 3.4. Relative contribution of the main groups of driving forces to the highest correlating variables and multiple regression models; correlation analysis and multiple regression models, grouped by scale level or year and independent from major kind of land use.

	Correlation Analysis				*Multiple Regression Analysis*			
Variables	*1984*	*1992*	*L1*	*L3*	*1984*	*1992*	*L1*	*L3*
Soil	-	0	-	0	0	0	0	0
Other bio-geophysical	0	+	+	0	0	0	0	0
Demographic	+	+	+	+	-	-	-	-
Policy	+	-	0	0	+	0	+	+

Table 3.5. Relative contribution of the main groups of driving forces to the highest correlating variables and multiple regression models; multiple regression models, grouped by major kind of land use and independent from scale level and year, unless otherwise noted.

Variables	*Pasture*	*Forest in 1984*	*Forest in 1992*	*Banana*	*Annuals in 1984*	*Annuals in 1992*
Soil	-	-	0	+	0	0
Other bio-geophysical	+	-	-	-	0	+
Demographic	-	+	0	-	-	0
Policy	+	+	-	0	+	-

Note: Plus signs indicate that a particular group of variables contributed more than expected based on the input percentage; minus signs indicate a less than proportional contribution; zeros indicate less than 10% deviation. For correlations, the five most explanatory variables were included; for multiple regressions, all significant variables were included ($P < 0.05$). L1 = level 1, grid size 2 km; L3 = level 3, grid size 7.5 km.

Temporal dynamics

Considering the multiple regression models, soil and biophysical variables tended to gain in importance between 1984 and 1992, while the importance of policy-related variables was diminished. This tendency could mainly be attributed to an interchange of highly correlated variables, *e.g.*, the presence of national parks and poor soil conditions. There was, however, one driving force and one major kind of land use that consistently changed between 1984 and 1992: the presence of IDA settlements (driving force) and annuals (major kind of land use). The percentage of a grid cell within recently (in 1980-1984) established IDA settlements lost significance between 1984 and 1992. The loss of importance is consistent with the land use changes that are expected to happen after the foundation of a settlement (Stoorvogel, 1995; Schipper, 1996): an initial dominance by annuals will gradually disappear and be replaced by a more diverse pattern in which differences from the area outside a settlement become less obvious. Of the most important major kinds of land use (banana, forest, pasture, annuals), regression models had the lowest coefficients of determination for annuals. In 1984, IDA settlements and soil conditions were the main factors, but the former lost importance and the models for 1992 were consequently less satisfying. The factors explaining this change are probably the diversity of the major types of land use and the fact that the area they occupied decreased drastically between 1984 and 1992. Phasing out of the subsidy system for basic grains after 1987 (Hansen-Kuhn, 1993, see Chapter 2) triggered almost complete disappearance of maize and rice.

Spatial dynamics

Differences in the contributing factors operating on different aggregation levels were small (Table 3.4). Typically the same factors were important at level 1 and level 3. However, regression coefficients belonging to the factors did change with scale of analysis. Often consistent changes were present from level 1 to level 3, thus indicating the presence of a scale effect. Whether factors gained or lost importance varied greatly among the major kinds of land use.

Individual cases

Figure 3.3 depicts the relationship between actual cover and cover predicted by the multiple regression equations applied to level 3 in 1984 for the most important kinds of land use. It shows that, on this aggregation level, the relationships between land use and its correlating variables were captured in a satisfactory manner. In Table 3.6 the main contributing variables in the multiple regression models applied to level 1 in 1984 are listed for the most important kinds of land use.

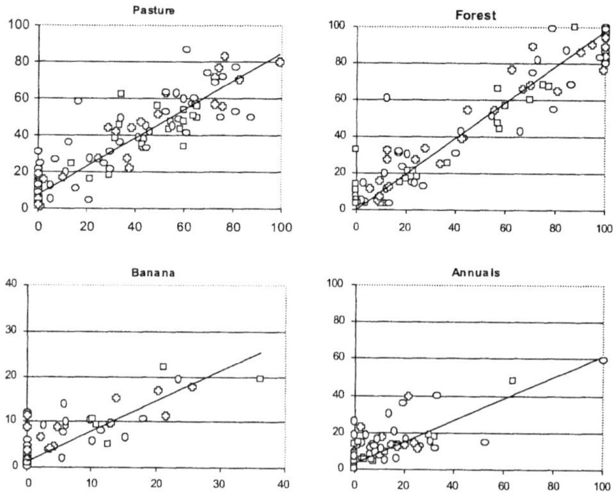

Figure 3.3. Relationship between actual cover (X-axis, % per grid cell) and cover calculated by multiple regression techniques (Y-axis, % per grid cell) for the major kinds of land use on level 3 in 1984. Lines indicate best linear relationship.

Table 3.6. Most important variables in the level 1 multiple regression models in 1984 for selected major kinds of land use, grouped by variable type.

Variable group	*Forest*	*Pasture*	*Annuals*	*Banana*
Soil	Shallow Poorly drained Sandy	Deep	Low fertility Leached but not: Poorly drained Shallow	Well drained High fertility but not: Leached Shallow
Other bio-geophysical	Steep slopes High altitude High rainfall	Steep slopes Stony surface		Low flooding risk Gentle slopes Higher altitude
Demographic		Near main roads	Higher rural density	
Policy	In natural parks In nature reserves	Outside national parks Inside old IDA settlements	Inside young and medium old IDA settlements	Not included in IDA settlements

Forest is still one of the dominant vegetation types in the AZ, despite the ongoing process of deforestation and agricultural expansion. Factors that explain the current distribution of forest clearly show the effects of agricultural expansion (as described by *e.g.*, Sader and Joyce, 1988). Forest is restricted to poor soils (shallow and poorly drained) in the national parks or to biophysically unfavorable conditions in mountainous areas (high rainfall, steep slopes). To some extent, a similar pattern was found in the distribution of annuals. Their presence correlated with leached soils of low fertility. However, poorly drained soils that are very shallow were apparently unsuitable for annuals. Farmers in the recently founded IDA settlements cultivated more annuals than those farming outside of them. The opposite could be observed in the case of banana. This crop requires well drained, highly fertile soils that are not leached and not too shallow (Delvaux, 1995). Flooding risk has to be low and the crop is restricted to flat or almost flat areas. Altitude contributed positively to the cover of banana, although the single correlation was a negative one. This points to the fact that *within* the area of gentle slopes with fertile soils, the higher areas (10 — 100 m) are preferred. The contributing biophysical factors display some characteristics of the distribution of pasture in the AZ: on steep slopes with a stony surface. The positive contribution of deep soils may seem illogical, but deep soils in the area coincide with the areas which are permanently flooded, making any other agricultural land use improbable.

Summarizing, multiple regression models seem to be able to adequately describe the land use distribution as it is known to exist in the AZ. Forest was largely restricted to the poor bio-geophysical conditions, coinciding with the presence of national parks. Banana plantations have claimed the best soils in flat, non-flooded areas. Annuals were thus restricted to the somewhat poorer circumstances, while pasture remained in the steep, stony, wet areas. It is assumed that no new driving forces of any great importance will appear in the future. It is thus justified to use the results of the multiple regression models in the allocation module of CLUE.

As no data were available for palm heart, an expert equation was constructed. Though only qualitative information on its current distribution and driving forces was available, an effort was made to quantify relations based on expert knowledge. The presence of palm heart is assumed to correlate with soil drainage, surface stoniness, flooding risk, altitude, and distance to the nearest market. A summarizing variable (PALMSUIT) was constructed with values ranging from 0 (all five variables unsuitable) to 4 (all suitable). The regression coefficient was set at 0.1 and the intercept at 40 (0.4%).

Regression hot spots

Based on the statistical analysis, so-called regression hot spot maps were constructed by subtracting the observed cover of 1984 from the predicted regression cover for the same year. The basic assumption is that competition is an important factor, that is not incorporated into the regression analysis. Residuals thus can be regarded as a measure of effects of competition. Examining regression hot spot maps, it is possible to get an idea of the potential changes that could occur, before running the allocation module. Values deviating from zero, *i.e.,* where actual and predicted cover differ, indicate areas where changes are most likely to take place. In Figure 3.4, the regression hot spot maps for banana and forest are given. Comparing both maps, it can be concluded that the potential dynamics for banana are much more geographically concentrated than forest dynamics. The pattern resulting from the allocation module might differ considerably from regression hot spots, because neither competition among major kinds of land use nor changes in the value of driving forces and total supply are taken into account.

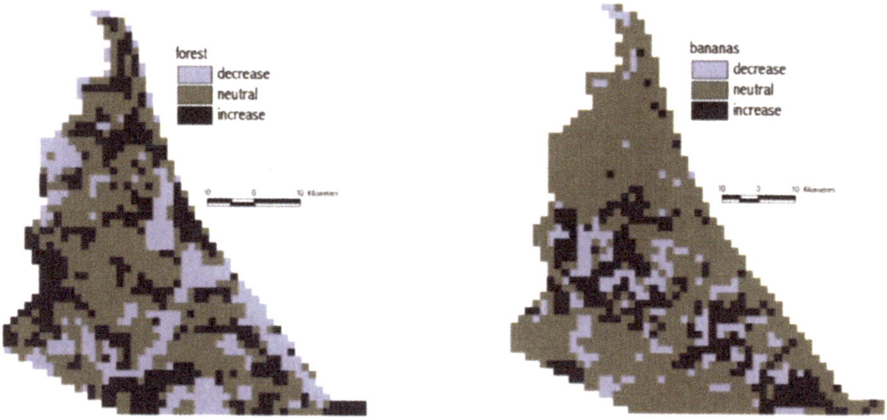

Figure 3.4. Regression hot spots maps for bananas (left) and forest (right). Green areas indicate areas where regression values exceed actual values and changes are to be expected. The opposite holds for red areas. In grey areas values are similar.

3.3.2 Allocation under given scenarios

The results of the base scenario for major kinds of land use are given in Figure 3.5, including a comparison with the initial 1984 situation.

Bottom two rows: Input data and results of base scenario. Left side: situation in 1984; right side: situation after 21 year (2005). Top: Annuals; bottom: palm heart. Palm heart did not exist in 1984.

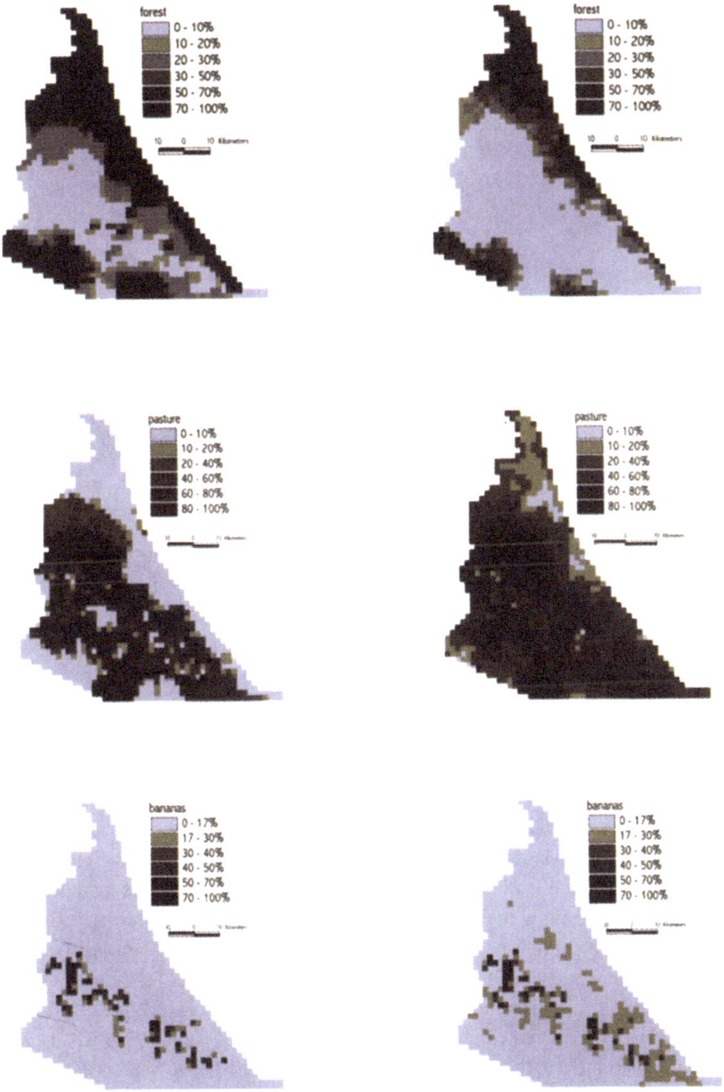

Figure 3.5. Input data and results of the base scenario. Left side: situation in 1984; right side: situation after 21 year (2005). Top: forest; middle: pasture; bottom: bananas. Light colors indicate a relatively low cover percentage; dark colors indicate a high cover percentage.

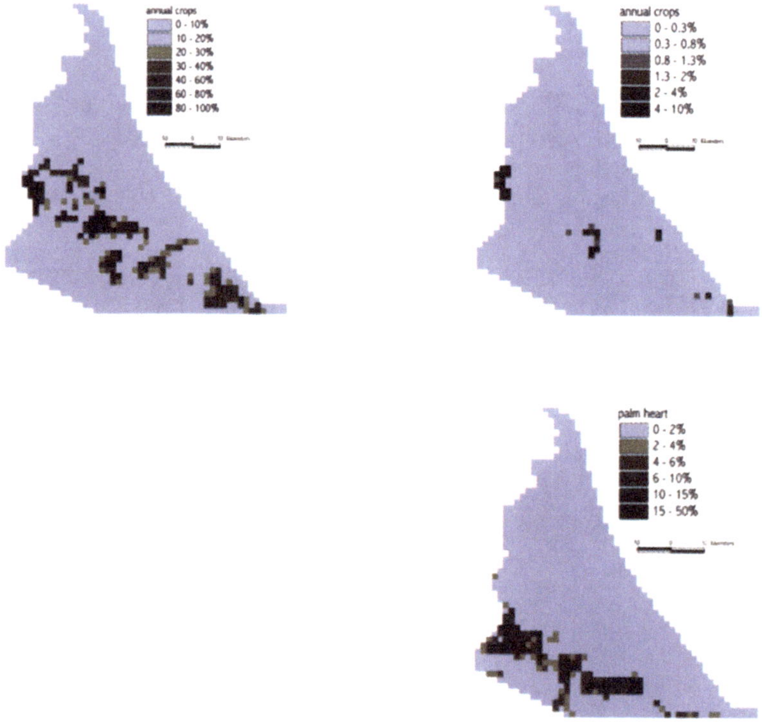

Figure 3.5 (continued). Input data and results of base scenario. Left side: situation in 1984; right side: situation after 21 years (2005). Top: annuals; bottom: palm heart. Palm heart did not exist in 1984. Light colors indicate a relatively low cover percentage; dark colors indicate a high cover percentage.

Base scenario

In the base scenario, forest will become restricted entirely to remote areas and to areas unsuitable for agriculture. Demand for agricultural land in potentially suitable areas for forest reduced natural cover in large areas to almost zero. Noteworthy is the persistence of forest in areas within the boundaries of national parks, even though these areas do not have a protected status in the base scenario. Pasture increasingly becomes the dominant land use in the entire zone. Besides some isolated pixels containing banana plantations and the already mentioned areas of forest, pasture becomes the dominant cover everywhere. Expansion of banana plantations will predominantly occur near areas where banana cultivation was already present. In particular new plantations are modeled close to Limón (in the South-eastern corner of the study area). The location of annual crops will be restricted to a few well-defined areas. Those areas do not necessarily correspond to the original 1984 areas. Palm heart was not present in the zone in 1984, but its projected presence is in the area just south of the main banana zone in the slightly higher locations.

Competition among major kinds of land use is an important factor in the allocation procedure. A comparison between Figure 3.4 (regression hot spots) and Figure 3.5 (changes between 1984 and 2005) illustrates that patterns can differ considerably. Forest is pushed away completely by agricultural land use. The matching patterns between regression hot spots and changes in the banana area indicate that in general banana plantations are strong competitors. The same holds for pasture (not shown). The process of deforestation and replacement of forest by (mainly) pasture was modeled accurately with CLUE. Figure 3.6 shows the strong correlation between the disappearance of forest and the increase of pasture. This observation confirms expert knowledge and is in agreement with conclusions drawn by Schipper *et al.* (1998).

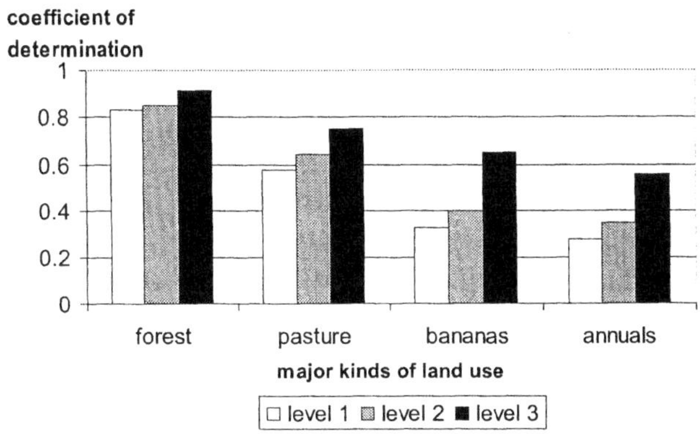

Figure 3.6. Relationship between the disappearance of forest and the increase of pasture area between 1984 and 1992, aggregated per administrative unit.

Other scenarios

Differences in land use patterns between the three demand-controlled scenarios (base, protection, and liberalization) were small. Despite large differences in changes in total area modeled in the market protection and market liberalization scenarios (Table 3.3), the overall patterns are much alike in terms of the major kinds of land use. Forest is consistently pushed out of agriculturally favorable areas and becomes restricted to the northern part of the region. Pasture increases in various places but predominantly in areas with high deforestation rates, and banana is present in the same areas in all scenarios. The location of areas with annual crops differed significantly among the various scenarios. In Figure 3.7, areas with annuals in the liberalization and base scenarios almost completely exclude each other. Again, competition with other major kinds of land use is the reason for this difference. Annuals have a preference for the same areas as banana and the stronger push of banana in the liberalization scenario causes the annuals to disappear completely from that area.

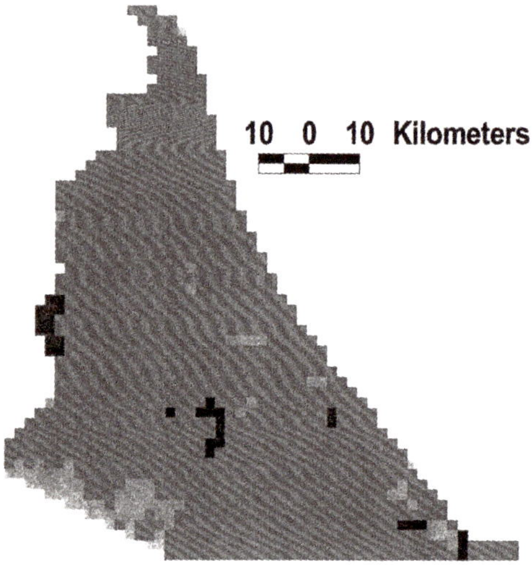

Figure 3.7. Differences in the allocation of annuals between the liberalization and base scenarios. Dark areas indicate the location of annuals in the base scenario; light gray areas indicate their location in the liberalization scenario.

Park protection scenario

Restricting land use allocation by protecting parks considerably influences land use patterns. Figure 3.8 shows the distribution of forest with and without the protection of national parks. Examining national aggregated numbers (Table 3.3), the pressure on forest outside parks is enhanced almost up to the point of complete disappearance. Only in the most remote area close to the border with Nicaragua does forest remain in reasonable quantities; the remainder of the AZ is almost exclusively devoted to agricultural uses.

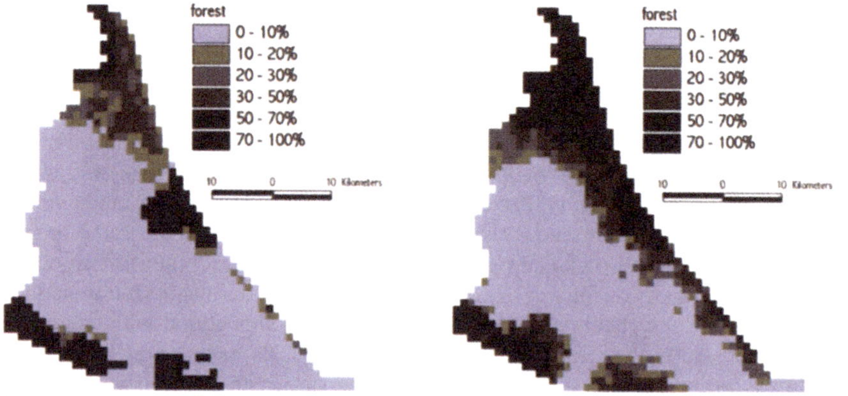

Figure 3.8. Distribution of forest cover in the liberalization scenario with (left) and without (right) protection of national parks, after 21 years of simulation. Light colors indicate a relatively low cover percentage; dark colors indicate a high cover percentage.

3.3.3 Validation

Figure 3.9 shows the aggregated results of the validation run of CLUE. CLUE proved to be capable of modeling land use patterns in 1992 based on the 1984 situation. The locations of pasture and forest were especially well modeled, with a correlation coefficient between actual and predicted land use of 0.87 and 0.95, respectively. For banana, the corresponding correlation coefficient was 0.74. The modeling of the location of annual crops was less successful, as reflected in a correlation coefficient of 0.36. For forest, pasture and annuals, the regression line between actual and predicted land use did not differ significantly from the 1:1 relationship. For banana, çover was overestimated in districts with little banana cultivation, and underestimated in districts in which a lot of banana was grown. From Figure 3.10, it can be concluded that the majority of the new plantations are located in or near areas where increases were modeled. However, the largest decreases in banana area were predicted incorrectly in areas where banana cover was close to 100% in both 1984 and 1992.

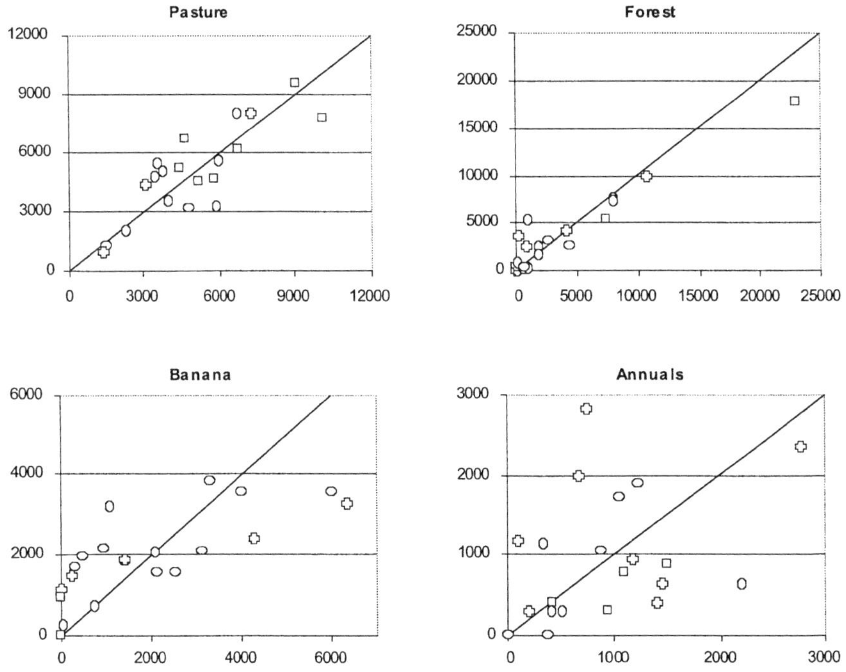

Figure 3.9. Modeled versus the actual cover in 1992 of major kinds of land use, aggregated per administrative unit. X axis: actual cover (ha); Y axis: modeled cover (ha). Lines are the 1:1 relationship (y = x).

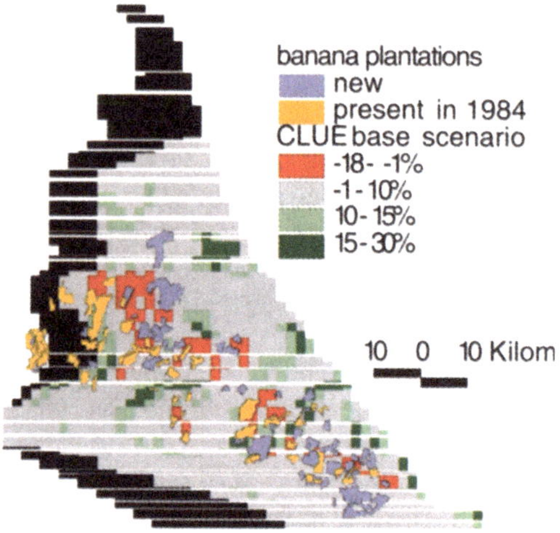

Figure 3.10. Comparison between the modeled increases in banana cover according to the base scenario (cover in 1992 minus cover in 1984) and the location of banana plantations as observed in 1984 and 1992 (based on Stoorvogel and Eppink, 1995).

3.4 Conclusions and discussion

3.4.1 General model performance

The spatial pattern of land use in the northern Atlantic Zone of Costa Rica was described in a satisfactory way by using the statistical techniques of the CLUE framework. Both socio-economic and bio-geophysical variables proved necessary to describe land use patterns. Using the results of this analysis it was possible to satisfactorily simulate land use pattern dynamics over a period of 20 years. Resulting maps showed patterns that could be explained with processes that are known to take place in the region. The competition for land among banana, pasture and forest was especially well modeled. The results indicate that the CLUE modeling framework can be successfully used on sub-national levels. Comparisons with the land use study based on linear programming techniques presented in Chapter 6 reveals interesting similarities. Under a similar scenario of excluding and including protection of national parks, it was concluded that extending the agricultural area into existing protected areas brings only marginal benefits, a point that is confirmed by this study (see Figure 3.8). A more elaborate comparison between the CLUE methodology and optimization techniques is given by Bessembinder (1997).

3.4.2 Model benefits and limitations

Two advantages of the CLUE methodology are its speed and data requirements. Although a relatively high amount of data is needed, data required to run CLUE are readily available in many parts of the world. Time-consuming and costly field work is not necessary, thus considerably accelerating the modeling procedure, which should normally take no longer than 4-6 months.

One of the main points of criticism against statistical modeling is its lack of explanatory power. Obviously, it is necessary to evaluate empirical relationships with expert knowledge of the area. Given the available amount of expert knowledge of land use and land use change dynamics in the AZ, the step from a statistical analysis to process-based rules is relatively small. It is not the *lack of explanatory power* but rather the *lack of knowledge* that limits the application possibilities of a statistical analysis. Even though by no means we wish to propagate the careless use of statistical relationships, we do want to stress the application value of statistical models, especially in combination with models that use process-based relationships. A true limitation of the CLUE modeling framework is its short time horizon, as it depends on a description of the (recent) past. Although a separate analysis of a second year enables a check on the stability of the relationships between land use and its driving forces estimated on the basis of the first year, uncertainty about future changes remains. This uncertainty is increased by incorporating economic variables like price changes or GDP growth. The current version of the CLUE modeling framework can therefore probably not be used for making extrapolations for more than 15 - 20 years into the future.

The observed differences between the modeled and the actual location of banana plantations in the AZ touch upon another limitation of the CLUE allocation procedure. The allocation algorithm was not designed to deal with extremely "clumpy" cover patterns. An inevitable effect of the large-scale production of banana in the AZ is that part of the production is located on less suitable soils, while better biophysical circumstances are present elsewhere in the region. Simulation of changes in allocation using CLUE will attempt to "correct" those apparent mis-allocations. Cells with high banana covers will be reduced and the remaining area of banana cultivation will be allocated in cells with similarly suitable biophysical conditions. The procedure works very well for covers that are omnipresent, like pasture, or for covers of which the pattern was well described by the statistical equations, like forest. Even though patterns of more local land uses (see Figure 3.10) might be captured reasonably, estimated cover percentages might not correspond to actually observed percentages.

3.4.3 Issues of scale

The present study has demonstrated how multiple scales can be incorporated in a quantitative modeling approach. Although the statistical analysis did not demonstrate dramatic differences between aggregation levels, the CLUE modeling framework offers tools to deal with multi-scale data sets. Not only can local variability in soil properties and the influences of a city or road be captured, but long term macro-economic changes and influences of the policies of other countries can be accounted for as well. Top-down modeling ensures that, on every scale level, variability from the higher levels is included in the lower ones, while bottom-up feedbacks in turn ensure that local constraints are not considered less important than global economic changes.

Comparing the CLUE model for the AZ presented here with the CLUE version for the whole of Costa Rica (Veldkamp and Fresco, 1996b) would allow interpretation over a still wider range of scales from 2 x 2 km to 45 x 45 km. A preliminary comparison between the present study and the national level study of Veldkamp and Fresco revealed similarities in the patterns of forest but differences in the case of pasture.

3.4.4 Application domain

The aim of the CLUE modeling framework is to understand relationships between the environment and land use, and the general or recurrent changes that are observed. CLUE demonstrates the spatially explicit effects of important processes: *e.g.*, the effects of accelerated deforestation, and the restriction of annual crops to poor soils by the strongly competitive banana plantations are convincingly shown in the presented maps. By analyzing countries with different geographical resources and in different phases of the process of land use change, an attempt can be made to draw conclusions on issues of global importance.

Finally, erosion, nutrient fluxes, changes in carbon pools, and loss of biodiversity are just a few of the possible effects that land use or land-cover change may have. Comprehension of the relationships between drivers and land use-change and the scale on which they operate could prove a powerful tool in a range of scientific fields. Past applications of CLUE proved the feasibility of linking its results with the modeling of nutrient balances (De Koning *et al.*, 1997).

Appendix 3.1 Independent variables used as input in the statistical analysis

Parameter abbreviation	Type	Description	Unit
DALF	Demographic	Average agricultural labor force	persons ha^{-1}
DRUR	Demographic	Average rural population density (districts)	persons ha^{-1}
DISDRUR	Demographic	Average rural population density (towns)	persons ha^{-1}
URBDEN	Demographic	Average urban population density	persons ha^{-1}
URBDIS	Demographic	Average distance to nearest city	m
DSROAD	Demographic	Average distance to nearest primary or secondary road	m
BSATHI	Soil	Soils with high base saturation; > 50% (between 25 and 100 cm)	%
CECHIGH	Soil	High cation-exchange capacity; > 24 meq/100 g soil	%
CECLOW	Soil	Low cation-exchange capacity; ≤ 16 meq/100 g soil	%
DEPDEEP	Soil	Deep soils; > 125 cm	%
DEPSHAL	Soil	Shallow soils; 0-25 cm	%
DEVWELL	Soil	Well developed and moderately leached soils	%
DEVNO	Soil	Not or slightly developed soils	%

Appendix 3.1 Continued

Parameter abbreviation	Type	Description	Unit
DRAIGOOD	Soil	Well drained soils	%
DRAIBAD	Soil	Very poorly or poorly drained soils	%
ECECHI	Soil	High acidity; > 2 meq/100 g soil and >2 meq KCl extraction Al+H/100 g soil	%
ECECLOW	Soil	Low acidity; < 2 meq/100 g soil	%
FERTHI	Soil	Fertile soils	%
FERTLOW	Soil	Infertile soils	%
PHHIGH	Soil	High pH; pH H_2O > 5.5	%
PHLOW	Soil	Low pH; pH H_2O < 4.5	%
TEXTCLAY	Soil	Clayey texture; SaCl, SiCl, Cl	%
TEXTSAND	Soil	Sandy texture; Sa, Lsa	%
TEXTWET	Soil	Wet texture; no texture (too wet)	%
SLOPSTEE	Other bio-geophysical	Steep slopes; steepness > 13%	%
SLOPFLAT	Other bio-geophysical	Flat area; steepness 0 - 2%	%
ALT	Other bio-geophysical	Average altitude	m
RAIN	Other bio-geophysical	Average yearly precipitation	mm
FLOODAL	Other bio-geophysical	Always flooded areas	%
FLOODLO	Other bio-geophysical	No flooding or low risk of flooding	%
STONNO	Other bio-geophysical	No stones to fairly stony on soil surface	%
STONYES	Other bio-geophysical	Stony to very stony on soil surface	%
DISRIVP	Other bio-geophysical	Average distance to nearest minor or major river	m
DISRIVS	Other bio-geophysical	Average distance to nearest river or gully	m
PARKNAT	Policy	Area within national park	%
PARKOTH	Policy	Area with other type of protective status	%
IDAYOU	Policy	Area within IDA settlement established after 1980	%
IDAMED	Policy	Area within IDA settlement established between 1970 and 1979	%
IDAOLD	Policy	Area within IDA settlement established between 1960 and 1969	%
IDANOT	Policy	Area outside IDA settlements	%

Appendix 3.2 Results of multiple regression analyses for 1984 and 1992 on aggregation level 1

Asterisks indicate significance level of individual variables: * = $P < 0.05$; ** = $P < 0.01$; *** = $P < 0.001$.

	ANN	maize	beans	rice	cassava	BAN	PAS	FOR	ANN	maize	beans	rice	cassava	BAN	PAS	FOR
driving force:	84	84	84	84	84	84	84	84	92	92	92	92	92	92	92	92
POPULATION:																
agricultural labor force	0.0080***	0.031***									-0.0010**					
rural density (districts)		-0.56***	0.0046**								0.020*					-0.39***
rural density (town + district)																
urban density																
distance to city				-0.0002***			0.00084***		-0.0004***			-0.0007***	-0.0000***			0.00096***
distance to road				0.00034***			-0.0016***					0.00026***			-0.0029**	
SOIL:																
base saturation high		-0.058***		0.033***		-0.12***						0.011**				
CEC high																0.48***
CEC low						-0.081*		-0.35***		-0.061*	-0.014*	-0.014*				-1.2***
shallow soils	-0.12***			-0.063***	-0.0049***	0.18***	-0.19***	0.27***							-0.17***	0.32***
no soil development				-0.034**		-0.043*										
well developed soils	0.13***	0.13***	0.0035**		0.018***	-0.21***			0.22***	0.18***			0.058***	-0.34***	0.17***	
bad drainage	-0.17***	-0.16***						0.38***	0.092***		0.022***		0.024***			
good drainage						0.19***								0.14***		
ECEC high																-0.68***
ECEC low																
fertility high	-0.093***	-0.080***									0.010**			0.36***		
fertility low					0.018***				0.13**	0.14***			0.036***		0.31***	-0.38***
pH high			0.0063**					-0.21***							0.62***	-0.77***
pH low									0.071**	0.046***	0.0084**		0.015***			

Appendix 3.2 Continued

driving force:	ANN 84	maize 84	beans 84	rice 84	cassava 84	BAN 84	PAS 84	FOR 84	ANN 92	maize 92	beans 92	rice 92	cassava 92	BAN 92	PAS 92	FOR 92
SOIL:																
texture clayey			-0.0046***		-0.0091***								-0.013**			
texture sandy						-0.29***		0.47***						-0.47***		1.2***
texture none																0.50***
OTHER BIO-GEOPHYSICAL:																
flat terrain	0.21***	0.21***		0.052***	0.012***	0.15***	-0.16***		0.14***	0.12***			0.022***	0.18***		-0.27***
steep slopes	-0.075**				-0.0065**			0.31***		-0.051*					-0.41***	
stony							0.30***	-0.24***	-0.15***	-0.091**			-0.043***		0.24*	-0.39***
flooding risk low	-0.06***			-0.060***			0.11***		0.078***	0.034***	0.019***		0.012***			
always flooded						-0.079***										
altitude		-0.0049***			-0.0006***	0.0043*	-0.022***	0.025***				0.0033***			0.082*	
precipitation	-0.0017*		-0.0001***				-0.011***	0.013***	0.0040**	0.0031***	0.00076**		0.00063**	-0.015***		
distance to river			-0.00011*	0.00023***	0.000030*		-0.0015***		-0.0005***		0.00003***	-0.00005**	-0.00006*	-0.00089**		0.0018***
POLICY:																
part of national park							-0.31***	0.33***		-0.050**	-0.015***		0.020***	0.12*		
part of other type of park							-0.28***	0.26***				0.014*			-0.25**	0.40***
part of recent IDA settlement	0.23***	0.19***	0.017***		0.0069***				0.13***	0.073***	0.015**			-0.17**		
part of older IDA settlement	0.16***	0.077***	0.0066***	0.056***	0.011***	-0.049*									0.14*	
part of old IDA settlement				-0.024***		-0.11***	0.22***			-0.026**	-0.0057*	0.0081***		-0.11***	0.13***	
intercept	13.8**	-4.4***	0.74***	6.5***	-0.65**	-19.8***	90.6***	-47.1***	-29.4***	-24.1***	-4.4***	1.2***	-5.0***	11.9***	29.7***	49.0
adjusted R^2	0.29	0.28	0.18	0.27	0.20	0.33	0.58	0.83	0.24	0.26	0.14	0.16	0.37	0.42	0.40	0.81

Appendix 3.3 Results of multiple regression analyses for 1984 and 1992 on aggregation level 3

Asterisks indicate significance level of individual variables: * = P < 0.05; ** = P < 0.01; *** = P < 0.001.

driving force:	ANN 84	maize 84	beans 84	rice 84	cassava 84	BAN 84	PAS 84	NAT 84	ANN 92	maize 92	beans 92	rice 92	cassava 92	BAN 92	PAS 92	NAT 92
POPULATION:																
Agricultural labor force																-0.0098*
rural density (districts)																
rural density (town + district)																
urban density																
distance to city							0.00030*						-0.00006*			0.00079***
distance to road								-0.00079*	-0.0013*		-0.00029*	0.00022*				
SOIL:																
base saturation high	-0.11*					-0.054*		-0.24**				0.032**	-0.021**	-0.39**		
CEC high						0.067**		0.32**						0.28**		0.44**
CEC low																
000shallow soils								0.75***				0.065***			-0.32***	3.2***
no soil development				-0.073***		-0.16***						-0.020*				
well developed soils									0.35***	0.24***		-0.044***	0.061***	-0.43***		0.29***
bad drainage					-0.019***	-0.13***					0.044***		0.041**			
good drainage											0.018*		0.026*			
ECEC high												0.099***				
ECEC low																
fertility high												0.031*		0.31**		2.3**
fertility low									0.46***	0.34***		0.11***	0.034**			2.7***
pH high							0.39*	-0.47***				-0.039**			0.91***	-1.1***
pH low	0.13**			0.062***	-0.010**								0.034***	0.31**		
texture clayey								0.45***								
texture sandy					-0.0078*		-0.46***	0.67***								-1.70*
texture none							-0.26*									

Appendix 3.3 Continued

driving force:	ANN 84	maize 84	beans 84	rice 84	cassava 84	BAN 84	PAS 84	NAT 84	ANN 92	maize 92	beans 92	rice 92	cassava 92	BAN 92	PAS 92	NAT 92
OTHER BIO-GEOPHYSICAL:																
flat terrain						0.27***			0.24**	0.18***	0.027**	-0.041***				
steep slopes					-0.021***				-0.26**	-0.12**						0.60***
stony							0.60***	-0.33**					-0.066***		0.60***	
flooding risk low				-0.074***			0.20***				0.018**	0.018**				
always flooded																
altitude							-0.044***	0.050***			0.0099**	-0.013***	-0.079*			-0.064*
precipitation	-0.004***				-0.00024**		-0.011***	0.011***					0.0015**			
distance to river							-0.0098**					-0.0002***	-0.00009**		0.0089**	
POLICY:																
part of national park								0.25***					-0.016**			
part of other type of park							-0.15*	0.23***								0.26*
part of recent IDA settlement	0.75***				0.019***		-0.45**				0.046***					
part of older IDA settlement				0.13*												-0.46*
part of old IDA settlement						-0.15***	0.26**								0.16*	
intercept	17.2**			5.5**	3.2***	-5.1*	80.4***	-72.6***	-38.6***	-15.7***	-4.7***	-7.4*	-8.7***	-36.3***	36.9***	-273.7***
adjusted R^2	0.56	n.s.	n.s.	0.30	0.46	0.65	0.78	0.91	0.70	0.64	0.35	0.58	0.69	0.61	0.71	0.95

[1] Here, broad classes of land use are meant, such as forest, pasture, annuals *etc*. See also the definition in the Appendix at the back of the book.

[2] In this case areas of individual crops are considered as major kinds of land use since biological, socio-economic as well as technological conditions are not further specified.

[3] The major kind of land use "pasture" was subdivided in the commodities "beef" and "cow milk" for demand purposes.

4 Spatial equilibrium modeling for evaluating inter-regional trade flows, land use and agricultural policy

PETER C. ROEBELING, HANS G.P. JANSEN,
AAD VAN TILBURG, and ROBERT A. SCHIPPER

Abstract

This chapter presents a Spatial Equilibrium Model (SEM) for analyzing the spatial patterns of agricultural supply, demand, trade and pricing in Costa Rica. The behavioral relations of producers and consumers are modeled, while simultaneously taking transaction costs and government policies into account. The SEM maps 17 major agricultural commodities across six planning regions and the rest-of-the-world, considered as a seventh region. The model is validated with 1995 data, and its results are used to assess spatial patterns of land use, trade flows and social welfare. The model's simulations show the potential effects of trade liberalization, changes in transport costs, technological progress in agriculture, and economic growth. Trade liberalization leads to increased welfare, mainly due to a rise in the consumer surplus resulting from lower import prices. Reductions in transport costs also have a positive welfare effect, as a result of increased domestic trade, more specialized regional production, and a growth in exports. Technological progress in agricultural production lowers unit production costs, and mainly favors production of export products, whose relative competitiveness is enhanced. Finally, economic growth increases domestic demand, leading to increased imports, fewer exports and a slightly enlarged domestic production. The study shows that on the basis of reliable data and sound econometric analysis, an agricultural sector model can be developed that both policy makers and research institutions could use to evaluate the effectiveness of alternative agricultural policy measures.

4.1 Introduction

This chapter is different from the other chapters in a number of respects. First, while the other chapters deal with land use issues on or below the regional level, the methodology presented here relates to the inter-regional allocation of land use. Second, the methodology used is an economic one with little involvement of other disciplines. Similar to the UNA-DLV methodology presented in Chapter 8, the methodology presented in this chapter is primarily intended to provide a means of supporting policy decisions, albeit on a different level.

Issues relating to the efficiency in the production, pricing and distribution of agricultural products are of primary importance in the development of agricultural markets and trade. An effective marketing system is one which efficiently links the various regions of surplus and deficit production within a country (and which takes foreign trade into account) in order to achieve the maximum benefit from regional comparative advantages.

Despite the availability of a considerable number of research methodologies for studying the trade of market commodities in geographically dispersed markets (*e.g.*, Samuelson, 1952; Takamaya and Judge, 1964; Takamaya and Judge, 1971), and of a clear theoretical exposition of the models of spatial equilibrium (Martin, 1981), the amount of applied research is much less.[1] This is particularly the case for developing countries where inadequate data often prevent successful development and application of Spatial Equilibrium Models (SEMs). Arguably, insights that result from spatial equilibrium analysis are even more important in developing as opposed to developed countries, given that resources in the former are more limited, making their optimal allocation particularly imperative.

For Costa Rica, previous research on the micro and meso levels in the Atlantic Zone (Jansen and van Tilburg, 1996) suggests that serious agricultural marketing deficiencies in the region result in sub-optimal marketing structure, conduct and performance. However, no information is available or has been generated on the trade flows between regions within Costa Rica. Proper analysis of inter-regional trade is important because several constraints may prevent an optimal flow between regions and, consequently, reduce consumer and producer welfare. While the size of trade flows is determined by conditions of supply (*e.g.*, costs of production) and demand (*e.g.*, purchasing power), these may be sub-optimal as a result of conditions related to transaction costs (*e.g.*, high transport costs) and government policies (*e.g.*, measures to prevent the free trade of basic staples for reasons of food security). Once bottlenecks in trade flows have been identified, measures can be taken to change particular conditions in order to increase national welfare. Therefore, it is important to design a macro-level theoretical framework in order to determine the optimal production, consumption and trading practices for the most important agricultural commodities in Costa Rica, against which current and future government policies can be evaluated.

Spatial Equilibrium Modeling is a well-known method of estimating the optimal allocation and trade flows of agricultural commodities. It examines the behavioral relations of producers and consumers, while simultaneously taking transaction costs and government policies into account. Moreover, it is a useful tool for the simulation and analysis of the short-term (1-5 years) effects of alternative policy measures on public welfare.

The objectives of this chapter are: (1) to model the actually prevailing spatial patterns of supply, demand, trade and pricing for the major agricultural commodities in Costa Rica; (2) to assess the degree to which current trade policies lead to sub-optimal welfare levels; and (3) to determine the welfare effect of future supply and demand developments as well as of possible infra-structural government policies. The SEM for Costa Rica developed in this chapter considers the 17 most important agricultural commodities in the country, the six planning regions as defined by the Costa Rican government, as well as the Rest-Of-the-World (ROW), which it interprets as a seventh region in order to take international trade into account.

The remainder of this chapter is structured as follows: the next section briefly discusses some of the major agricultural policies over the last decades that are relevant to the simulations performed. The third section justifies the regional analysis and selection of commodities. The fourth section provides a short description of the SEM in which

[1] Examples include Martin and Zwart (1975), Pieri *et al.* (1977) and Krishnaiah and Krishnamoorthy (1988).

aggregate welfare is maximized subject to supply, demand and resource restrictions. Moreover, results of model parameter estimations are presented, including demand and supply elasticities as well as transport costs. In the fifth section, results of the base run as well as of a number of policy simulations are discussed. The base run model specification reflects the prevailing situation in terms of regional equilibrium in commodity supply and demand, corresponding prices, domestic trade flows, and international trade. The base model is validated against actual 1995 data, and used as the basis for policy simulations. The latter consider such variables as government trade policies, infra-structural development, technological progress in agricultural production, and changes in demand resulting from income and population growth. Each scenario is evaluated in terms of changes in welfare, land use and trade patterns. Finally, the last section provides a summary of the main results followed by some concluding observations.

4.2 Main agricultural policies in Costa Rica after 1980

This section focuses on the agricultural policies in Costa Rica that are relevant to the model simulations in Section 4.5. A more extensive overview of economic and agricultural policies is given in Chapter 2. Structural adjustment programs were introduced in Costa Rica after 1980, as it was realized that the size of the domestic market is too small to serve as a base for rapid and sustained growth in the agricultural sector. Adjustment measures mainly consisted of lowering trade barriers, financial sector reform, and reform of the state sector. These measures have had clear positive effects on both economic growth and employment (Chapter 2; Schipper *et al.*, 1998). The structural adjustment measures in the agricultural sector resulted in a much higher degree of integration into the world market (Pomareda, 1996; SEPSA, 1997). The system of guaranteed producer prices and consumer subsidies, implemented by the National Production Council (CNP, *Consejo Nacional de Producción*) was gradually phased out, while production of agricultural export crops was promoted through tax reductions for exports to new markets, reduced import taxes for inputs such as agrochemicals and agricultural equipment, and credit on favorable terms for companies engaged in export activities (Mora Alfaro *et al.*, 1994). The consequences included a strengthening of the comparative advantage for traditional export crops (*e.g.*, banana, coffee, sugar-cane) over basic food crops, as well as the promotion of non-traditional export crops (*e.g.*, pineapple, palm heart, flowers, ornamental plants, roots and tubers). In the latter case, an incentive system centered around tax rebates, called the "Export Tax Credit" (CAT, *Certificado de Abono Tributario*, initiated in 1984 and phased out in 1999), played an important role in boosting earnings from non-traditional export crops, despite a lax accounting system which left it open to abuses. Overall, the dependence of total export earnings on banana and coffee cultivation decreased as a result of the stimulation of non-traditional export crops (González, 1994). Exports in general are further stimulated by the exchange rate policy, which uses a system of mini-devaluations to maintain the competitive position that Costa Rica holds with its main trading partners.

At present the jurisdiction of the CNP, as far as agricultural policy is concerned, is largely restricted to regulating imports and exports of basic food grains, for which it has the exclusive authority to issue permits. Imports of many agricultural commodities other than basic grains are regulated by a variable tariff system. In addition, the CNP is involved in determining the maximum prices of certain staple foods according to the quality specifications to which they conform.

In general, structural adjustment in Costa Rica has been qualified as incomplete (Hausmann, 1998), and considerable trade barriers still exist (Table 4.1). Import tariffs have their legal base in the so-called Import Duties Law (DAI, *Derecho Arancelario de Importaciones*) and Law No. 6946. The DAI defines commodity-specific preferential import taxes, which are calculated on the basis of *c.i.f.* import prices. DAI tax levels differ by country of origin; the figures presented in Table 4.1 are those that apply to countries outside Central America. Law No. 6946 defines relatively minor non-preferential import taxes, which are calculated on the basis of the *c.i.f.* import prices of all commodities, independent of the country of origin.

4.3 Regional analysis and commodity selection

For the analysis in this chapter, Costa Rica is divided into six planning regions including the *Central*, *Pacífico*, *Chorotega*, *Brunca*, *Norte* and the *Atlántica* regions. These planning regions correspond to those distinguished by the Costa Rican government during the period 1986-1988 (Figure 4.1). Distinguishing between these regions is important for a number of reasons. First, since regional agro-ecological conditions tend to differ significantly, they determine to a large extent the commodities that can be produced in each region. In addition, bio-physical factors are important determinants of production technologies and corresponding yield levels. Second, demand for agricultural products shows significant inter-regional variation, mainly due to regional differences in per capita household income, household size, degree of urbanization, and consumer preferences (Geurts *et al.*, 1997).

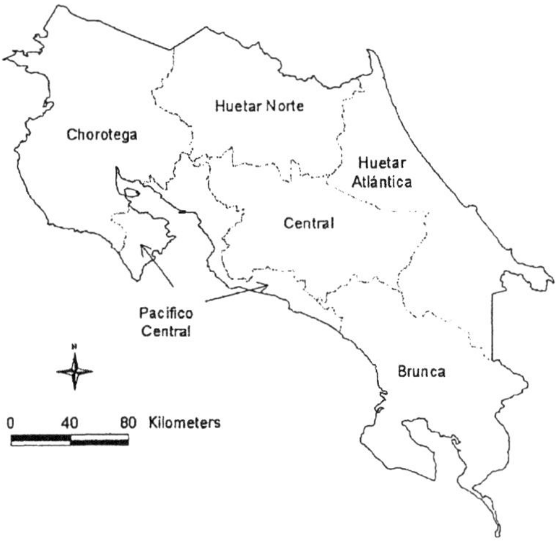

Figure 4.1. Costa Rica: Planning regions 1986-1988.

Table 4.1. Import prices and export prices of agricultural commodities as of 1995 (\$ kg^{-1}), along with the import tariffs on them[1]

	Cassava	*Maize*	*Mango*	*Melon*	*Orange*	*Beef*	*Plantain*	*Beans*	*Rice*	*Potato*	*Onion*	*Sugar*	*Milk*
p^{export} f.o.b.	0.33	0.15	0.59	0.40	0.09	1.66	0.21	0.35	0.30	0.17	0.20	0.028	0.17
$p^{Costa\ Rica}$	0.19	0.18	0.55	0.57	0.10	1.55	0.20	0.53	0.24	0.47	0.31	0.027	0.27
p^{import} c.i.f.	0.35	0.16	0.62	0.42	0.10	1.85	0.22	0.48	0.33	0.33	0.25	0.029	0.19
p^{import} +tax	0.40	0.18	0.73	0.49	0.11	2.20	0.27	0.58	0.45	0.49	0.38	0.044	0.39
DAI[2]	14%	14%	18%	18%	18%	18%	18%	20%	35%	45%	52%	55%	102%
Law 6946	1%	1%	1%	1%	1%	1%	1%	1%	1%	1%	1%	1%	1%
Total	15%	15%	19%	19%	19%	19%	19%	21%	36%	46%	53%	56%	103%

Notes: [1] All monetary units in this study are in \$, at the average 1994-1996 exchange rate of \$1 = ¢181.

[2] DAI, *Derecho Arancelario de Importaciones*: import tariffs, which were obtained as unpublished information from the Ministry of Trade and Commerce (1998).

In this study 17 major agricultural products (fifteen crops, along with beef and milk) are considered, and have been selected on the basis of their relative importance (at the national level) in terms of cultivated area and value of production. The crops included in the study are: basic grains (rice, maize and beans), traditional export crops (coffee, banana and sugar), non-traditional export crops (plantain, palm heart, mango, melon, pineapple and cassava) and fruits and vegetables (orange, onion and potato).

4.4 Methodology

4.4.1 Specification of the Spatial Equilibrium Model

In this section a short description of the SEM is presented. Each of the regions included in the SEM may be a producer of a commodity, a consumer of that commodity, or a combination of the two. While it is assumed that in principle each region can trade any commodity with any other region (including the ROW region), actual production of a given commodity in a given region is limited by the bio-physical conditions that prevail in that region. Given a set of user-specified restrictions (including those determined by both bio-physical limitations and socio-economic policies), a SEM allocates resources over a designated area in an efficient manner by maximizing a welfare function that consists of the sum of the domestic consumer and producer surplus of commodities for all regions, plus exports minus imports and transportation costs. The model simulates competitive market equilibrium for commodities in regions, where commodity prices are equal to their marginal costs.

A SEM can be considered as a particular type of sector model (Hazell and Norton, 1986) in which a spatial dimension is introduced on both the supply and demand sides. The mathematical representation of the relevant part of the SEM used in this study is shown in Appendix 4.1. Regional domestic prices are determined endogenously on the basis of consumer behavior (as expressed in the demand functions that underlie the demand elasticity estimates) and producer behavior (as expressed in the production functions that underlie the supply elasticity estimates), assuming competitive market clearing processes and taking into account inter-regional transport costs. As Costa Rica is a "small country," such determination of domestic prices takes place within the limits as set by exogenously given f.o.b. (free-on-board) export prices and c.i.f. (cost, insurance and freight) import prices, assuming perfectly elastic export demand and import supply. However, exogenous export and import prices are not a necessary feature of a SEM (see also Section 4.5.1). Furthermore, unlike in a general equilibrium model, incomes are kept exogenous. In addition, data limitations prevented the inclusion in the model of interdependencies among commodities on the supply side of the agricultural sector (as captured by cross-price supply elasticities) and product substitution on the demand side (as represented by cross-price demand elasticities).

4.4.2 Estimation of regional demand elasticities

Geurts *et al.* (1997) used budget data for nearly 4000 households from the 1987-1988 National Household and Income and Expenditure Survey (DGEC, 1988) to construct demand models, from which expenditure and own-price elasticities for 24 major food categories were calculated. Each food category consisted of a variety of agricultural products, and estimates were made on a nation-wide basis. Since regional elasticities are required for the SEM, the same data set was used to obtain domestic regional elasticity estimates for the 17 agricultural commodities included in the SEM, and for each of the six planning regions covered.

Specification and estimation of demand models

The specification of the demand models for the estimation of regional expenditure and own-price elasticities was based on the model presented in Geurts *et al.* (1997), as follows:

$$\ln exp_c = \alpha_c + \beta_c \ln x + \gamma_c(\ln x)^2 + \delta_c \ln p_c + \varphi_c(\ln p_c \,.\, \ln x) + \lambda_c \ln N + \kappa_c \ln CPI \qquad \textbf{(1)}$$

Per capita expenditure (exp_c) on agricultural commodity c was hypothesized to depend on (1) per capita total monthly consumption expenditure (x) as a proxy for total income; (2) own-price (p_c); (3) the number of household members per household (N); and (4) the general monthly consumer price index (CPI) as a proxy for the general price level. Per capita expenditure can be expected to be negatively influenced by household size, as larger households normally have lower per capita income, as well as expenditure, and may be more efficient in their use of foods. The monthly overall consumer price index (the CPI for food products only was unavailable) was included for two reasons, the first of which is to capture the effects of other prices on the demand for a particular category whose share of total expenditure is assumed to be small. In this way, a potential source of missing variable bias is eliminated (Deaton and Case, 1988). The second reason to include the CPI as an explanatory variable in the model is its traditional role of deflating nominal economic variables. A quadratic logarithmic expenditure term was included to allow for the possibility that commodities may be considered luxury, necessity or inferior goods by earners of different levels of income (Timmer, 1981). Finally (even though not used in this chapter), the interaction term of per capita expenditure with unit price allows price elasticities to vary according to total expenditure level.

Estimation results

Based on the econometric estimation of equation (1), own-price elasticities of demand were estimated for each of the 17 commodities and 6 regions included in the SEM, as follows:

$$\varepsilon^D = \delta + \varphi \ln x - 1 \qquad \textbf{(2)}$$

Results are shown in Table 4.2. A t-test was used to determine whether the elasticity estimates are significantly different from zero (Lizano, 1994). All own-price elasticities have the expected negative sign, with the exception of the positive demand elasticities for palm heart in the *Norte* and *Atlántica* regions. In the SEM model these own-price elasticities were assigned their corresponding national own-price elasticity values.

Similarly, equation (1) is used to estimate the expenditure elasticities for the 17 commodities and 6 regions included in the SEM, as follows:

$$\eta = \beta + 2\gamma \ln x + \varphi \ln p \qquad \textbf{(3)}$$

Results are shown in Table 4.3. All expenditure elasticities have the expected positive sign, with the exception of the negative expenditure elasticities for plantain in *Brunca* and *Atlántica*, melon in *Brunca*, pineapple in *Norte* and cassava in *Brunca* and *Norte*. For use in the demand shift scenario as assessed by the SEM model, these expenditure elasticities were assigned their corresponding national values.

4.4.3 Estimation of regional supply elasticities

The estimation of regional supply elasticities requires time series data for the prices and production of the respective commodities. Such data are not readily available in Costa Rica, and construction of the necessary data base turned out to be a tedious and time consuming exercise. Nevertheless, time series data could be obtained from various sources, either on a yearly basis (18 years or longer) or on a monthly basis (120 months or longer). Price data refer to the average annual or monthly price, while production data represent total annual or monthly production.

Yearly regional production and producer price data for coffee and sugar were obtained from the Costa Rica Coffee Institute (ICAFE, *Instituto Costarricense del café*) and the Consortium of Sugar Cane Processors (LAICA, *Liga Agrícola Industrial de la Caña de Azúcar*), respectively. Regional production and national price data for basic grains were available from the CNP, while regional banana production data were obtained from the National Banana Corporation (CORBANA, *Corporación Bananera Nacional*). Export prices for banana were taken from the FAO statistical database. For non-traditional export crops, fruits and vegetables, regional production and national price data on a monthly basis could be constructed from data available at the national wholesale market, called the National Center for Supply and Distribution of Food Products (CENADA, *Centro Nacional de Distribución de Alimentos*), where an estimated 60% to 70% of the total national production of these products is traded. Finally, yearly production and price data for beef and milk were obtained from Montenegro and Abarca (1998).

Specification and estimation of supply models

Supply response models for annual crops were based on the standard Nerlove model (Askari and Cummings, 1976), which includes the effects of price expectations and adjustment lags in production on the supply of agricultural goods. In the Nerlove

Table 4.2. Regional and national own-price elasticities of demand for the 17 major commodities

	Central	*Pacífico*	*Chorotega*	*Brunca*	*Norte*	*Atlántica*	Costa Rica
Rice	-0.83**	-0.89**	-0.99**	-0.77**	-0.85**	-0.99**	-0.86**
Maize	-0.93**	-0.97**	-0.89**	-1.09**	-0.81**	-0.84**	-0.94**
Beans	-0.93**	-0.98**	-0.67**	-0.89**	-0.87**	-0.93**	-0.89**
Coffee	-0.84**	-0.92**	-0.81**	-0.78**	-0.90**	-0.82**	-0.86**
Banana	-0.68*	-0.98**	-0.42	-0.53	-0.51	-0.61	-0.71**
Sugar	-0.88**	-1.07**	-1.10**	-0.94**	-0.95**	-1.05**	-0.99**
Plantain	-0.91**	-0.91**	-0.85*	-0.83*	-0.77*	-0.82*	-0.83**
Palm heart	-1.13	-1.22	-1.17	0.25	0.32	-1.11	-1.19**
Mango	-0.67*	-0.85*	-0.64*	-0.91*	-0.56*	-0.61*	-0.97**
Melon	-0.81**	-1.42**	-1.02**	-0.21**	-0.92**	-0.91**	-0.81**
Pineapple	-0.81**	-0.69*	-1.27	-0.95*	-0.14	-0.95*	-0.74**
Cassava	-0.75**	-0.74**	-0.53	-0.39	-0.56**	-0.43	-0.59**
Onion	-0.88**	-0.95**	-0.74**	-0.89**	-0.66**	-0.81**	-0.85**
Orange	-0.73**	-0.73*	-0.95*	-0.42	-0.08	-0.44	-0.74**
Potato	-0.86**	-0.85**	-0.82**	-0.82**	-0.75**	-0.80**	-0.82**
Beef	-0.92**	-0.93**	-0.91**	-0.94**	-0.78**	-0.90**	-0.90**
Milk	-0.85**	-0.98**	-0.83**	-0.89**	-0.93**	-0.83**	-0.88**

Significance level: ** (*) significantly different from zero at the five (ten) percent level according to the t-test.

Table 4.3. Regional and national expenditure elasticities for the 17 major commodities

	Central	*Pacífico*	*Chorotega*	*Brunca*	*Norte*	*Atlántica*	Costa Rica
Rice	0.21*	0.24*	0.25*	0.37**	0.13	0.39*	0.26**
Maize	0.26*	0.18	0.43*	0.41*	0.02	0.15	0.26**
Beans	0.19*	0.09	0.15	0.19	0.04	0.32*	0.17**
Coffee	0.22*	0.28*	0.29*	0.25*	0.12	0.36*	0.26**
Banana	0.26*	0.19	0.32	0.23	0.12	0.21	0.22**
Sugar	0.20*	0.23*	0.37**	0.31*	0.16	0.50**	0.29**
Plantain	0.31*	0.21	0.29	-0.02	0.18	-0.04	0.17**
Palm heart	1.86	1.73	1.88	0.33**	0.28**	2.09	0.41**
Mango	0.15	0.23	0.14	0.54	0.56	0.63	0.67**
Melon	0.42	0.94	1.27*	-6.63	0.94	0.95	0.36**
Pineapple	0.25	0.15	0.36	0.03	-0.30	0.09	0.08**
Cassava	0.13	0.06	0.35	-0.05	-0.02	0.17	0.08**
Onion	0.39*	0.22	0.39*	0.44*	0.14	0.50*	0.35**
Orange	0.65**	0.72*	0.62	0.62	0.18	0.46	0.60**
Potato	0.27*	0.32*	0.44*	0.36*	0.30*	0.36*	0.33**
Beef	0.58**	0.69**	0.63**	0.48*	0.35*	0.50**	0.55**
Milk	0.51**	0.45*	0.61**	0.60**	0.59**	0.56**	0.55**

Significance level: ** (*) significantly different from zero at the five (ten) percent level according to the t-test.

model, actual production (Q_t) is considered a function of lagged deflated prices (P_{t-1}), lagged production (Q_{t-1}) and non-price variables.

The supply response model used in this chapter is an adaptation of the standard Nerlove model, since it includes additional explanatory variables, such as the lagged deflated prices of competitive crops ($P_{c,t-1}$), the lagged deflated prices of major inputs ($P_{i,t-1}$) for inputs i, and a time trend (T) to correct for the possibility of technical change, improvements in infrastructure or other structural factors. This resulted in the following supply response model to be estimated for annual crops, as presented by Roebeling *et al.* (1999b):

$$Q_t = c_0 + c_1 Q_{t-1} + c_2 P_{t-1} + c_3 P_{c,t-1} + c_4 P_{i,t-1} + c_5 T + v_t \qquad \textbf{(4)}$$

where the term c_0 represents the intercept and v_t is the disturbance term. Model coefficient estimates provide information on the area adjustment coefficient ($1-c_1$) as well as on short-term (c_2) and long term supply response ($c_2/(1-c_1)$).

Often the size of the harvested or planted area is preferred as a proxy for production, since the latter is influenced by factors that cannot be controlled by the producer (Rao, 1989). However, since data on harvested or planted area in the regions were not available, regional production data were used as the dependent variable in equation (4). The assumption that farmers consider only last year's prices for this year's planting decisions, is restrictive but convenient because of its simplicity. Moreover, in the case of time series with a limited number of observations, the inclusion of two or more lags may lead to insufficient degrees of freedom. Commodities for which supply response was estimated on the basis of monthly data (due to a lack of a sufficiently long time series of yearly observations), a seasonal dummy was introduced in the above specification to correct for seasonalities in production, as well as an auto-regressive term to correct for serial correlation of residuals by taking into account slowly moving influences (Maddala, 1992).

Supply response models for perennial tree crops require a slightly different approach, since planting and harvesting decisions reflect two different moments in time. Stryker (1990) and Frimpong-Ansah (1992) suggest that production of perennials in a given time period is also determined by the so-called "normal" production (N_t). Normal production in this context refers to the level of production that can be expected from a given planted area. On the basis of time series data about planted area, yield levels over the lifetime of the tree, as well as the average tree life, the "normal" production of the planted area is calculated. This leads to the following specification of the supply response model for perennial crops (Roebeling *et al.*, 1999b):

$$lnQ_t = c_0 + c_1 lnQ_{t-1} + c_2 lnP_{t-1} + c_3 lnP_{c,t-1} + c_4 lnP_{i,t-1} + c_5 lnN_t + c_6 T + v_t \qquad \textbf{(5)}$$

The short term price elasticity (c_2) reflects adjustments in the application of variable production factors, while the long term price elasticity ($c_2/(1-c_1)$) reflects the area adjustment coefficient (Nerlove, 1958). Although these price elasticities do not capture the effect of prices on planting decisions (since these are incorporated in the normal production variable), the long term price elasticity does give an indication of farmer behavior in maintaining and improving the planted area (Frimpong-Ansah, 1992).

Estimation results

Supply elasticities were estimated for all commodities included in this study for each of the relevant planning regions as well as for the nation as a whole (Table 4.4). They were calculated on the basis of the short term supply response or own-price coefficient ($dq/dp=c_2$ in equations 4 and 5). Supply elasticities are defined as follows:

$$\varepsilon_s = \frac{dq}{dp} * \frac{\bar{p}}{\bar{q}} \qquad \textbf{(6)}$$

where $\bar{p}$ represents average own-price and $\bar{q}$ average production.

Table 4.4. Regional and national own-price elasticities of supply for the 17 major commodities

	Central	*Pacífico*	*Chorotega*	*Brunca*	*Norte*	*Atlántica*	Costa Rica
Rice	-	0.54**	0.74*	0.18	0.90**	0.26	0.82**
Maize	0.54**	0.61	0.35**	0.37*	0.24**	0.64	0.56**
Beans	0.20	0.25*	0.33**	0.40*	0.11	-	0.18
Coffee[1]	0.08	0.08	0.05	-	-	0.18*	0.08**
Banana	-	-	-	1.07**	-	1.78**	1.75**
Sugar	0.11**	0.39**	0.13**	0.49**	0.20**	-	0.29**
Plantain	n.a.	n.a.	-	-	0.63	0.53	0.53*
Palm heart	-	-	-	-	n.a.	n.a.	n.a.
Mango	n.a.	n.a.	n.a.	n.a.	n.a.	-	n.a.
Melon	0.44*	0.76**	0.55**	n.a.	-	n.a.	0.65**
Pineapple	n.a.	n.a.	-	n.a.	0.51**	n.a.	0.51**
Cassava	0.36**	-	-	-	0.23**	0.21**	0.24*
Onion	0.23**	-	0.68**	-	-	-	0.22**
Orange	n.a.	n.a.	n.a.	n.a.	n.a.	n.a.	n.a.
Potato	0.12	-	-	-	-	-	0.12
Beef	n.a.	n.a.	n.a.	n.a.	n.a.	n.a.	0.46**
Milk	n.a.	n.a.	n.a.	n.a.	n.a.	n.a.	0.56**

Significance level: ** (*) significantly different from zero at the five (ten) percent level according to the t-test. Significance level according to original slope coefficients for price parameter.

Notes: [1] Long term supply elasticities.

"n.a." : supply elasticities could not be calculated due to too few observations or absence of regional production data.

"-" : no supply elasticity is determined as production is not possible in the region for agro-ecological reasons.

All supply elasticity estimates have the expected positive sign, and most are significantly different from zero according to the standard t-test. In some cases, regional supply elasticities could not be estimated due to a limited number of observations or a lack of data. For such cases it is assumed that the regional supply elasticity equals the corresponding national estimate.

4.4.4 Estimation of inter-regional transport costs

Transportation costs determine to a significant extent the comparative advantage of a particular region in the production of a specific commodity. Given wholesale prices as determined in properly functioning markets, farm gate prices depend mainly on transport costs, which thus are an important determinant in the economics of production in a given region. In this research, transportation costs between regions as well as to major export harbors were calculated on the basis of an adaptation of regression models estimated for the *Atlántica* region as described in Jansen and Stoorvogel (1998).

In the original models used by Jansen and Stoorvogel (1998), transportation costs were hypothesized to depend on the geographical distances between markets and farms (Dis_n) and the quality of the road infrastructure (n). These models were estimated by using farm-level data on the transportation costs of commodities from farms to farmers' markets or to the national wholesale market (CENADA). Using a GIS, these survey data were combined with geographical data on the approximate location of sample farms and the distances. Calculations were then made for four road types, and used to econometrically estimate a number of alternative regression models to assess the influence of road type on transportation costs. This resulted in the following preferred transportation cost model, where all estimated coefficients are significantly different from zero at the 1% level or better:

$$UC = 0.0028 + 0.00022\,(Dis_1+Dis_2) + 0.00055\,(Dis_3+Dis_4) \qquad \textbf{(7)}$$

$$(N=56,\ R^2=0.72)$$

where UC represents the unit transport cost in \$ kg^{-1}; Dis_n is the distance on road type n (in kilometers); and n represents road quality (n = 1, 2, 3, 4, from best to worst type of road). For the SEM, the model as depicted by equation (7) needed to be adjusted in several ways in order to obtain inter-regional transport costs for the considered commodities. First, the SEM considers inter-regional transport flows, taking into account only the best type of road, since inter-regional road connections are all qualified as type 1 roads. Second, distances between regions are calculated as the average distance between the geographical centers of each pair of regions, while distances to export harbors were calculated as the average distance from each region's geographical center to the nearest export harbor (Table 4.5). Distances are based on a digital road map for Costa Rica (Figure 4.2).

Finally, for products that permit bulk transport (rice, maize, beans, banana and sugar), variable transport costs were estimated at \$ 0.066 10^{-3} kg^{-1} km^{-1} (Schipper *et al.*, 1998). To summarize, total transport costs per kilogram product between regions were calculated as the sum of fixed costs (\$ 2.8 10^{-3} kg^{-1}) and variable transport costs (\$ 0.22 10^{-3} kg^{-1} km^{-1} or \$ 0.066 10^{-3} kg^{-1} km^{-1}), where the latter are determined as the multiplicate of distance and variable transport costs per kilogram.

Table 4.5. Distances between regional centers and to export harbors (in km)

	Central	*Pacífico*	*Chorotega*	*Brunca*	*Norte*	*Atlántica*	World
Central	0.0	110.0	261.0	189.0	129.3	150.7	130.3
Pacífico	110.0	0.0	159.0	299.0	143.3	260.7	102.5
Chorotega	261.0	159.0	0.0	450.0	209.3	411.7	159.0
Brunca	189.0	299.0	450.0	0.0	318.3	339.7	299.0
Norte	129.3	143.3	209.3	318.3	0.0	219.7	181.5
Atlántica	150.7	260.7	411.7	339.7	219.7	0.0	88.0
World	130.3	102.5	159.0	299.0	181.5	88.0	0.0

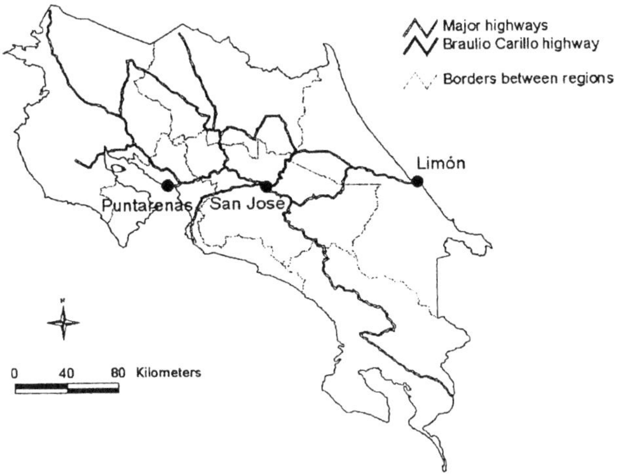

Figure 4.2. Costa Rica: main roads connecting the centers of the planning regions.

4.5 Model results

4.5.1 Model validation

Model calibration was performed with the data for the year 1995, the most recent year for which a complete set of agricultural statistics for each commodity could be constructed. The model's base run is based on 1995 data for production, consumption, imports and exports; corresponding prices in regional markets as well as in the relevant world markets; transport costs; own-price elasticities of supply and demand; and prevailing national trade policies. The latter encompass both tariff and non-tariff measures, including import taxes for a number of products, and export quotas for basic grains, potato, onion and milk.

Commodities for which Costa Rica engages in international trade, supply of imports as well as demand for exports are all treated as completely elastic (*i.e.*, infinitely large elasticities in absolute terms). The underlying assumption that neither Costa Rican import demand nor export supply will significantly affect international prices is justified given the relatively limited quantities traded.[2] The implication is that, even though exports contribute to domestic producer surplus, surplus accrued by consumers in foreign countries is disregarded. Similarly, while imports contribute to domestic consumer surplus in Costa Rica, possible contributions to the surplus of foreign producers are disregarded as well. Finally, total land use is limited to the total agricultural area as determined by the CNP (CNP, 1990).

The model is validated by comparing base run results with the actual situation in 1995 in terms of area allocation, agricultural production levels, and production value, including product prices. In general, base run results are very close to the actual 1995 situation (Roebeling *et al.*, 1999b). The total area devoted to the 17 commodities in the base run is about 4% lower than in the actual situation, due to the slightly higher yield levels used in the base run. With the exception of melon and cassava, production of individual commodities in the base run never deviates more than 10% from the actual 1995 levels. Total production value in the base run exceeds the actual 1995 value by about 6%. This difference can be explained by the larger importance of export production in the base run for a number of commodities, leading to higher domestic prices.[3]

Imports and exports of basic grains, potato, onion and milk were all subject to government trade regulations in 1995, either directly by imposing import and export quotas or indirectly through subsidies or tariffs. The latter often are determined in a rather *ad-hoc* way. The base run therefore assumes that no exports of these products take place, while import prices equal world market prices plus import tariffs. For all other commodities, simulated exports in the base run situation generally exceed actual values, suggesting that there may have been trade barriers in effect in 1995 that are not considered by the model. In the case of coffee, for example, it is well known that, by law, part of the harvest below a specified quality has to be directed to the national rather than the world market.

On the other hand, import tariffs on virtually all agricultural commodities are such that imports are limited to an absolute minimum. Imports that were allowed in 1995 were mostly determined by government regulations (*e.g.*, the CNP for basic grains), guided by temporary shortages in the national market due to seasonal fluctuations in supply and/or demand (*e.g.*, potato and onion). Import restrictions, in combination with a slight overestimation of exports, result in levels of domestic demand in the base run situation that are somewhat below actual 1995 demand.

4.5.2 Base run results

The regional and national land use patterns for Costa Rica determined in the base run are given in Table 4.6. While in the *Chorotega* and *Norte* regions virtually all of the available agricultural area is utilized by the 17 commodities included in this study, in the *Central* region they cover 83% of the available area, whereas in the *Pacífico*,

[2] While this is true even for coffee, Costa Rica's share in the total world exports of banana is about 15%, making export demand for Costa Rican bananas less than infinitely elastic. Nevertheless, for simplicity's sake, an infinitely elastic demand in the world market was also assumed for banana. In contrast, export demand for banana is considered to be inelastic in Chapter 7.

[3] In the calculation of production values, all production was valued against regional commodity producer prices.

Brunca and *Atlántica* regions they account for between 60% and 70% of the total available agricultural area. The *Chorotega* and *Central* regions together account for over half of the total cultivated area in Costa Rica.

According to the base run, the major part of agricultural land in all regions is used for beef and milk production, with pastures covering 87% of the cultivated area in Costa Rica. Consequently, crop production covers only 13% of the cultivated area, and is concentrated on traditional export crops (mainly coffee and banana) and/or basic grain production (mainly rice and beans). Most important non-traditional export crops include plantain, mango, pineapple and cassava, while orange is an important crop for the domestic market (consumed mainly in the form of juice).

Basic grain production accounts for only 3.6% of the national agricultural land use. Relatively important regions for basic grain production include *Chorotega*, *Brunca*, *Norte* and (only for rice) *Pacífico*. The regions *Chorotega*, *Brunca* and *Pacífico* are the most important rice producing regions and contain 92% of the total rice area, while the production of beans is largely concentrated in the *Norte* region which includes over 60% of the total bean area. Maize production is relatively unimportant in all regions.

Table 4.6. Base-run regional and national land use and production values

	Central		*Pacífico*		*Chorotega*		*Brunca*		*Norte*		*Atlántica*		Total	
	ha 10^3	\$ 10^6	ha 10^3	\$ 10^6	ha 10^3	\$ 10^6	ha 10^3	\$ 10^6	ha 10^3	\$ 10^6	ha 10^3	\$ 10^6	ha 10^3	\$ 10^6
Rice	0.0	0.0	9.1	8.6	19.5	17.9	12.2	11.2	3.5	3.4	0.0	0.0	44.3	41.1
Maize	1.1	0.3	1.2	0.3	5.0	0.3	4.1	1.1	2.3	0.6	0.6	0.2	14.2	3.8
Beans	2.1	0.8	1.5	0.6	3.1	1.3	7.7	2.7	22.9	7.9	0.2	0.1	37.5	13.3
Coffee	94.9	411.0	9.8	42.7	1.6	7.0	0.0	0.0	0.0	0.0	0.3	1.5	106.7	462.3
Banana	0.0	0.0	0.0	0.0	0.0	0.0	2.4	25.6	0.0	0.0	46.9	535.1	49.2	560.7
Sugar	10.6	19.9	4.8	9.3	16.8	31.6	2.4	4.6	6.1	11.5	0.0	0.0	40.8	76.9
Plantain	0.0	0.1	0.0	0.0	0.0	0.0	0.0	0.0	0.7	3.3	5.7	23.3	6.4	26.6
Palm heart	0.0	0.0	0.0	0.0	0.0	0.0	0.0	0.0	2.1	2.7	2.3	3.1	4.4	5.8
Mango	5.4	6.0	1.0	1.1	0.1	0.2	0.3	0.3	0.0	0.0	0.0	0.0	6.7	7.6
Melon	0.1	0.7	2.6	28.4	0.3	3.5	0.1	0.6	0.0	0.0	0.0	0.2	3.1	33.3
Pineapple	0.2	2.1	0.0	0.0	0.0	0.0	0.0	0.3	6.4	54.5	0.0	0.1	6.6	57.0
Cassava	0.2	0.9	0.0	0.0	0.0	0.0	0.0	0.0	6.3	27.0	0.0	0.1	6.5	28.0
Onion	0.7	4.8	0.0	0.0	0.0	0.0	0.0	0.0	0.0	0.0	0.0	0.0	0.7	4.8
Orange	11.4	7.7	0.0	0.0	0.3	0.2	0.8	0.7	7.5	3.4	1.6	0.7	21.5	12.8
Potato	2.2	20.9	0.0	0.0	0.0	0.0	0.0	0.0	0.0	0.0	0.0	0.0	2.2	20.9
Beef	311.5	25.7	90.7	7.4	746.8	71.8	227.9	18.3	333.1	24.0	247.3	20.3	1957.2	168.0
Milk	140.4	60.8	26.2	10.3	50.8	18.2	23.6	8.6	72.5	27.4	38.8	14.7	352.2	140.1
Total	580.8	561.7	147.0	108.6	844.3	152.5	281.4	73.9	463.3	165.8	343.7	599.4	2660.5	1662.4
Available[1]	697.2	-	217.2	-	844.3	-	464.8	-	463.3	-	504.6	-	3190.1	-

[1] Source: CNP, 1990.

Traditional export crops are relatively important, accounting, on average, for 7.4% of total regional land use. Geographical diversification of coffee and banana production is relatively limited, due to the specific agro-ecological requirements of these crops. Coffee production is concentrated in the elevated and therefore relatively cool *Central* region, where 90% of the total coffee area is situated. Banana production is also highly concentrated, with over 95% of the total banana production area located in the hot and humid lowlands of the *Atlántica* region. Sugar production, on the other hand, is geographically more diversified, with the major production regions in *Chorotega* and *Central.*

Non-traditional export crops account for just over 2% of total agricultural land use. The most important production regions include *Norte*, *Atlántica*, *Central* and *Pacífico*, each of which specializes in one or more non-traditional export crops. The *Atlántica* region specializes in plantain production and is responsible for nearly 90% of the total national plantain area. Mango production mainly takes place in the *Central* region, where over 80% of the total mango area in Costa Rica is located. Finally, production of both pineapple and cassava is concentrated in the *Norte* region, containing over 95% of the total area of each crop.

Milk and beef production is predominant in all regions, responsible for about 87% of total national land use. Of the total national pasture area, about 85% is devoted to beef production and only 15% to milk production. Pasture for beef production is most important in the *Chorotega* and *Brunca* regions where, respectively, 94% and 90% of the regional pasture area is for beef production. On the other hand, pasture for milk production is most important in the *Central* and *Norte* regions where, respectively, 31% and 18% of the regional pasture area is for dairy cattle. By far the most important beef producing region is *Chorotega*, which includes 38% of the total national pasture area used for beef production. The *Central* region is the most important milk producing region, accounting for 40% of the national pasture area for milk production purposes.

National and regional values of production (calculated as regional production times regional price) are also presented in Table 4.6. Comparisons of regional land use and value of production reveal that regions which account for a large share of total cultivated area in Costa Rica do not necessarily account for a large share of total national agricultural income. For example, while the *Chorotega* region contains 32% of the total cultivated area in Cost Rica, it contributes only 9% to national agricultural income. On the other hand, the *Atlántica* and *Central* regions generate the lion's share of national agricultural income: together these regions are responsible for 70% of national agricultural income while occupying only 35% of the total cultivated area.

Basic grain production generates less than 4% of total agricultural income on almost 4% of the total cultivated area. Rice and bean production account for, respectively, 71% and 23% of agricultural income obtained from basic grain production, while revenues from maize production are relatively minor. Traditional export crops are the major source of agricultural income, generating 66% of total national agricultural income while occupying just over 7% of the total cultivated area. Banana and coffee are the most important traditional export crops, accounting for 51% and 42% of total traditional export crop earnings, respectively. Non-traditional export crop production generates

12% of total agricultural income, while occupying just over 2% of the total cultivated area. Pineapple, melon, cassava and plantain prove to be the most profitable non-traditional export crops, respectively accounting for 29%, 17%, 14% and 13% of the total earnings from non-traditional export crops. Finally, beef and milk together generate 19% of the total agricultural income, of which 55% is generated by beef production. Even though beef production is the third most important source of agricultural income (after banana and coffee), it occupies the major share of the total cultivated area (87%).

Values of inter-regional product flows are shown in Table 4.7, representing product shipments from supply to demand regions and valued at regional supply prices. *Central* and *Atlántica* are important export regions (together these regions are responsible for 76% of total exported production value), as they are the major suppliers of coffee (*Central*) and bananas (*Atlántica*). Major supply regions for the domestic market include the *Norte*, *Brunca* and *Chorotega* regions, with basic grains, beef and milk as the most important traded commodities. Supply from the ROW region (*i.e.*, imports) is negligible for the considered commodities, due to relatively high import taxes and other trade barriers.

The ROW and *Central* regions constitute the major demand regions, absorbing 76% and 17% of the value of total national production, respectively. The *Central* region is the major domestic demand region, responsible for nearly 70% of the value of national demand. This is explained by the fact that over 60% of the Costa Rican population lives in the *Central* region (DGEC, 1997). Moreover, average per capita income in the *Central* region is about 55% higher than in other regions of Costa Rica (Geurts *et al.*, 1997). This wealth results in greater consumption, as well as a greater demand for higher quality goods in the *Central* region in comparison with other regions in Costa Rica.

Table 4.7. Base run value of product flows between supply and demand regions ($ 10^6)

Demand regions	Supply regions							
	Central	*Pacífico*	*Chorotega*	*Brunca*	*Norte*	*Atlántica*	ROW	Total
Central	141.7	13.9	27.1	28.6	46.6	22.5	0.0	280.5
Pacífico	2.5	14.2	0.0	0.0	0.6	0.9	0.0	18.1
Chorotega	2.7	0.2	18.6	0.0	2.1	0.0	0.7	24.3
Brunca	8.6	0.0	0.0	19.5	0.3	1.6	0.0	30.0
Norte	7.6	0.0	0.6	0.0	21.9	0.0	0.0	30.2
Atlántica	3.3	2.7	0.0	0.0	1.0	13.6	0.2	20.8
ROW	395.2	77.7	106.7	25.9	93.2	560.8	0.0	1259.4
Total	561.7	108.6	152.5	73.9	165.8	599.4	1.0	1662.4

In general, all regions primarily produce for the export market, in the second place for the *Central* region, and in the third place for other regions in Costa Rica.

On average, 66% of the total national production value is exported, while about 21% is directed towards the *Central* region. All regions are self-sufficient for at least 65% of regional commodity demand value, apart from the *Central* region which is self-sufficient for only 50%. This difference is not only explained by the fact that the *Central* region is responsible for the lion's share of domestic demand, but also by the fact that transport costs from other regions to the *Central* region are relatively low because of the latter's geographical location. Product flows between regions other than the ROW and the *Central* region are rather insignificant, constituting about 2% of the total trade flow value of nearly $1.7 10^9.

4.5.3 Policy simulations

In this section the SEM is used to evaluate a number of hypothetical policy measures in terms of their effect on welfare, land use and trade patterns.[4] These hypothetical policy measures were chosen on the basis of issues that are widely considered to be important in the Costa Rican agricultural policy arena (see also Section 4.2 and Chapter 2). The resulting simulations can be divided into three broad groups. First, ample attention is given to an assessment of the potential effects of trade liberalization measures. Second, as in many other developing countries, marketing of agricultural products in Costa Rica is hampered by relatively high transportation costs (Jansen and Van Tilburg, 1996). Consequently, a number of simulations are carried out to analyze the effects of changes in transportation costs. Finally, continuing economic development and technological progress ensure that demand for, as well as supply of, most agricultural commodities will keep on shifting in the future (future perspective scenarios), both of which may have important effects on land use and trade flows which in turn influence welfare.

The results of each model simulation are compared to either the base run simulation (trade liberalization scenarios and transport costs scenarios) or to a situation of complete trade liberalization (future perspective scenarios), while focusing on changes in land use patterns (Table 4.8), value of production, trade flows (Table 4.9), and shifts in producer and consumer surplus (Table 4.10). In all simulations it was assumed that the maximum available area for cultivation equals the cultivated area that was determined in the base run.

Trade liberalization

As a member of the World Trade Organization and as a signee to its various international agreements, Costa Rica has committed itself to free trade in basic food grains and eventually in all agricultural commodities. In Costa Rica, rice and several non-basic grain commodities, like potato, onion, sugar and milk, are heavily protected by import taxes (Table 4.1). This policy simulation therefore allows an assessment of the likely effects on welfare, agricultural land use and trade flows (both between regions and internationally) of partial trade liberalization (*i.e.*, free trade in basic grains only) and complete trade liberalization (*i.e.*, free trade in all agricultural commodities). Trade liberalization refers to the lowering or abolishment of tariff and non-tariff

[4] The outcomes of the scenarios depend on the set of export and import prices used. As world market prices fluctuate considerably, the results of trade liberalization scenarios will depend on these prices. However, though different world market prices would lead to different results, welfare effects can be expected to show similar tendencies.

Table 4.8. Simulation results: optimal land use distribution (ha 10^3)

	Base	Trade liberalization		Transport costs			Technological progress (supply shift)				Demand shift
		Basic grains	All products	Variable	Fixed	Braulio	Rice	Beans	Cassava	Beef	
Rice	44.3	50.3	50.4	44.5	44.4	44.3	54.5	50.5	50.5	50.3	52.5
Maize	14.2	14.2	14.3	14.2	14.2	14.2	14.3	14.3	14.3	14.2	14.6
Beans	37.5	36.6	36.7	37.6	37.5	37.5	36.7	42.3	36.7	36.8	36.6
Coffee	106.7	106.7	106.7	106.8	106.7	106.2	106.7	106.7	106.7	106.7	106.7
Banana	49.2	49.2	49.3	49.6	49.4	49.3	49.3	49.3	49.3	49.2	49.2
Sugar	40.8	40.8	40.8	40.9	40.9	40.8	40.8	40.8	40.8	41.0	41.0
Plantain	6.4	6.4	6.4	6.5	6.4	6.4	6.4	6.4	6.4	6.4	6.8
Palm heart	4.4	4.4	4.4	4.4	4.4	4.4	4.4	4.4	4.4	4.4	4.5
Mango	6.7	6.7	6.8	6.8	6.7	6.6	6.8	6.8	6.8	6.7	6.8
Melon	3.1	3.1	3.1	3.1	3.1	3.1	3.1	3.1	3.1	3.1	3.1
Pineapple	6.6	6.6	6.6	6.7	6.6	6.6	6.6	6.6	6.6	6.6	6.6
Cassava	6.5	6.5	6.5	6.6	6.5	6.5	6.5	6.5	7.4	6.5	6.5
Onion	0.7	0.7	0.7	0.7	0.7	0.7	0.7	0.7	0.7	0.7	0.7
Orange	21.5	21.6	21.7	21.7	21.7	21.7	21.7	21.6	21.7	21.7	24.7
Potato	2.2	2.2	2.1	2.2	2.2	2.2	2.1	2.1	2.1	2.1	2.1
Beef	1957.2	1952.5	1996.8	1954.8	1956.8	1959.2	2003.5	2000.0	1997.3	2146.6	1988.7
Milk	352.2	352.2	307.6	353.5	352.3	350.7	307.4	307.6	307.6	307.0	309.7
Total	2660.5	2660.9	2660.9	2660.6	2660.6	2660.4	2671.6	2669.7	2662.4	2810.0	2661.0

Table 4.9. Simulation results: values of production and international trade flows related to agricultural commodities ($ 10^6)

	Base	*Trade liberalization*		*Transport costs*			*Technological progress (supply shift)*				*Demand shift*
		Basic grains	*All products*	*Variable*	*Fixed*	*Braulio*	*Rice*	*Beans*	*Cassava*	*Beef*	
Central	561.7	561.7	542.7	566.2	562.2	540.4	542.5	543.0	543.1	544.0	546.5
Pacífico	108.6	111.7	109.4	109.3	108.8	108.8	112.8	109.8	109.3	112.7	110.8
Chorotega	152.5	160.5	155.8	154.7	153.1	153.2	159.5	156.4	155.8	156.1	158.7
Brunca	73.9	77.0	75.6	75.0	74.1	74.1	91.3	76.9	75.6	81.4	79.0
Norte	165.8	166.1	158.7	169.8	166.1	166.5	159.3	174.0	176.5	185.8	162.1
Atlántica	599.4	599.4	595.9	606.2	602.9	597.1	595.9	596.6	596.0	601.0	600.8
Total	1662.4	1676.4	1638.1	1681.3	1667.2	1640.2	1661.3	1656.7	1656.2	1,681.0	1657.8
Export	1259.4	1,275.7	1,285.0	1275.7	1264.0	1241.8	1292.9	1285.2	1301.8	1315.2	1139.3
Import	1.0	2.7	42.0	0.9	1.0	1.6	42.2	40.8	42.0	42.1	148.4
Trade balance	1258.4	1,273.0	1,243.0	1274.8	1263.0	1240.2	1250.6	1244.4	1259.8	1273.1	990.9

Table 4.10. Simulation results: producer and consumer surplus related to agricultural commodities ($\$10^6$)

	Base	*Trade liberalization*		*Transport costs*			*Technological progress (supply shift)*				*Demand*
		Basic grains	*All products*	*Variable*	*Fixed*	*Braulio*	*Rice*	*Beans*	*Cassava*	*Beef*	*shift*
Producer surplus	1090.6	1096.1	1096.0	1114.2	1093.4	995.6	1099.0	1111.5	1099.4	1138.4	1100.8
Consumer surplus	218.3	210.6	255.6	218.8	218.1	216.1	255.6	256.4	255.6	256.1	1223.1
Economic surplus	1308.9	1306.7	1351.6	1333.0	1311.5	1211.7	1354.7	1368.0	1354.9	1394.5	2323.9

(*i.e.*, the setting of quotas *etc.*) measures. Under free trade conditions, producers face (f.o.b.) world market export prices, while consumers are confronted with (c.i.f.) world market import prices (Table 4.1).

Land use patterns in both trade liberalization simulations (Table 4.8) show only minor differences from the base run situation. This conformity is understandable because trade liberalization mainly affects import prices, and therefore exerts its main effect on the domestic consumption side, which is relatively unimportant (only 24% of the total national production value in the base run is for internal consumption). Partial trade liberalization results in increased rice and (to a minor extent) maize production for export purposes, at the expense of bean and beef production. Rice exports increase, as export quotas are abandoned while export prices exceed the domestic prices set by the CNP. In the case of beans, the import price is lower than the domestic price, leading to the importation of beans. In the case of complete trade liberalization, domestic prices for onion, potato and (especially) milk products considerably exceed their respective export prices, providing a strong incentive for imports of these commodities, at the expense of domestic production. Trade restrictions cause inflated opportunity costs of land, since land use is not determined by international comparative advantage but rather by artificially high domestic prices. Under a regime of complete trade liberalization, one may therefore expect a decrease in the production of commodities that were previously protected by import barriers.

Total value of agricultural production increases under partial trade liberalization, while complete trade liberalization leads to a net decrease in agricultural income (Table 4.9). Both forms of trade liberalization lead to increased levels of exports and imports, but the latter is considerably larger under complete trade liberalization. Under partial trade liberalization, the largest gains in agricultural income are obtained in *Chorotega*, *Brunca* and *Pacífico*, regions that have the largest comparative advantage in rice and (to a minor extent) maize production. Production is oriented towards the export market, at the expense of production for domestic purposes. The overall result is increased export earnings, even though import expenditures also rise, due to the increased bean imports. Complete trade liberalization leads to a net decline in agricultural income, as domestic prices and production for beans, onion, potato and milk decrease under the influence of the lower import prices. These tendencies are especially strong in the *Norte* and *Central* regions, the main producers of these commodities. An increasing portion of the total demand for these products is met by imports, while the agricultural area that no longer needed to satisfy domestic demand is allocated to the production of exports, which in turn leads to higher export earnings.

The small share that the consumer surplus contributes to the total economic surplus generated through agriculture in Costa Rica (16%) has two main causes (Table 4.10). First, a major part of total agricultural production is exported. Even though such exports can be expected to generate surplus for foreign consumers, this surplus is not considered in the objective function of the SEM. Second, for most commodities, demand is relatively elastic while supply is relatively inelastic, resulting in a relatively small consumer surplus and a relatively large producer surplus.

Partial trade liberalization results in small gains in agricultural producer surplus, since higher export prices for rice and (to a minor extent) maize lead to increased production for export. This growth occurs at the expense of the domestic consumption of these commodities, leading to a decline in the consumer surplus. This decline is partially compensated by a rise in domestic consumption of imported beans, as a result of lower import prices. Under complete trade liberalization, the net effect on producer surplus is small since gains obtained through free trade in basic grains are absorbed by a decline in the production of other commodities (particularly milk, potato and onion), the imports of which increase (in turn resulting in a decrease in domestic prices). Lower import prices permit higher levels of consumption through the substitution of imports for domestic products, and as a result consumer surplus increases in the complete trade liberalization scenario.

Transport costs scenarios

The cost of transportation is generally an important determinant of farm production decisions and, as a result, of aggregate land use, which in turn influences aggregate welfare and trade flows. Even though virtually all major inter-regional Costa Rican road connections are paved, the quality of the pavement is usually poor, leading to high variable transport costs (Hausmann, 1998). Moreover, high vehicle import and road taxes, as well as high insurance costs, also lead to high fixed transport costs. Besides road conditions, transportation costs are obviously influenced by geographical distance. The highway between the harbor city of Limón in the Atlantic Zone and the capital of San José, passing the *Braulio Carillo* National Park (Figure 4.2), constitutes the country's main trunk road over which a large part of both agricultural and non-agricultural commodities travel to and from the Central Valley. This road is often closed, since it suffers from frequent land slides, and traffic is forced to look for alternative routes, which have transport costs that are some 2.7 times higher (estimated on the basis of a digital road map). A number of policy runs were carried out simulating the effects of a 20% reduction in variable or fixed transport costs, as well as a possible closure of the *Braulio Carillo* highway (assuming a duration of one year for modeling purposes). Such closure of the *Braulio Carillo* highway is a realistic possibility, either because of major landslides and/or a major reconstruction.

Changes in land use relative to the base run as a result of reductions in the cost of transportation are not dramatic (Table 4.8), since transport costs are already low when compared to total production value. Decreases in transport costs lead to a decline in pasture area for beef production, while favoring crop production and the raising of dairy cows. Exports of both traditional and non-traditional crops increase, while growth in basic grain production is much lower. On the other hand, closure of the *Braulio Carillo* highway leads to a reduction in (export) crop area in favor of pastures for beef production. Response reactions are larger for the decline in variable transport costs, in comparison with a decline in fixed transport costs, as variable transport costs account for between 73% (between *Central* and *Pacífico*) and 92% (between *Brunca* and *Chorotega*) of total transport costs. Closure of the *Braulio Carillo* passway has a much stronger effect on land use than the decline in either variable or fixed transportation costs, because of the considerable increase in the costs of transportation to the major export harbor (Limón).

Reductions in transport costs result in higher agricultural incomes in all regions, as well as increases in both exports and inter-regional transport flows (Table 4.9). While imports are negatively affected by a decrease in variable transport costs, they show an increase as a result of a decrease in fixed transport costs, even though in both cases the total volume of trade (*i.e.*, the sum of inter-regional and international trade) increases. Closure of the *Braulio Carillo* highway would result in a decline in national agricultural income due to reduced exports and lower inter-regional transport flows. This decline in exports, combined with increased imports, results in a deteriorating foreign trade balance.

Since a reduction in variable transport costs mainly favors export crops, regions with a relatively distant location from export harbors (*i.e.*, *Brunca* and *Norte*) benefit most from a reduction in variable (*i.e.*, distance related) transport costs. On the other hand, a decline in fixed transport costs benefits all regions equally, and therefore the region that houses the major export harbor (Limón in the *Atlántica* region) shows the largest relative decline in transport costs, and thus the largest response reaction. Inter-regional transport flows increase in both the reduced variable and the reduced fixed transport cost simulation, as lower transport costs allow for higher levels of regional specialization in agricultural production. The latter also causes a decrease in imports, even though the overall effect on imports of a reduction in fixed transport costs is positive, since prices of imported commodities in the region that houses the major export harbor fall relatively more than prices of commodities obtained from other regions (because inter-regional transport costs are largely determined by variable transport costs).

Closure of the *Braulio Carillo* passover in the road connecting the *Atlántica* region with the *Central* region leads to lower levels of agricultural income in these regions, while income in the other regions increases. Exports from the *Central* region decrease as a result of increased transport costs to the major export harbor (Limón) and because of less inter-regional trade with the *Atlántica* region. Trade between the Atlantic and other regions decreases, while trade among the latter increases. On the other hand, imports into the *Atlántica* region increase as a result of the isolating effect that a closure of the *Braulio Carillo* exerts on this region. On the national level, lower exports and higher imports result in reduced gains from trade.

Economic surplus (or welfare) increases as a result of decreases in transport costs, while reduced transport options result in a decrease in economic surplus. The increase in producer surplus exceeds that in consumer surplus. The former results from the downward shift of the supply curves, even though part of the gains are lost due to commodity price decreases. Since most demand elasticities are high (in absolute terms) relative to supply elasticities, price decreases are relatively minor, with a correspondingly small increase in consumer surplus. Closure of the *Braulio Carillo* passover generates opposite effects. The *Central* region purchases less from the *Atlántica* region and more from other regions, but at higher costs. Total consumption decreases in all regions, in the *Central* region because of higher prices and in the other regions because of larger shipments to the *Central* region. Production in other regions increases in order to meet the higher requirements of intra-regional trade as well as *Central* region demand, thus partially compensating for the decline in producer surplus.

Future perspectives: supply shifts

Technological change (*e.g.*, improved varieties and/or improved crop husbandry methods) results in the increased productivity of production systems and implies a downward shift of the supply curve, which results in lower unit production costs. A number of commodities that form part of our analysis are included in the mandate of international agricultural research centers. These include rice, beans and cassava; pastures are also subject to considerable international research. Consequently, for each of these commodities as well as for beef, model simulations were carried out assuming a 20% downward shift of the supply curve. Contrary to the previous simulations, supply shift simulation results are based on, and evaluated relative to, the complete trade liberalization situation as presented in the previous section, since supply response reactions can be expected to be relatively minor under the present restrictive trade policy regime. That is, currently existing export barriers would force any production increase resulting from a productivity increase to be absorbed by the domestic market, whose limited size would lead to artificially low prices.

A downward shift in the supply curve for a particular commodity leads to an increase in the cultivated area of that commodity as a direct result of lower unit production costs. Thus, production is increased through increases in both cultivated area and productivity. No significant trade-offs were observed of those commodities that did not experience similar technological change. Because of the sheer size of the pasture area in Costa Rica, technological progress in the beef production sector has the largest impact on total land use (in terms of absolute area changes), followed by the rice and bean sectors (Table 4.8). The absolute area changes caused by productivity gains in maize and cassava are relatively minor, mainly due to their initially much smaller areas. Supply response to technological change in a relative sense is highest for products that face favorable export markets (particularly cassava) and/or for products that are initially imported (particularly beans). Exports take place at fixed export prices without influencing regional price levels, and substitution of imports by domestic production generates producer surplus while lowering import expenditures.

All supply shifts lead to an increase in total agricultural income (Table 4.9). Productivity improvements in the beef sector in particular lead to relatively large increases in agricultural income, due to the ample pasture area. Productivity gains in rice, beans and cassava lead to similar growth in agricultural income, where production increases are either dedicated to exports (rice, cassava and beef) or used as a replacement for imports (beans). The international trade balance improves, while domestic consumption levels are maintained as regional product prices are held constant. In the case of productivity gains in rice, the resulting additional production of rice in the *Pacífico*, *Brunca* and *Norte* regions is mostly shipped to the *Central* region. This lowers the demand pressure exerted by the *Central* region on the *Chorotega* region for this commodity, thus enabling *Chorotega* to fully exploit its comparative advantage in the production of rice for the export market. A similar situation holds for productivity increases in cassava and beef, for which the *Norte* region has a comparative export advantage, while consumption requirements in the *Central* region can be satisfied by production increases in *Brunca* and the *Central* region itself. Finally, a productivity increase in bean cultivation causes bean imports to be substituted by domestic production in the *Norte* and *Brunca* regions.

For each of the commodities analyzed in this section, a downward shift of the supply curve results in an increase in the economic surplus which can be mainly attributed to a rise in producer surplus (Table 4.10). In line with the growth rates in production values (Table 4.9), productivity improvements are largest for beef. The increase in producer surplus stems from increased export production and/or the substitution of domestic production for imports, while production for internal consumption is maintained or increased. In combination with regional prices that either remain constant or decrease, consumer surplus remains either constant or increase. Shifts in the supply curves of rice and cassava have no effect on domestic consumer surplus, since additional production is exported while domestic consumption and prices remain constant. On the other hand, a shift in the supply curve of beans positively affects domestic consumer surplus, since imports are substituted by domestic production at prices that are below import prices. Similarly, a downward shift in the beef supply curve also results in a (small) rise in consumer surplus. In the major beef cattle raising region (*Chorotega*), productivity improvements in beef production permit a reduction in the pasture area while maintaining beef production, thus liberating agricultural land for the production of sugar cane for the domestic market at lower domestic prices.

Future perspectives: demand shift

In addition to population growth, demand for agricultural commodities is significantly influenced by income (Geurts *et al.*, 1997), thus making income growth an important determinant of future shifts in commodity demand. Unlike population growth, income growth does react to policy measures in the short-to-medium term, and scenario analysis based on income growth is therefore important. A model simulation was undertaken that assumes a "normal" rise in average per capita real income over the next 10 years (2.5% y^{-1}), in addition to the estimated annual population growth rate (2.0% y^{-1}) for the same period. Demand curve shifts were calculated by region for each commodity on the basis of 1995 demand data, which in turn were calculated from 1987-88 survey data obtained from DGEC (1988) using the population and income growth rates, as well as the regional expenditure elasticities for the commodities concerned (Section 4.4.2).

Similar to the supply shift simulations analyzed in the previous section, demand shift simulations are based on, and compared to, a situation of complete trade liberalization. If currently existing trade barriers in general, and restrictions on imports of many agricultural commodities in particular, continue to prevail in the future, commodity prices can be expected to increase significantly as a result of increased demand. Even though production increases could be achieved by expansion of the agricultural area, the agricultural frontier has been virtually reached in many parts of Costa Rica, at least in the areas with reasonable road and marketing infrastructure (Quesada Mateo, 1990). In addition, government policies favor continued protection of land that is potentially valuable for agriculture.

Since the agricultural area is limited to the cultivated area of the base run situation, total land use remains unchanged (Table 4.8). Increasing demand over time causes the area allocated to the cultivation of oranges, rice and pasture for milk production to increase, at the expense of the pasture area for beef production. Even though growth in real income results in increased consumption levels, its effect on consumption patterns is relatively minor. This can be explained by the expenditure elasticities, which are generally low. On the other hand, the expenditure elasticity estimates used in this study may not be valid much beyond the simulated period (*i.e.*, beyond the year 2000), since the country's demand structure can be expected to change in the long run as a result of continued economic development.

Continued population growth combined with rising incomes results in increases in demand, which, in turn, lead to four general developments. First, production of commodities with relatively unfavorable import prices increases, leading to the increased importance of rice and (to a lesser extent) maize in the overall cropping pattern. Similarly, the bean area decreases slightly as a consequence of more favorable import prices.

Second, even though domestic income growth increases domestic demand and decreases total export earnings (Table 4.9), production of the most profitable traditional as well as non-traditional export crops is hardly affected (Table 4.8). Traditional export crops (mainly banana and coffee) are highly profitable and remain responsible for the lion's share of agricultural income, despite the increasing prices of other crops. Moreover, increased domestic demand for these commodities hardly affects the total demand for these crops. This also holds for non-traditional export crops like pineapple, melon and cassava. In addition, since the share of the total production of non-traditional export crops that is exported is lower than for traditional export crops, the area of some of the former (orange, plantain, palm heart) increases as a result of increased domestic demand and unfavorable import prices. Milk, potato, rice, beans and plantain are all crops for which imports increase significantly as a result of demand growth, while possible previous exports of these commodities disappear as regional prices (increased by export transport costs) exceed world market prices due to higher domestic demand pressure.

Third, demand shifts lead to an increase in the importance of crop production in general (Table 4.8), at the expense of pastures for beef production, since returns on the latter are low (Jansen *et al.*, 1997b). Higher demand and subsequent higher prices lead to a corresponding increase in production values (Table 4.9). The largest growth in agricultural income occurs in regions that in the base run exhibited the lowest agricultural income (*Chorotega*, *Brunca* and *Norte*), since these are the regions where a large part of the area devoted to beef production gets converted into crop land.

Finally, with rising income levels, exports diminish while imports increase, resulting in significant deterioration of the foreign trade balance.

Demand increases resulting from growth in population and real incomes have a positive effect on both producer and consumer surplus (Table 4.10). However, the increase in producer surplus is much lower than the rise in consumer surplus, a fact that is not surprising given the demand-driven character of increased agricultural production. The relatively large gain in consumer surplus is a result of the considerable upward shifts of the demand curves. Due to area limitations, options for increasing the producer surplus are limited to the substitution of crop production systems at the expense of pasture-based beef production. The increase in producer surplus fully stems from the (limited) increase in crop production combined with higher (regional) equilibrium prices. The latter show a gradual increase due to increased domestic demand which in turn negatively affects exports. On the other hand, domestic prices are kept under control by world market import prices.

4.6 Summary and conclusions

In this chapter a Spatial Equilibrium Model (SEM) for Costa Rica was developed in order to model spatial patterns of agricultural supply, demand, trade and pricing, as well as to assess agricultural policy effectiveness in terms of social welfare. The SEM developed in this chapter considers the 17 main agricultural commodities in Costa Rica, its six planning regions as well as the Rest-Of-the-World (ROW) which it interprets as a seventh region to take international trade into account. A SEM uses econometric estimates about the behavioral relations of producers (supply response curves) and consumers (demand response curves), while simultaneously considering inter-regional transaction costs and government policies. Supply functions implicitly take into account factor markets (as they reflect marginal production costs) as well as bio-physical factors within each of the supply regions. However, a SEM is a static sector model, that does not take into account the process of change from one market equilibrium to a new market equilibrium, nor the transfers of production factors to and from other sectors in the economy.

The model was validated by comparing base run results with the actual situation in 1995 concerning land use, production quantities, value of production and product prices. Model specification and data reflected the actual 1995 situation quite well. Base run results show that about 87% of the total cultivated area is dedicated to pasture for livestock production, the remainder being used for traditional export crop production (using more than half of the total crop area), basic grains and non-traditional export crops (28% and 16% of the total crop area, respectively). Pasture is predominant in all regions, with *Chorotega* (beef) and *Central* (milk) being the most important production regions. Production of traditional export crops is concentrated in the *Central* and *Atlántica* regions, while non-traditional export crops are more widely dispersed. Basic grains are concentrated in the western and northern regions of Costa Rica. Traditional export crops account for almost two-thirds of the total agricultural production value, while the shares of livestock and non-traditional export crop production are only 19% and 12%, respectively. Over two-thirds of the total agricultural production value is exported, while about one-fifth is shipped to the *Central* region. Imports are negligible (due to relatively high import duties), as are product flows between regions other than the ROW and the *Central* region.

Policy scenarios include various trade liberalization measures, reduced transport costs, and shifts in supply and demand. While the first two scenarios are compared to the base run situation, the latter two scenarios are assessed relative to the outcome of the complete trade liberalization scenario, since effects of supply or demand shifts are expected to be low under the restrictive trade policy regime represented in the base run.

Trade liberalization simulations include partial (*i.e.*, free trade in basic grains only) and complete (*i.e.*, free trade in all commodities) liberalization scenarios, and refer to the abolishment of tariff and non-tariff measures regulating the respective commodities. The increase in producer surplus is relatively small, while the effect on consumer surplus is ambivalent. Higher domestic prices for basic grains result in a decline in consumer surplus in the partial simulation, while complete trade liberalization largely favors consumers as a result of favorable import prices. The net effects on the trade balance are slightly positive for the partial liberalization and slightly negative for the complete liberalization scenario. In agreement with economic theory, trade liberalization scenarios result in higher welfare. This increase in welfare insofar as it concerns the 17 agricultural commodities considered in the present study (over 3% in the complete liberalization scenario) does, however, not take into account the transition costs resulting from shift towards the new market equilibrium.

Transport cost scenarios include a 20% decline in variable and/or fixed transport cost, and a closure of the *Braulio Carillo* highway. While variable transport costs refer to costs t^{-1} km^{-1}, such as costs of petrol (including tax) and vehicle maintenance, fixed transport costs include vehicle depreciation, import and periodic tax levies, and insurance costs. Reductions in variable transport costs have a larger (positive) impact on social welfare than reductions in fixed transport costs, as variable transport costs account for about 75% of total transport costs. Of course, such welfare gains (nearly 2% in the variable transport cost scenario) should be compared with the costs for such improvements. The analysis in this study provides a clear justification for the past decision to construct the *Braulio Carillo* highway and for the costs involved in its maintenance. Closure of this highway would result in a decrease in producer surplus of nearly 10%, besides a reduction of 1% in consumer surplus. This result stems from the fact that such a closure would lead to a sizable increase in the transport costs required to reach the export harbor Limón, with obvious consequences for Costa Rica, which exports about 75% of its agricultural production value.

Advances in the production technology involved in producing a particular commodity lead to a reduction in unit production costs. A number of commodities included in our analysis are covered by the mandate of international agricultural research centers, such as rice, beans, cassava and beef (through pasture research). For each of these commodities model simulations were carried out assuming a 20% downward shift of the supply curve. A downward shift of the supply curve for a particular commodity leads to an increase in the cultivated area of that commodity, while domestic prices remain the same as they are largely determined by world market prices in a free trade environment. Production expands as a result of increases in both cultivated area and productivity, with no significant trade-offs with other commodities. Production increases are allocated to either exports (rice, cassava and beef) or used for import substitution (beans). Gains in producer surplus vary between 0.3% for rice and 3.9% for beef, while consumer surplus stays about constant. Because of the large size of pasture areas in Costa Rica, technological progress in the beef production sector has the largest absolute impact on land use and producer surplus, followed by technological progress in the bean sector.

In agreement with projections of the World Bank (World Bank, 1997) and the Inter-American Development Bank (Hausmann, 1998), a demand shift simulation is performed on the basis of a "normal" per capita real income growth rate of 2.5% per year for the period 1995 to 2005, and of the expected 2.0% annual population growth rate for the same period. Land allocated to commodities important for domestic consumption (orange, rice and milk) increases at the expense of pasture area for beef production, in view of the latter's relatively low returns. Demand increases are projected to be mainly met by increased imports and reduced exports, although the area allocated to the most profitable traditional as well as non-traditional crops is hardly affected. Domestic production also shows a small increase. In turn, the increase in consumer surplus exceeds the increase in producer surplus. The foreign trade balance for agricultural products can be expected to deteriorate significantly as a result of continuing economic growth, and should be offset by an increase in the export of industrial products or services. It is expected that the growth of industrial production will strengthen the foreign trade balance, relative to services and export earnings from tourism, banana and coffee.

In conclusion, this study has resulted, for the first time in Costa Rica, in the development and application of an agricultural sector model that can be used by policy makers to evaluate the likely outcomes of alternative policy measures affecting trade liberalization, improvement of the transport infrastructure, technological progress in agriculture, and economic growth. Before the SEM could be constructed, validated and used for scenario analysis, considerable effort was spent on the construction of a previously non-existing data base, needed to estimate regional supply elasticities for individual crops. In addition, regional demand elasticities for individual agricultural products were estimated using data from the latest national household expenditure survey. Finally, data on road infrastructure were collected and stored in a GIS and used as input in an econometric estimation of a transport cost model. The successful sequence of data collection followed by the use of solid econometric methods to estimate both supply and demand responses, combined with the use of a GIS and econometrics to estimate transport cost models and the use of all these building blocks to construct a SEM to analyze the effects of different (policy) scenarios, is still quite unique for (small) developing countries.

Appendix 4.1 Mathematical formulation of the Spatial Equilibrium Model

The following is a mathematical representation of the SEM developed in this study, whereas Tables A4.1 to A4.4 provide a description of the notation used. The objective function (in \$ 10^3 y^{-1}) to be maximized is given by the sum of consumer and producer surplus resulting from domestic demand and supply of commodities, plus exports earnings, minus imports costs, and minus total transport costs involved in trade flows of commodities between supply and demand regions. Demand and supply functions are assumed to be linear, resulting in a quadratic objective function.

$$\text{Max } NSW = \sum_{c} \sum_{j \neq ROW} \left[\frac{1}{2\varepsilon_{cj}^{D'}} Q_{cj}^{D^2} - \frac{\bar{q}_{cj}(1-\varepsilon_{cj})}{\varepsilon_{cj}^{D'}} Q_{cj}^{D} \right] -$$

$$\sum_{c} \sum_{i \neq ROW} \left[\frac{1}{2\varepsilon_{ci}^{S'}} Q_{ci}^{S^2} - \frac{\bar{q}_{ci}(1-\varepsilon_{cj})}{\varepsilon_{ci}^{S'}} Q_{cj}^{S} \right] +$$

$$\sum_{c} \left[p_c^X X_c - p_c^M M_c \right] - \sum_{c}\sum_{i}\sum_{j} t_{cij} T_{cij} \qquad \textbf{(1)}$$

where $\varepsilon_{cj}^{D'} = \varepsilon_{cj}^{D} \dfrac{\bar{q}_{ci}}{\bar{p}_{cj}}$ for all c, j **(2)**

and $\varepsilon_{ci}^{S'} = \varepsilon_{ci}^{S} \dfrac{\bar{q}_{ci}}{\bar{p}_{ci}}$ for all c, i **(3)**

Subject to:

- supply restriction per supply region per commodity (t y^{-1}):

$$Q_{ci}^{S} \geq \sum_{j} T_{cij} \qquad \text{for all } c, i \qquad \textbf{(4)}$$

- demand restriction per demand region per commodity (t y^{-1}):

$$Q_{cj}^{D} \leq \sum_{i} T_{cij} \qquad \text{for all } c, j \qquad \textbf{(5)}$$

- commodity balance per commodity, balancing commodity supply and demand (t y^{-1}):

$$\sum_{j} Q_{cj}^{D} = \sum_{i} Q_{ci}^{S} \qquad \text{for all } c \qquad \textbf{(6)}$$

- commodity export balance per commodity (t y^{-1}):

$$X_c = \sum_{i \neq ROW} T_{j=ROW,\, ci} \qquad \text{for all } c \qquad \textbf{(7)}$$

- commodity import balance per commodity (t y^{-1}):

$$M_c = \sum_{j \neq ROW} T_{i=ROW,\, cj} \qquad \text{for all } c \qquad \textbf{(8)}$$

- land resource restriction per supply region (ha y^{-1}):

$$\sum_{c} Q_{ci}^{S} / y_{ci} \leq l_i \qquad \text{for all } i \qquad \textbf{(9)}$$

Table A4.1. Superscripts

Indices	*Description*
X	exports
M	imports
S	supply
D	demand

Table A4.2. Subscripts

Indices	Description
i	supply regions: *Central, Pacífico, Chorotega, Brunca, Norte, Atlántica,* and Rest Of the World (ROW)
j	demand regions (same as supply regions)
c	commodities: rice, maize, beans, coffee, banana, sugar, plantain, palm heart, mango, melon, pineapple, cassava, onion, orange, potato, beef, and milk

Table A4.3. Variables

Variables	Description	Unit of measurement
NSW	net social welfare	$ 10^3 y^{-1}
Q^S_{ci}	supply quantity of crop c in region i	t y^{-1}
Q^D_{cj}	demand quantity of crop c in region j	t y^{-1}
T_{cij}	trade flow of crop c from region i to region j	t y^{-1}
X_c	export quantity of crop c ($X_c = Q^D_{c,j=ROW}$)	t y^{-1}
M_c	import quantity of crop c ($M_c = Q^S_{c,i=ROW}$)	t y^{-1}

Table A4.4. Parameters

Parameters	*Description*	*Unit of measurement*
t_{cij}	transport costs of commodity c from region i to region j	$ kg^{-1}
$\bar{p}_{ci} = \bar{p}_{cj}$	actual equilibrium price of commodity c in supply (i) and demand (j) regions	$ kg^{-1}
p^X_c	actual export price of commodity c	$ kg^{-1}
p^M_c	actual import price of commodity c	$ kg^{-1}
$\bar{q}_{ci}$	actual equilibrium production of commodity c in supply regions i	t
$\bar{q}_{cj}$	actual equilibrium consumption of commodity c in demand regions j	t
ε^S_{ci}	supply elasticity of commodity c in supply regions i	-
ε^D_{cj}	demand elasticity of commodity c in demand regions j	-
y_{ci}	yield of commodity c in supply region i	t ha^{-1}
l_i	land availability in supply region i	ha

5 Technical Coefficient Generators for quantifying land use systems

HUIB HENGSDIJK, BAS A.M. BOUMAN, ANDRÉ NIEUWENHUYSE, ROBERT A. SCHIPPER, and JANETTE BESSEMBINDER

Abstract

Many approaches to land use modeling employ linear programming techniques. Technical Coefficient Generators are expert systems designed to "generate" the technical coefficients of land use systems, herds and feed supplements used in linear programming models. The generated coefficients represent inputs and outputs of the production systems. This chapter describes two generic Technical Coefficient Generators, PASTOR (Pasture and Animal System Technical coefficient generatOR) and LUCTOR (Land Use Crop Technical coefficient generatOR), both of which quantify land use systems based on the integration of systems-analytical knowledge and expert knowledge. PASTOR quantifies pasture land use systems, herds and feed supplements in livestock production, while LUCTOR is concerned with annual, perennial and timber systems and managed natural forest. The main inputs quantified are labor requirements, fertilizers, biocides and associated costs. Outputs consist of yield and associated sustainability indicators: changes in soil N, P and K stocks (Δ stock); N losses via leaching, volatilization and (de)nitrification; quantities of the active ingredients applied in biocides and a so-called biocide index. PASTOR and LUCTOR are illustrated with data from the northern Atlantic Zone of Costa Rica.

5.1 Introduction

A Technical Coefficient Generator is an expert system for quantifying the input-output structure of actual and alternative land use systems. Land use systems refer to any type of land use under specific biophysical (*e.g.*, soil, climate) and technological (*e.g.*, management) conditions, that requires inputs and produces outputs, both of which are so-called technical coefficients (Fresco *et al.*, 1992).[1] The term "technical coefficient" is derived from terminology used in the literature on linear programming, a much-used technique in quantitative land use modeling (Hazell and Norton, 1986). For each land use system, *e.g.*, cropping, timber plantation, animal husbandry, a unique combination of inputs results in a unique combination of outputs. Inputs may include external nutrients (*e.g.*, fertilizer), biocides, labor use and agricultural implements. Typically, outputs are production items in physical or financial terms, but may also include indicators related to natural resource use, such as changes in soil stocks (*e.g.*, nutrients, organic matter), and to waste loss and environmental emissions, such as nutrients, biocides and trace

[1] A complete definition of land use system is given in the section 'Concepts and definitions employed in land use analyses' at the end of this book.

or greenhouse gases. As such, technical coefficients include indicators that are required in the quantification of socio-economic and biophysical sustainability trade-offs within and among land use systems. Technical Coefficient Generators are in principle designed to generate the technical coefficients of production systems required as the inputs for the optimization models used in land use studies (see Chapters 6-8). However, they are also useful as stand-alone tools to study the biophysical and economic viability of alternative land use systems on the basis of field experimentation (Hengsdijk *et al.*, 1999).

Building upon experiences gained in previous phases of REPOSA (Jansen and Schipper, 1995; Stoorvogel *et al.*, 1995) and in related studies in The Netherlands (Habekotté, 1994), Europe (De Koning *et al.*, 1995) and West Africa (Hengsdijk *et al.*, 1996), two Technical Coefficient Generators were constructed for the northern part of the Atlantic Zone of Costa Rica (AZ): PASTOR (Pasture and Animal System Technical coefficient generatOR) for the region's cattle systems, and LUCTOR (Land Use Crop Technical coefficient generatOR) for its cropping systems. The value of PASTOR and LUCTOR in land use studies involving linear programming techniques has been demonstrated both on the regional and farm levels (Bouman *et al.*, 1998c; Sáenz *et al.*, 1998; Bouman and Nieuwenhuyse, 1999; Schipper *et al.*, 1998), and will also be illustrated in Chapters 6, 7 and 8 of this book. This chapter explains the underlying, generic concepts of PASTOR and LUCTOR, describes their functioning and data requirements, illustrates some results, and discusses the benefits of the presented approach.

Both PASTOR and LUCTOR are available on the CD-ROM that comes with this book, accompanied by exercises illustrating how they function and how the generated technical coefficients can be used in optimization models.

5.2 Concepts involved in the generation of technical coefficients

5.2.1 Type of land use systems

A land use system is defined as a combination of a *land unit* with a *land utilization type*; it is characterized by its specific sets of inputs and outputs, *i.e.,* its technical coefficients, and possibly by land improvement systems such as irrigation or drainage (after FAO, 1976, and Driessen and Konijn, 1992). Here, the term "land use system" is synonymous with the term "production activity" as defined in production ecology: the cultivation of a crop or rotation of crops in a particular physical environment, that is characterized in terms of its inputs required and its outputs produced (van Ittersum and Rabbinge, 1997). *A land utilization type* is a specific kind of land use under well-described biological (*e.g.,* crop, variety, product), socio-economic (*e.g.,* labor use) and technological (*e.g.,* management, sequence operations, use of inputs) conditions (after Fresco *et al.*, 1992, and Jansen and Schipper, 1995). A *land unit* is defined as a physical area of land that is uniform in its characteristics and qualities (after FAO, 1983). Pasture land use systems come under the more general category labeled *livestock systems,* which also include herds that graze upon the pastures and the addition of any feed supplements (Bouman *et al.*, 1998a; Fresco *et al.*, 1992).

Land use studies are usually concerned with *i*) descriptions and explanations of current land use (*e.g.,* Alexandratos, 1995), *ii*) evaluations of the magnitude of the problems involved (*e.g.,* Penning de Vries *et al.*, 1995), *iii*) explorations of technical options that may contribute to solutions (*e.g.,* Rabbinge and Van Latesteijn, 1992) or *iv*) predictions about the possible effects of policy intervention on land use (*e.g.,* Hengsdijk *et al.*, 1998b). Typically, the goals along with the spatial and temporal scales of analysis differ for each type of land use study. The combination of both (*i.e.,* goals and scale) determines to a large extent which type of land use systems are to be included in the analysis.

In *biophysical explorative studies*, that look some 20-30 years ahead (*e.g.,* as in Chapter 7), sustainable land use options may be explored under what are mostly biophysical boundary conditions. Such studies require *alternative* land use systems that are technically feasible and sustainable from a biophysical point of view, but most likely not yet widely practiced. Such systems use inputs more efficiently than current systems due to supposed future efficiency gains in agricultural production (De Wit *et al.*, 1987). The biophysical sustainability of alternative land use systems is mainly operationalized in terms of a zero change in soil nutrient stock: all nutrients withdrawn from the system (resulting from both product removal and unavoidable losses) are balanced by various external inputs (*e.g.,* fertilizer, natural deposition). This implies that the productivity of such alternative land use systems, as determined by nutrient stocks in the soil, is maintained over time. Furthermore, these types of land use systems may include crop or grass species that are currently not (yet) grown, or production techniques that are not widely used (*e.g.,* mechanization or the use of less biocides), but that are potentially interesting alternatives. Disregarding such options in long-term explorative studies would overlook possible prospects for agricultural development in *e.g.,* "labor extensive" or "environmentally friendly" directions.

In combined *biophysical socio-economic* land use studies, the emphasis is often on identifying the possible effects of policy intervention on the performance of so-called representative farm types (*e.g.,* Hengsdijk *et al.*, 1998b). Therefore, besides alternative land use options, *actual* land use systems that characterize current farm conditions are included. Often, though not necessarily, such actual land use systems are unsustainable because they deplete the soil nutrient stock (*e.g.,* as is the case in the AZ, see Chapter 2), whereas the alternative ones are (biophysically) sustainable by definition. With a time horizon of 0-5 years, however, future efficiency gains in agricultural production are presumably less pronounced, and land use systems should represent actual systems and incorporate changes in production techniques that can be expected to be realized only in the short-term. Combined biophysical socio-economic land use studies with a long time horizon (20-30 years), can use the same alternative systems studied as in the more biophysical oriented studies discussed above. Analyzing *actual* and sustainable *alternative* land use systems together or alternately in land use models under various biophysical sustainability restrictions quantifies the trade-offs between the biophysical and economic dimensions of sustainability, including the scope for arriving at positive, rather than negative, trade-offs (so-called win-win situations, Bouman *et al.*, 1999a). In this book, examples of long-term studies of land use options are described in Chapter 6 and 7, and Chapter 8 reports on a relatively short-term study.

5.2.2 Quantifying technical coefficients

In both LUCTOR and PASTOR, the so-called "target oriented" approach (Van Ittersum and Rabbinge, 1997) is used to quantify *alternative* land use systems: target production levels are predefined and the various combinations of inputs required to realize these target levels are subsequently quantified. For example, the target production levels for crops and pastures may vary from maximum (*i.e.,* potential), close-to-actual to very low yields, resulting in simulated high and low external input levels (*e.g.,* fertilizers, biocides) for the first and the last case, respectively. Substitution between different types of inputs is reflected by changes in labor and capital inputs (De Wit, 1979). This implies that production techniques can be quantified using either herbicides or manual weeding methods, or using either manual or mechanized field preparation methods. Data for quantification of alternative production systems are derived from knowledge of the ecological processes involved, survey data on crop husbandry practices of the most efficient farmers in the region, relevant scientific literature and expert knowledge (Bouman *et al.*, 1999a). For quantification of *actual* production systems a descriptive approach is used. Primary data regarding inputs and physical production are obtained from field surveys and agricultural statistics, while the remaining data gaps are estimated using standard agronomic and animal husbandry data, supplemented by expert knowledge.

Land utilization types were defined for crops (LUCTOR) and pastures (PASTOR). The distinction between *land units* is based on the diagnostic land qualities of soil fertility and drainage conditions, and on the diagnostic land characteristics of slope and stoniness (see Chapter 2). Weather is not a diagnostic characteristic, since it is considered sufficiently homogeneous in the AZ. The diagnostic land qualities and characteristics in the AZ determine the range of possible crops, the maximum yield level, the suitability for mechanization, the costs for field preparation and the nutrient recovery rates. Three major land units were distinguished, each of which were subdivided into mechanizable and non-mechanizable sub-units: Soil Fertile Well drained (SFW), Soil Infertile Well drained (SIW) and Soil Fertile Poorly drained (SFP) (see Chapter 2).

Both PASTOR and LUCTOR generate three categories of technical coefficients: *i)* input requirements in physical and economic terms, *i.e.* labor, fertilizers, biocides, implements and costs, *ii)* physical production (*i.e.,* crop yield, meat, milk) and *iii)* biophysical sustainability indicators. The first two categories are universal, but the use of sustainability indicators depends on the biophysical sustainability issues relevant to each specific case study. Based on the relevant issues in the AZ (Chapter 2), the following technical coefficients of biophysical sustainability were implemented: the change in soil nutrient stock (Δ stock) for nitrogen (N), phosphorus (P) and potassium (K); N losses to the environment; the volatilization and denitrification/nitrification of N as proxy for N-related greenhouse gas emissions; and the use of biocides, expressed as the total amount of active ingredients used (BIOA) and by an ordinal biocide index (BIOI; Jansen *et al.*, 1995) that takes into account not only the active ingredients used but also their degree of toxicity and their persistence in the environment:

$$BIOI = \frac{1}{Y} \sum_{\substack{\text{biocides,}\\ \text{applications}}} Q*f*TOX*\sqrt{DUR}$$

Where: Y = duration of land use system (y); Q = biocide quantity (kg); f = fraction active ingredient in biocide; TOX = toxicity of active ingredient (World Health Organization (WHO) code), and DUR = persistence of active ingredient in environment (d). The WHO codes, indicating the toxicity of the active ingredients, are translated into numbers (see Table 5.1). Attaching different values to different toxicity levels allows us to differentiate more or less different types of active ingredients according to their toxicity (Chapter 6).

Table 5.1. Levels of toxicity used in calculating biocide indices, along with their corresponding WHO-codes.

WHO code	Description	Toxicity parameter
Ia	Extremely hazardous	9
Ib	Highly hazardous	7
II	Moderately hazardous	5
III	Slightly hazardous	3
V	No acute risk	1

In a later phase, emission/sequestration of carbon (C) was incorporated as an additional sustainability indicator to monitor the fate of the greenhouse gas CO_2 (Plant and Bouman, 1999; Bouman *et al.*, 1999c). Biophysical sustainability indicators are calculated by bookkeeping the biocides and nutrients in the system. Nutrient efficiencies and loss fractions are based on a combination of systems-analytical and expert knowledge. Input and outputs are expressed per hectare per year.

Input costs consist of two types: the costs of current inputs (*e.g.*, seeds, fertilizers and biocides) and the costs of services either from movable capital items (*e.g.*, implements) or immovable items (*e.g.*, on-farm post-harvest processing units, drainage canals). Input costs of movable inputs are based on rent prices. To calculate the cost of immovable inputs it is implicitly assumed that the scale on which such inputs are used is economically optimal. These costs are expressed as an annuity factor to take the investment costs of materials with a life span exceeding one year into account. Annuity costs are calculated using the capital recovery factor (Gittinger, 1982) with a discount rate specified by the user.

5.2.3 Complementary information sources

The calculation of technical coefficients is based on standard data regarding agronomic and animal husbandry relationships, empirical data and systems-analytical knowledge of the physical, chemical, physiological and ecological processes involved. In situations where data are incomplete or lacking, or where processes are poorly understood, expert knowledge is used as a complementary information source. For example, process-based models predicting the complex interactions between pests and crops and their effect on yields are not yet sufficiently developed for the useful generation of technical coefficients (Kropff *et al.*, 1995). This is due to the stochastic and location-specific nature of crop-pest complexes, which make their effects on yields highly diverse and difficult to model. Crop experts with years of location-specific field experience, on the other hand, are often able to make reliable predictions with sufficient accuracy for use in the generation of technical coefficients. In the development of PASTOR and LUCTOR, teams of experts were frequently consulted because of their knowledge about the livestock and cropping systems in the AZ, which resulted in the well-considered and much-discussed formulation methods of quantifying technical coefficients.

5.3 PASTOR

PASTOR (Bouman *et al.*, 1998a) contains separate modules for the calculation of technical coefficients for pasture, herd and feed supplement systems.

5.3.1 Pasture

The pasture module can model three pasture land utilization types: *i*) fertilized, improved pastures (grasses), *ii*) grass-legume mixtures, and *iii*) unfertilized natural pastures that represent the *actual* systems in the AZ. The first two types are *alternative* systems that are sustainable in the sense that their soil nutrient balances are in equilibrium (zero Δ nutrient stock); the third type may be unsustainable in terms of soil nutrient stock, depending on management characteristics such as stocking rate and feed supplementation practices. All pasture land utilization types are characterized by their management: botanical composition (species), stocking rate, weeding technique and production level as determined by the rate at which fertilizers are applied. The pasture land utilization types are combined with land units to form pasture land use systems. Table 5.2 gives an example of an implementation for the AZ as developed in a regional land use study (Bouman *et al.*, 1998c).

Table 5.2. Definition criteria and options for the pastures in the northern Atlantic Zone of Costa Rica used in PASTOR.

Definition criterion	*Maximum number of options*
Botanical composition	6 (Improved grasses *Cynodon nlemfuensis*, *Brachiaria brizantha*, and *Brachiaria radicans*; grass-legume mixtures *B.brizantha-A.pintoi* and *B.humidicola-A.pintoi* mixture; "Natural" which represents a mixture of the naturalized and native grasses *Ischaemum ciliare*, *Axonopus compressus* and *Paspalum* spp.)
Land unit	3 (Fertile well drained, fertile poorly drained, infertile well drained)
Stocking rate	21 (From 1 to 6 animal units per ha, in steps of 0.25. For the grass-legume mixtures and the natural pasture, stocking rates only varied from 1-3.25)
Weeding manner	3 (Only herbicides, only manual, mixed herbicides and manual)
Fertilizer application	11 (From 0 to 100% required to reach maximum attainable production in steps of 10%)

For fertilized pastures, technical coefficients are calculated with a pre-defined allowable loss of soil nutrient stock, *i.e.,* the maximum quantities of N, K and P that are allowed to be removed from the soil stock are pre-defined by the user. For sustainable *alternative* pastures, these losses in soil nutrient stock are zero. The procedure for calculating technical coefficients for fertilized pastures is rather complex and involves a number of steps (Figure 5.1). For each grass species, upper and lower production boundaries are estimated for each land unit in the study area in terms of biomass and contents of metabolizable energy (ME), crude protein (CP) and phosphorus (P). The upper boundary corresponds to the maximum attainable production with no nutrient constraints (Bouman *et al.*, 1996), whereas the lower boundary corresponds to the minimum production level attained on exhausted land units where the grass just manages to survive. On the basis of the maximum attainable production, PASTOR calculates the attainable feed (*i.e.,* biomass and amounts of ME, CP and P) produced as a function of a range of (user-defined) stocking rates. With increasing stocking rate, less of this biomass can be eaten by the grazing cattle because of trampling and deposition of faeces and urine (Van der Ven, 1992). N, P and K soil nutrient balances are calculated using an adapted version of the model presented by Stoorvogel (1993). The calculations are based on estimates/calculations for all inputs, namely atmospheric deposition, fixation by micro-organisms, weathering, manure and urine (from the grazing stock), and all outputs, namely the attainable amount that may be removed by grazing and losses by erosion, leaching, volatilization, denitrification/nitrification, and fixation (only for P). A negative balance (*i.e.,* loss of soil nutrient stock) indicates the amount of fertilizer that is needed to sustain the attainable amount of biomass that may be removed by grazing. Next, a user-defined range of fertilizer application levels is specified, ranging from 0-100% of the amount needed to sustain this amount of attainable feed. Gross fertilizer input is calculated from the applied net amount by taking account of loss fractions specified per nutrient type. For each fertilizer application level, the actual amount of feed is calculated from the total amount of nutrients available (from fertilizer, manure and all other external sources) and non-linear energy and nutrient concentrations in the pasture biomass as functions of nutrient availability. For example, with 0% fertilizer application, the amount of feed can not be higher than the amount that is produced using only external inputs from atmospheric deposition, fixation by micro-organisms, weathering, faeces and urine. In the case of 100% fertilizer application, the amount of feed equals the maximum attainable production. In a last step, the amount of available feed at the various fertilizer application rates is compared to the uptake capacity of the cattle at the various stocking rates. Since cattle can not remove more feed than their intake capacity, the actual amount of pasture removed by grazing is limited to the intake capacity of the grazing cattle. Any over-production of pasture is recycled into the soil. When pasture production is not sufficient to sustain cattle intake (*e.g.,* as in the combination of relatively low fertilizer application rates with high stocking rates), a feed supply shortage is obtained. In a whole cattle production system, this shortage should be filled by feed supplements, thus constituting an additional source of external nutrients to the cattle.

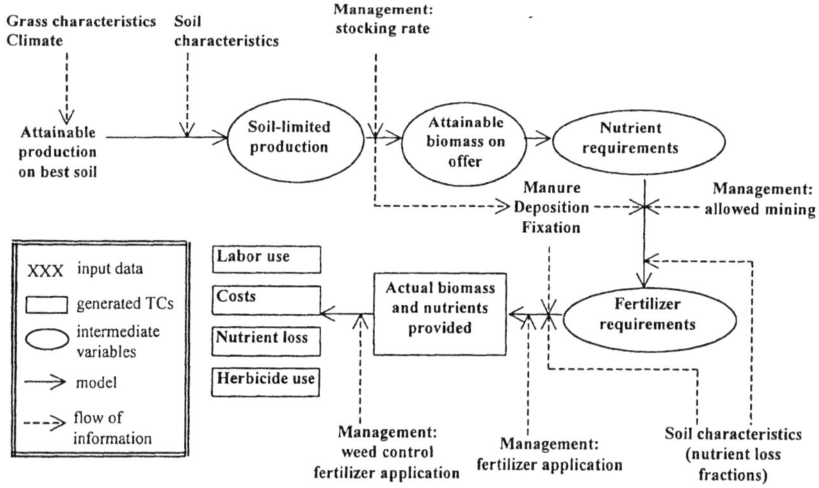

Figure 5.1. Schematic representation of the procedure for calculating technical coefficients for fertilized, alternative pastures.

For unfertilized pastures (*i.e.,* grass-legume mixtures and natural pastures), the calculation procedures are relatively simple. Since no fertilizer is applied, actual feed on offer is specified by the user as function of a range of feasible stocking rates. In the case of grass-legume mixtures, the soil nutrient balance model takes account of the additional input of N by the legume. The soil nutrient balance is merely the result of bookkeeping all nutrient inputs and outputs, and may be in equilibrium, as it is in grass-legume mixtures, or have a negative (*i.e.,* stock depleting) value, as it does in most *actual* grass-only systems. For all pastures (*i.e.,* fertilized or unfertilized), costs and labor requirements are related to material inputs such as fences, tools and herbicides, as well as to operations such as establishment, weeding, fertilizer application (if any) and maintenance. Different modes of weeding may be specified by using different combinations of herbicides and mechanical weeding techniques (manual).

For the AZ, the following pastures are included in PASTOR. Fertilized pastures which include three improved grass species: *Cynodon nlemfuensis*, *Brachiaria brizantha* and *Brachiaria radicans*. For each species, technical coefficients can be generated for the combination of 21 stocking rates from 1 to 7 animal units ha^{-1} (AU; 1 AU = 400 kg live weight), in steps of 0.25 AU ha^{-1}, with 10 fertilizer application levels ranging from 0 to 100% (in steps of 10%) of the amount required to sustain maximum attainable production. Unfertilized pastures include three options: "natural", which stands for current mixtures of native and naturalized species (*Ischaemum ciliare*, *Axonopus compressus* and *Paspalum spp.*), and two grass-legume mixtures: *Brachiaria brizantha* with *Arachis pintoi* and *Brachiaria humidicola* with *Arachis pintoi*. Technical coefficients can be generated for 10 stocking rates ranging from 1 to 3.25 AU ha^{-1} in steps of 0.25 AU ha^{-1}. At stocking rates higher than 3.25 AU ha^{-1}, the grass-legume mixtures have proven to be non-persistent (Ibrahim, pers. comm.). Natural pastures generally have negative changes in soil N (and K) stock, whereas in grass-legume mixtures, yield levels and legume N contributions are balanced in such a way that the soil N balance is in equilibrium (although there still can be negative changes in soil K stock). Pasture characteristics have been compiled on the basis of previous studies (Vicente-Chandler *et al.*, 1974; Salazar, 1977; Veldkamp, 1993; Ibrahim, 1994; CIAT, 1995; Hernández *et al.*, 1995; Nieuwenhuyse, 1996), unpublished experimental data of the Ministry of Agriculture and Livestock of Costa Rica and expert knowledge. Production potential and the possibility of growing certain pastures vary with land unit: the fertilized *Cynodon nlemfuensis* and *Brachiaria brizantha* grow on well-drained land (SFW, SIW); fertilized *Brachiaria radicans* on poorly drained land (SFP); natural pastures on all land; *Brachiaria brizantha* with *Arachis pintoi* on fertile, well-drained land (SFW); and *Brachiaria humidicola* with *Arachis pintoi* on infertile, well-drained land (SIW). Management parameters are derived from field surveys (Van Loon, 1997). In all pastures, weeds are combated by a combination of manual weeding and herbicides, except in grass-legumes where only manual weeding can be used.

5.3.2 Herd

The herd module in PASTOR can quantify technical coefficients for breeding, fattening and double-purpose systems. A breeding system is defined as a system where calves are bred and subsequently sold at a certain age or live weight. No animals are bought externally. A fattening system is defined as a system where young animals are bought, fattened for a period of time, and sold afterwards. No animals are bred internally. A double-purpose system is managed the same way as a breeding system, the difference being that, besides meat, milk is also sold. For all types, the modeled herds are "stationary", which means that there are no changes in herd size and composition over the year(s) (Upton, 1989; 1993). Production and feed requirements of the herd are computed, based on specified herd structure characteristics, target growth of the animals and target buying/selling strategy, and total composition. The (stationary) composition of the herd, *i.e.*, the number and type of animals per age class, is calculated using the method presented by Hengsdijk *et al.* (1996). The production of the herd is obtained by adding together the user-specified target live-weight gains and milk production for all animals in the herd, using the user-defined buying/selling strategy. Feed requirement calculations are based on the equations presented by the National Research Council (NRC, 1989, 1996). Calculations are performed for each animal in the herd according to sex and age group, and for females according to stage of pregnancy and lactation. They are then added together to obtain total herd requirements. The costs and labor requirements of herds are related to construction, buying and maintenance of corrals, feed troughs, various equipment, vaccinations, assistance at birth and animal health care. The costs and labor requirements are quantified for each of these items and operations, and added together to obtain herd totals.

For the AZ, breeding and fattening herds are generated, each with a low and a high growth rate representing *actual* and *alternative* herds, respectively. Currently in the AZ, the growth rates of calves in their first year are relatively low because of the poor quality of natural pasture and limited use of good quality feed supplements (Hernández *et al.*, 1995; Jansen *et al.*, 1997b; Van Loon, 1997). Based on the field observations of well-managed herds (Van Loon, 1997), first-year growth rates of *actual* herds are set at 0.65 kg head^{-1} d^{-1} for males and 0.52 kg head^{-1} d^{-1} for females in breeding systems, and at 0.5 kg head^{-1} d^{-1} for males and 0.4 kg head^{-1} d^{-1} for females in fattening systems. *Alternative* herds are modeled with first-year growth rates of 1.0 kg head^{-1} d^{-1} for males and 0.9 kg head^{-1} d^{-1} for females in breeding systems, and of 0.9 kg head^{-1} d^{-1} for males and 0.8 kg head^{-1} d^{-1} for females in fattening systems (using calves of eight months and older). These herds are based on the assumption that higher growth rates can be sustained through the use of improved pastures and high-quality feed supplements (van der Grinten *et al.*, 1992; Hernández *et al.*, 1995; Coopemontecillos). Input parameters on herd characteristics, management and selling strategies are derived from field surveys (Van Loon, 1997), representative of the situation of well-managed herds in the area.

5.3.3 Feed supplements

The feed supplement module converts data on supplements into feed characteristics (metabolizable energy, crude protein and phosphorus), costs and units of required labor. For the AZ, these supplements included: green rejected bananas, sugar cane molasses, two types of chicken-dung based concentrates, and a P mineral salt. Nutrient and energy concentrations are taken from Vargas (1984) for bananas and molasses, and from supplier's information for the concentrates and the salt.

5.4 LUCTOR

5.4.1 Crops and technology

LUCTOR (Hengsdijk *et al.*, 1998a) generates technical coefficients for annual cropping systems, perennial cropping systems, timber plantations and managed natural forests. These systems are characterized in terms of the complete operation sequences involved and all the quantified inputs and outputs of these operations (Stomph *et al.*, 1994). For annual cropping systems, periods are defined for each well-defined operation (*e.g.*, field preparation, sowing, *etc.*) to take into account the timeliness of operations and to identify labor peaks. For perennial cropping systems and timber plantations, no such periods are identified since these systems require different operations throughout the entire year, as a results of the relatively uniform climatic conditions in the AZ, and since such operations typically occur simultaneously. Therefore, labor requirements for these systems are spread evenly over the year.

Actual and alternative cropping systems are characterized by environmental and management criteria. The most important criteria and their corresponding options are shown in Table 5.3 and discussed below. Based on user-defined environmental and management options, LUCTOR calculates physical input requirements and total costs of input use of each unique land use system, as well as the associated indicators of natural resource use and emissions to the environment.

Table 5.3. Definition criteria and options for cropping systems in the northern Atlantic Zone of Costa Rica

Definition criterion	*Maximum number of options*
1. Land unit	3 (Fertile poorly drained, fertile well drained, infertile well drained)
2. Crop type	12 (Black beans, cassava, maize-grain, maize cobs, export pineapple, local pineapple, palm heart, plantain, banana, teak, melina and managed natural forest)
3. Yield level	11 (10 Target yields for alternative systems, 1 yield level for actual systems)
4. Mechanization level	2 (Low and high)
5. Crop residue strategy	2 (Harvesting, left in the field)
6. Herbicide level	2 (Low and high)
7. Pesticide level	2 (Low and high)

Crop

LUCTOR generates land utilization types for the following crops: palm heart, black beans, cassava, maize (grain and fresh cobs), melina and teak tree, banana, pineapple (both for export and local markets), plantain, and managed natural forest. These crops are chosen based on the expert knowledge about their biophysical and economic potential in the AZ. For maize and pineapple, two types of crops are considered since their marketable products have different economic values and market outlets. In addition, since their crop characteristics and growth cycles are distinct, their input-output relations differ as well. For all other crops, only one single crop type is defined.

Identification of feasible crop-land unit combinations is based on a qualitative land evaluation. The fertile, well drained land unit (SFW) is suitable for all crops, although palm heart, plantain, timber, and pineapples for export all require the construction of a drainage system. The poorly drained land unit (SFP) can be used in its natural state for natural forest management, and is suitable after the construction of a drainage system for both banana and plantain. The infertile land unit (SIW) is unsuitable for banana, plantain, beans and maize, mainly because of high soil acidity. Costs of construction and maintaining drainage systems for crop-land unit combinations are included in the description of the relevant land use systems. The land characteristics of slope and stoniness determine the feasibility of mechanized cropping systems. Land units having either slopes of more than 25% and/or more than 1.5% stones are considered unsuitable for cropping systems that require machinery. Due to the erosion hazard, teak plantations are feasible only on land with slopes of less than 25% (Chavarria and Valerio, 1993).

Yield level

Ten target yields are defined for the alternative land use systems. The maximum target yield level, being the maximum attainable production without nutrient constraints

(Bouman *et al.*, 1996), is reduced in 10% steps, so that the lowest yield is 10% of the maximum attainable production. The maximum attainable production level takes into account the quality characteristics required by some crops. For example, most cassava cultivated in the AZ is for export, a market which demands relatively small tubers, which are harvested before the maximum crop biomass is attained. Yields may include as many as three product qualities for annuals and two for perennials, all of which may have their own price and market outlets.

Mechanization level

Mechanization levels refer to soil preparation operations and the application of biocides (subdivided into pesticides and herbicides). Other mechanized field operations are limited in view of the high rainfall intensities in the AZ, the high risk of soil compaction, and the characteristics of certain crops (*i.e.,* narrow passage in perennials). In the high mechanization option, field preparation is mechanized while biocides may be applied with a boomspray or spray plane (for pesticides only) instead of a backpack sprayer.

Crop residue strategy

Crop residues may either be left in the field after harvesting or be collected and used for *e.g.,* fodder purposes. Both options influence the labor requirements as well as the nutrient relationships of cropping systems.

Herbicide and pesticide levels

Biocides are divided into herbicides and pesticides, the latter including fungicides, insecticides and nematicides. In the low herbicide option, herbicides are completely substituted by manual weeding, which requires more labor and which reduces the emission of active ingredients into the environment. In the low pesticide option, insecticides and fungicides are reduced by a crop-dependent percentage to a level lower than the high pesticide option. It is assumed that with better crop monitoring and hygienic measures - both of which require additional labor - the use of these pesticides can be reduced. Lower pesticide use reduces emissions of active ingredients but may also lower yields, since yield losses are, in general, inevitable when insecticide and fungicide use is lowered. The extent of these yield losses is estimated on the basis of expert knowledge.

5.4.2 Alternative and actual cropping systems

For the quantification of *alternative* cropping systems, yield levels are based on available field experiments and on discussions with field experts. Furthermore, these systems attempt to maintain soil nutrient balances of N, P and K in equilibrium; this requirement implies that the annual nutrient uptake and losses due to erosion, leaching, volatilization, denitrification and fixation (only for P) are replenished with nutrients from natural sources (atmospheric deposition, crop residues and fixation by micro-organisms), in addition to a certain amount of fertilizer that is calculated by LUCTOR. In case of black beans, the additional input of fixed N by the crop is taken into account. The procedure of determining fertilizer requirements is straightforward and is based on the same bookkeeping procedure used in PASTOR. Loss fractions by type of nutrient are based on a combination of systems-analytical and expert knowledge. For some perennial and timber systems, nutrient balances may show a positive result, *i.e.,* nutrients in the soil are enriched. In these systems, nutrient turnover in the different years of the crop cycle (*i.e.,* the time during which the land is planted with a crop) is taken into account. Nutrients in crop residues left in the field after harvesting as well as nutrients in the standing crop are discounted in the following year. At the end of a crop cycle, a large flush of nutrients from decomposing crop residues is released, and is available at the start of a new crop cycle. In such situations the inputs of nutrients may exceed the sum of the crop uptake and nutrient losses, thus resulting in positive changes in soil nutrient stock. In Figure 5.2 the procedure for calculating nutrient cycles in actual and alternative perennial and timber systems is presented.

Although yield levels of alternative cropping systems are evenly distributed over their range, other inputs and outputs are not; a practice that is justified since higher yield levels are usually associated with higher crop nutrient concentrations (Van Keulen and de Wolf, 1986). In this way, non-linear (*i.e.,* diminishing return) relationships are determined between fertilizer requirements and yield levels. It is assumed that the use of all insecticides and fungicides decreases proportionally with diminishing yield levels because a number of fungal diseases and insects pests require less effort to be controlled under less favorable growing conditions (De Wit, 1994). Finally, it is assumed that inputs in alternative cropping systems are applied in a more technically efficient manner than in actual cropping systems, which may be expressed in: *i*) crop characteristics that are geared towards higher yields than in the actual systems (*i.e.,* higher harvest indices); *ii*) a shift in the distribution of quality class towards a higher fraction of prime quality as a result of better crop management (*i.e.* in fruits such as pineapples); *iii*) higher planting densities, and *iv*) higher frequencies of fertilizer applications.

For *actual* cropping systems, the calculation procedures are to a large extent similar to those for alternative systems. However, in the case of actual cropping systems, empirical data on yield and use of inputs such as nutrients, labor and biocides are used to determine associated biophysical sustainability indicators. Any missing values are estimated using agronomic knowledge and expert judgement. Unlike the approach for alternative systems - where nutrient balances are in equilibrium by design - nutrient balances of actual cropping systems are simply the result of summing all outputs (nutrient losses) and inputs (nutrient gains). Actual cropping systems do not necessarily have lower yields than alternative cropping systems. However, alternative cropping systems, at least theoretically, can be practiced without depleting soil nutrient stocks, while most actual cropping systems deplete the soil nutrient stock and are, therefore, not sustainable in the long run.

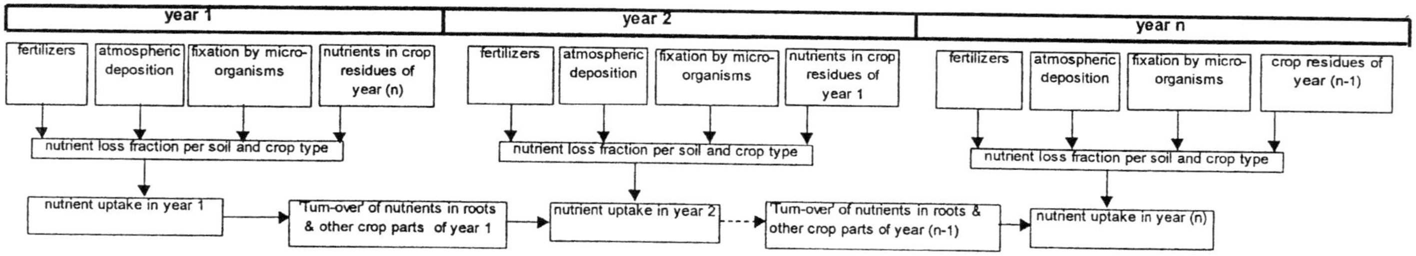

Calculation of nutrient balances of actual cropping systems:

For each year i: (fertilizers + deposition + fixation + residues) * (1 - nutrient loss fraction) = available nutrients in year i

nutrient balance in year 1 = available nutrients in year 1 - (nutrient uptake in year 1 + nutrient losses due to erosion)

nutrient balance in year 2 = available nutrients in year 2 - (nutrient uptake in year 2 + nutrient losses due to erosion - 'turn-over' nutrients of year 1)

⋮

nutrient balance in year (n) = - available nutrients in year (n) - (nutrient uptake in year (n) + nutrient losses due to erosion - 'turn-over' nutrients in year (n-1))

Calculations of nutrient balances of alternative cropping systems:

For each year i: (deposition + fixation + residues) * (1 - nutrient loss fraction) = available nutrients in year i

Note: Nutrient balances of alternative systems are predefined and set at 0, instead the required amount of fertilizers is calculated to obtain such 'closed' balances.

nutrient requirements in year 1 = ((nutrient uptake in year 1 + nutrient losses due to erosion) - available nutrients in year1) / (1 - nutrient loss fraction)

nutrient requirements in year 2 = ((nutrient uptake in year 2 + nutrient losses due to erosion) - available nutrients in year 2 - 'turn-over' nutrients of year 1) / (1 - nutrient loss fraction)

⋮

nutrient requirements in year (n) = ((nutrient uptake in year (n) + nutrient losses due to erosion) - available nutrients in year (n) - 'turn-over' nutrients of year (n-1)) / (1 - nutrient loss fraction)

Figure 5.2. Procedure of calculating technical coefficients for the nutrient dynamics of perennial and timber systems (n=length of crop cycle in years).

5.5 Examples

In this section, some of the land use systems generated with PASTOR and LUCTOR are illustrated. The technical coefficients of these systems were used in different optimization models to explore land use options in the AZ, and the results are presented in Chapters 6, 7 and 8.

Pasture

Table 5.4 presents some technical coefficients of four of the different pastures generated with PASTOR: two natural pastures and two fertilized improved grasslands with *Cynodon nlemfuensis* (Estrella), all on a fertile well drained soil (SFW).

Table 5.4. Technical coefficients (all annual values) of four pastures (two natural pastures and two fertilized improved Estrella pastures) generated by PASTOR.

	Natural		*Estrella*	
Stocking rate (AU ha^{-1}):	1	2	2	2
N-fertilizer level (%)[1]	0	0	0	100
Outputs				
Supplied dry matter (kg ha^{-1})	4566	4995	3150	4646
Supplied metabolizable energy (ME in Mcal ha^{-1})	7762	8492	5741	11 424
Supplied crude protein (CP in kg ha^{-1})	274	300	250	548
Supplied ME by pasture minus ME eaten by cattle	1834	-3366	-6117	-434
Supplied CP by pasture minus CP eaten by cattle	0	-248	-298	0
Δ soil N stock (kg ha^{-1})	-16	-8	0	0
Inputs				
Nitrogen fertilizers (kg N ha^{-1})	0	0	0	106
Herbicides (kg a.i. ha^{-1})	0.75	0.75	1.5	1.4
Labor requirements (d ha^{-1})	1.4	1.4	3.0	3.5
Total costs (\$ ha^{-1})	28	28	52	258
Fertilizer costs (\$ ha^{-1})	0	0	0	215

[1] Percentage of N-fertilizer level required for maximum attainable production

The two natural pastures represent current pastures in the AZ and differ only in stocking rate (1 AU and 2 AU ha^{-1}, respectively). At a stocking rate of 1 AU ha^{-1}, the natural pasture supplies sufficient crude protein (CP) and metabolizable energy (ME) to support a stocking rate of one AU ha^{-1} as indicated by the technical coefficients quantifying the supplied ME and CP minus the consumed ME and CP. Since PASTOR limits the amount of feed eaten from a pasture at a given stocking rate to the minimum of the CP-intake requirements of cattle or the calculated amount of CP provided for by the pasture, there is even a surplus of ME. At a stocking rate of 2 AU ha^{-1}, natural pastures are not able to supply sufficient ME and CP to feed the grazing cattle, a fact indicated by the negative ME and CP balances.

As a consequence, feed supplements are required to maintain the given stocking rate. Both natural pastures deplete the soil N stock, indicating that, in the long run, the production level of both pastures can not be maintained. The depletion rate is smaller with the higher stocking rate because more nutrients are added to the system via an external supply of feed supplements that is subsequently excreted in urine and faeces.

The improved Estrella pastures differ only in N-fertilizer level: one is unfertilized, and the other receives 100% of the fertilizer needed to sustain the attainable feed on offer. For both systems, a zero change in soil N stock was predefined (*i.e.,* soil N balance in equilibrium). Table 5.4 shows that the unfertilized Estrella is not able to supply sufficient amounts of CP and ME to maintain a stock of 2 AU ha^{-1}, and additional feed supplements are required. The amount of dry matter supplied by this pasture is even less than that of the natural pastures. The 100% fertilized Estrella supplies sufficient CP but still shows a (small) deficit of 400 Mcal ME. The higher weed suppressing capacity of fertilized Estrella compared to its unfertilized variant is reflected in a lower use of herbicides. The higher costs for Estrella compared to natural pastures is caused by the higher costs of planting material, and of the labor needed for planting and management, while the higher costs of the fertilized Estrella compared to its unfertilized variant is due to fertilizer costs.

Crops

Table 5.5 shows the technical coefficients of cassava (*Manihot esculenta* Crant) and banana (*Musa AAA*) land use systems, both on a fertile well drained soil (SFW) with a high use of biocides and low use of mechanization.

Table 5.5. Technical coefficients (all annual values) of four cropping systems generated by LUCTOR

	Cassava		*Banana*	
Type of system	*Actual*	*Alternative*	*Alternative*	*Alternative*
Herbicide level	*High*	*High*	*High*	*Low*
Outputs				
Prime quality product (kg ha^{-1})	5100	12 750	65 277	65 277
Second quality product (kg ha^{-1})	2550	6375	11 519	11 519
Third quality product (kg ha^{-1})	850	2125	0	0
Δ soil N stock (kg ha^{-1})	-54	0	0	0
Inputs				
Nitrogen fertilizers (kg N ha^{-1})	0	290	713	713
Biocides (kg a.i. ha^{-1})	1.2	2.2	56	53
Biocide index (ha^{-1})	4007	4051	2563	476
Labor requirements (d ha^{-1})	34.1	67.4	193	210
Total costs ($ ha^{-1})	118	621	9528	9487
Fertilizer costs ($ ha^{-1})	0	465	1868	1868
Biocide costs ($ ha^{-1})	50	63	1684	1644

For cassava, an actual and an alternative system are shown, with weeds combated in both systems by herbicides. Three product qualities are distinguished, each having their own price and market outlet. The difference between the systems is the much higher attainable yield and the predefined zero soil N depletion rate of the alternative system. As a consequence, the alternative system requires a large amount of nitrogen fertilizer to sustain this production level. As shown in Table 5.5, the costs for fertilizer determine to a large extent the total costs of the alternative cassava system. The higher demand for labor is caused by the higher labor requirement for harvesting due to the higher yield. Even though only annual labor requirements are given in Table 5.5, LUCTOR also calculates labor requirements for annual cropping systems on a monthly basis in order to identify labor peaks during the year.

For banana, only two product qualities exist. The two (alternative) production systems differ in their use of herbicides: one with and one without the use of herbicides, indicated in Table 5.5 as "high" and "low" respectively. Yield levels, and thus required N fertilizers, are the same since it is assumed that herbicides can be completely substituted by manual weeding without affecting yield. The use of biocides is almost the same in both systems, since banana requires substantial amounts of fungicides, nematicides and insecticides. Since herbicides account for only a small part of total biocide use in banana cultivation, the total amount of active ingredients (in biocides) is only marginally smaller in the zero-herbicide system than in the high-herbicide system. However, the type of herbicide used in banana (paraquat) is very persistent and has a high impact on the total biocide index (see Section 5.2.2). Therefore, the biocide index is much lower in the zero-herbicide system than in the high-herbicide system. Finally, the additional labor required for manual weeding is expressed in the higher labor requirements of the zero-herbicide system. Since labor costs are not considered to be part of the total costs, the banana system without herbicides has lower costs (because it does not use herbicides). The costs of labor are taken account of in the linear programming models that make use of LUCTOR see Chapters 6-8). The much higher costs of both banana systems as compared to the cassava systems are caused by higher costs of establishment and of post-harvest processing, and by the use of more fertilizer and biocides. Even though the amount of biocides used in both cassava systems are a fraction of the amount used in both banana systems, the biocide indices of both cassava systems are much higher due to the more frequent use of the herbicide paraquat, which has a high impact on the biocide index.

5.6 Conclusions

Both PASTOR and LUCTOR have been successfully used to systematically generate the necessary input data for various land use studies of the AZ in Costa Rica (Bouman and Nieuwenhuyse, 1999; Bouman *et al.*, 1998c; Sáenz *et al.*, 1998; and see also Chapters 6, 7 and 8). Since both PASTOR and LUCTOR are highly generic and modular, their parameters can easily be adjusted to reflect such location-specific conditions as those shown in the case study of the Aranjuez watershed (Hengsdijk, 1999; Sáenz *et al.*, 1999; Section 10.6), to incorporate new information (such as the design of new production systems, *e.g.*, precision agriculture technologies for banana production, presented in Chapter 9), or to analyze how uncertainty and variability in underlying data affect the various dimensions of land use systems.

Both PASTOR and LUCTOR were used to train groups of potential users in Costa Rica, such as extensionists of the Ministry of Agriculture and Livestock and researchers of several universities. This knowledge transfer, however, has not been a one-way street, since development as well as application of the Technical Coefficient Generators in land use studies resulted in fruitful discussions about — and adaptation of — expert-based assumptions. Generic expert systems, such as PASTOR and LUCTOR, thus stimulate field experts (which often are also users) to be explicit about their knowledge, and to make that knowledge transparent and open to critical review and discussion. The advantage is that such important knowledge is not left unused. Considered in this way, both PASTOR and LUCTOR as applied in the AZ studies are also important tools to store, order and integrate agro-ecological information that is currently not readily available or accessible.

Both PASTOR and LUCTOR allow the quantification and analysis of the different dimensions of land use systems, *i.e.*, in economic terms (*e.g.*, costs), social terms (*e.g.*, labor), production terms (*e.g.*, yields), in terms of emissions to the environment (*e.g.*, use of biocides) and in terms of natural resource use (*e.g.*, use of soil nutrients). Such explicitly described dimensions of land use systems, recorded in their own physical and monetary units, stimulate a more fact-driven discussion on land use-related objectives. In this way land use systems can be analyzed in the context of their resource use efficiency and the possible trade-offs that may exist between economic and environmental objectives (Hengsdijk *et al.*, 1999).

6 Integrated biophysical and socio-economic analysis of regional land use

ROBERT A. SCHIPPER, BAS A.M. BOUMAN, HANS G.P. JANSEN, HUIB HENGSDIJK, and ANDRÉ NIEUWENHUYSE

Abstract

A methodology is presented with which regional land use options can be explored in order to provide policy support. This methodology is called SOLUS (Sustainable Options for Land USe) and integrates a linear programming model called REALM (Regional Economic and Agricultural Land-use Model) with Technical Coefficient Generators for crops (called LUCTOR) and for livestock activities (called PASTOR) used to quantify the technical coefficients of land use systems, and with a geographic information system (GIS). SOLUS is implemented for a case study in the northern Atlantic Zone of Costa Rica, with the objectives of *i*) analyzing regional land use practices while taking into account economic and environmental objectives and restrictions, and *ii*) evaluating economic and agrarian policies that influence the agricultural sector. Salient characteristics of the SOLUS methodology are the geographic explicit delineation of land and labor resources, the incorporation of endogenous prices of outputs and labor, and the variation of output prices according to quality of roads and distance to markets. The latter two aspects are related to the size of the region, while endogenous prices and wages are necessary because the supply originating in the region is capable of influencing prices and wages. Seven policy scenarios are studied that address policy and sustainability issues relevant to the region: technological change, zero soil nutrient depletion, limiting biocide use, taxing biocides, forest conservation, lowering interest rates and increasing real wages. It is shown that the SOLUS methodology is a suitable tool for the analysis of policy options in order to support policy decisions, as well as to analyze future land use options in view of their effects on income and the environment.

6.1 Land use analysis

6.1.1 Modeling the agricultural sector

Before embarking on a discussion of land use analysis, it is useful to present a skeleton model of the agricultural sector (Moll and Schipper, 1994). In Figure 6.1, the right-hand side of the model is structured according to the various actors in the agricultural sector and their occupations; policies and policy measures affect operators in markets, as well as institutions and infrastructure. Together, these actors determine the socio-economic environment in which farm households, or primary producers in general, operate. On the left-hand side, natural resource endowments and technological possibilities determine the types and technology levels of crop, pasture and forestry land use systems, and of animal and fishery production systems. These systems utilize land, either directly or indirectly. Farm households make a selection of the mentioned systems on the basis of their resources and preferences, guided by the socio-economic environment. The total output of the agricultural sector (in terms of primary products) depends on the actual selection of systems by farm households. This output consists of a) the types and quantities of products, and b) negative or positive effects on the sustainability of agricultural production, thereby affecting future production possibilities. Farm households are thus the final decision-makers in agricultural production, but their behavior is influenced by the biophysical environment on the one hand and by the socio-economic environment on the other.

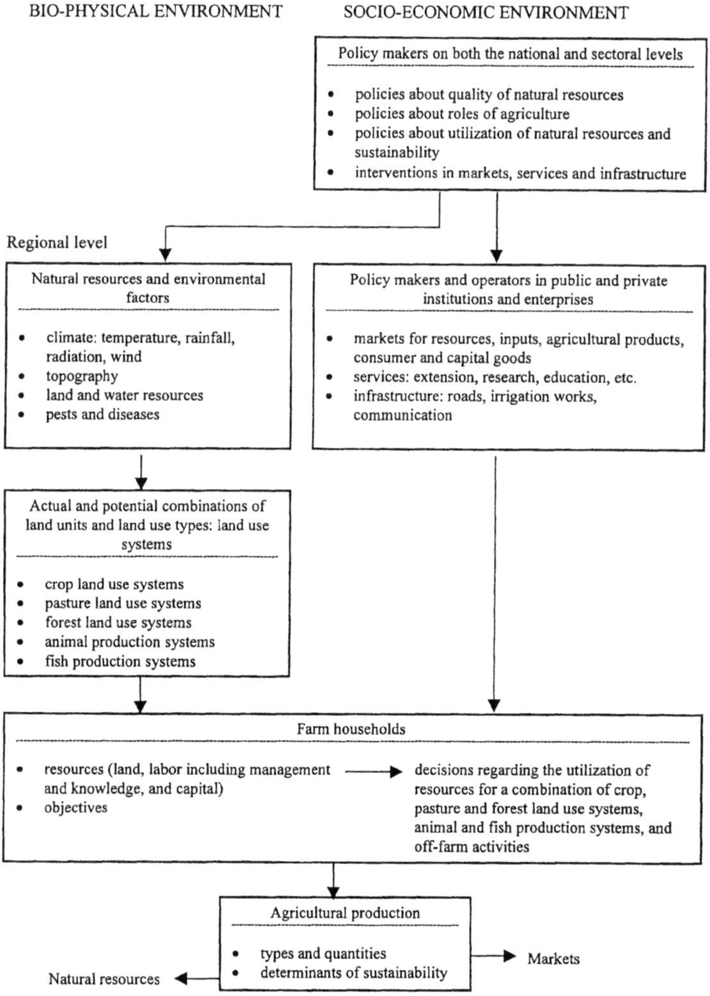

Figure 6.1. Skeleton diagram of the agricultural sector (source: Moll and Schipper, 1994).

6.1.2 SOLUS: a framework for land use analysis

Land use analysis attempts to survey present and future land use in a specific area. It involves a description and diagnosis of the current land use and farming systems, followed by an exploration of future land use options. As such, land use analysis forms part of more general procedures for land use planning (Schipper, 1996). Several procedures for land use planning have been designed, for example FAO (1993) and Fresco *et al.* (1992). The latter emphasize the importance of analyzing land use on different levels (land use system, farm, sub-regional, regional and national), as well as the fact that land evaluation and farming systems analysis are two complementary tools for land use planning. Land evaluation is a multidisciplinary tool for assessing the suitability of land for different uses, while farming systems analysis diagnoses the present farming and land use practices, and presents insights into possible improvements in the existing ways of farming. Because they are complementary, both approaches are useful for land use analysis. Based on this insight, Fresco *et al.* (1992) developed the LEFSA (Land Evaluation and Farming Systems Analysis) sequence.

Using concepts from the LEFSA sequence and tools and ideas from systems analysis, a general methodology for land use analysis has been developed, entitled SOLUS (Sustainable Options for Land USe) by an interdisciplinary team of scientists in REPOSA[1] (Figure 6.2). The core of the methodology consists of a (multiple goal) linear programming model, two Technical Coefficient Generators, and a Geographic Information System (GIS). The linear programming model selects land use systems and other production activities by determining which ones are optimal for the attainment of a specific goal, for instance the maximization of economic surplus in the agricultural sector (*e.g.*, as elaborated in this chapter), or some farm household utility function (as elaborated in Chapter 8). The linear programming model may also be of the multiple-goal type in which the optimal means of achieving several different goals can be found, *e.g.*, the maximization of employment in the agricultural sector or the minimization of certain environmental effect indicators (*e.g.*, Jansen *et al.*, 1995, and Chapter 7 of this book). Optimizations are performed under restrictions, which may be absolute, *e.g.*, no more land can be used than is available in the area, or normative, *e.g.*, limitations may be imposed on certain sustainability parameters. The maximization of a specific objective function under a set of coherent restrictions is called a scenario. Trade-offs between economic and sustainability objectives are quantified by running the linear programming model for different scenarios and/or by generating and offering different land use systems. Land use systems are generated by Technical Coefficient Generators, and are quantified in terms of their technical coefficients, *i.e.*, their inputs and outputs including yields, costs, labor use, and sustainability indicators. GIS plays an important role in archiving and manipulating geo-referenced input data and in presenting spatial output results. There is a semi-automated flow of data between the GIS, the Technical Coefficient Generators, and the linear programming model (Bouman *et al.*, 1998b; Stoorvogel, 1995). The process of interaction with stakeholders and interest groups, problem definition and scenario formulation in the implementation of the methodology is discussed in more detail in Section 7.2 of this book (see Figure 7.1).

[1] Research Program on Sustainability in Agriculture, a collaboration between Wageningen Agricultural University (WAU), the Ministry of Agriculture and Livestock of Costa Rica (MAG), and the Tropical Agricultural Research and Higher Education Center (CATIE).

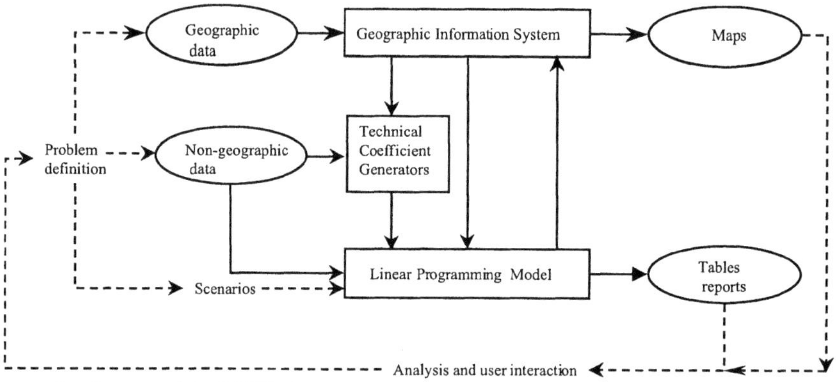

Figure 6.2. Structure of the SOLUS framework. Grey boxes are models/tools; ovals are data; blank names are activities; drawn lines are flows of data; dotted lines are flows of information.

The concept of SOLUS builds upon extensive experience with the use of linear programming models in land use studies (*e.g.*, De Wit *et al.*, 1988; Rabbinge and van Latesteijn, 1992; Erenstein and Schipper, 1993, Veeneklaas *et al.*, 1994; Kuyvenhoven *et al.*, 1995; Kruseman *et al.*, 1995; Stroosnijder *et al.*, 1995; Bakker *et al.*, 1998). In Costa Rica, SOLUS evolved from the USTED (*Uso Sostenible de Tierras En el Desarrollo*; Sustainable Land Use in Development) methodology. USTED operated initially on the sub-regional level. In particular, it was developed for a settlement (Schipper *et al.*, 1995; Schipper, 1996; Stoorvogel *et al.*, 1995). Thereafter, it was gradually scaled-up and used on the county level (Jansen *et al.*, 1997a; Jansen and Stoorvogel, 1998). SOLUS operates in particular on the (sub-)regional level, *e.g.*, as presented in this chapter (Schipper *et al.*, 1998; Bouman *et al.*, 1998c, 1999a).

SOLUS is designed to explore land use practices in order to support agricultural policy design. Explorations can be made either with a relatively long time horizon (20-30 years) or with a short-to-medium time frame (0-5 years). The linear programming model of the SOLUS framework needs to be constructed specifically for the purpose of each study, and the Technical Coefficient Generators are used to produce relevant land use systems. For instance, in Chapter 7, a long-term study is presented that explores the biophysical "outer bound" potentials of agriculture in the AZ study area. For this study, the Technical Coefficient Generators PASTOR (Pasture and Animal System Technical coefficient generatOR) and LUCTOR (Land Use Crop Technical coefficient generatOR) (Chapter 5) were used exclusively to generate alternative land use systems based on a target-oriented approach. A multiple-goal linear programming model was developed called GOAL-AZ (General Optimal ALlocation for the AZ), that largely excludes socio-economic factors. On the other hand, the studies presented here and in Chapter 8, take as their point of departure that, beside biophysical conditions, land use is, to a significant degree, determined by socio-economic conditions. Therefore, the linear programming models developed for such short-to-medium term analysis take into account the mechanisms involved in the product and labor markets (thus allowing for endogenous pricing), while PASTOR and LUCTOR are used to generate both actual and alternative land use systems.

6.1.3 Levels of analysis

An important aspect of the SOLUS framework is the differentiation between levels of analysis (land use system, farm, sub-region, region and nation), in a way that is comparable to the LEFSA sequence (Fresco *et al.*, 1992). Given a certain set of objectives, levels of analysis are related to levels of decision making. Addressing several levels of analysis at the same time gives rise to aggregation issues. In the context of land use analysis three points are at stake (Erenstein and Schipper, 1993):

1. Land use on the regional level is often considered without detailed information regarding the behavior of the farm households responsible for the actual use of land.
2. Aggregation bias, as individual farmers have resources at their disposal in different proportions from the aggregated resources of a region.
3. Variables that are exogenous on the micro level become endogenous on higher levels.

Aggregate decision problems

Aggregate decision problems involve choices on least two levels (Erenstein and Schipper, 1993). On the macro level, a policy maker tries to decide how best to allocate (financial) resources, given the existence of *i*) more than one objective, and *ii*) uncertainty about the exact outcome, for example with regard to land use. On the micro level, farmers have their own decision problems. They have to decide how best to respond to the policy environment, given their own objectives, possibilities, and resources. However, it is not known beforehand on the macro level what the response on the micro level will be. It is this lack of knowledge that causes uncertainty about policy response at the macro level. In order to solve the macro or policy problem, uncertainty surrounding micro responses has to be reduced. In other words, some means of simulating the probable response of farmers is required to evaluate the likely effects of a policy decision. In this context, multi-level models have been proposed. Hazell and Norton (1986), following Chandler and Norton (1977) as well as Chandler *et al.* (1981), outline the principles of such models, involving (interdependent) constrained optimization on different levels. However, such models are not directly workable (Hazell and Norton, 1986). In practice, efforts are concentrated on simulating producer decisions by building models that reflect their constraints, opportunities and objectives. Such models are then solved under varying assumptions about the ways in which the policy environment affects producers. However, agricultural producers differ widely in their resource endowments and production possibilities. Therefore, an adequate investigation of producer response to policy measures requires representative farms to be modeled. Simulation of the probable response of farmers is further complicated by the fact that farmers usually have a number of objectives and preferences. This precludes, for example, the establishment of profitability as a sole choice criterion (Diltz, 1980). A farm household may strive to achieve a number of objectives. It may attempt to meet the present subsistence requirements of the family (either by on-farm production or by making purchases), to provide funds for the family's emergency or short-term educational expenses, to maximize the long-term profitability of the farm, or to sustain the (natural) resource base.

Finally, it must be stressed that since it is the farmer who decides on, and is responsible for, the actual use of land, finding the "*optimal*" cropping patterns from a policy viewpoint may not be very useful, unless ways are found to induce farmers to adopt those cropping patterns.

Aggregation bias

On the level of the region or of an entire sector, an aggregation bias arises because farms are not identical in terms of their resource endowments. Ideally, to completely avoid any aggregation bias, a model should be constructed for every individual farm. These individual models could, at least theoretically, be linked together to form a sector model. Since such a composite model is practically impossible due to data, manpower, financial and computer limitations (Jansen and Stoorvogel, 1998), two modeling approaches may be considered:

1. The aggregate regional model: this involves the aggregation of a region's resources and the modeling of all aggregated variables as if they formed a single large farm. An elaboration of this approach is given in this chapter and in Chapter 7.
2. The representative farm model: this involves classification of farms into a smaller number of representative groups, mostly on the basis of relative factor endowments (Jansen and Stoorvogel, 1998) or according to the most limiting resource (Sheehy and McAlexander, 1965). A model is constructed for each representative farm group. These farm models are then aggregated to a sector model using the number of farms in each group as weights. To limit aggregation bias, this procedure places a high demand on the proper definition of the representative farms and weighting procedures. An example of this approach is given in Chapter 8.

Both approaches overstate resource mobility by enabling farms to combine resources in proportions that are not available to them individually. Both approaches implicitly assume that all farms (in each group) have equal access to the same technologies of production. Therefore, the value of the objective function (in a maximization problem) of an aggregated model is, in general, higher than that of the objective function of a disaggregated one (Hazell and Norton, 1986; Erenstein and Schipper, 1993). In order to minimize aggregation bias, farms are to be classified into groups or regions defined according to requirements of homogeneity.

Exogenous variables on the farm level becoming endogenous on the regional/sector level

In the transition from farm-level to regional or sector-level analysis, there is an aggregation problem with respect to the nature of the variables. Variables that are exogenous on the micro level may be endogenous on the meso or macro level. Product prices, for instance, are normally considered as a given for individual producers, but may be variable across a region as a whole. The same applies to factor prices. The entire service sector is normally considered as given for individual producers, but it is a variable when considered regionally. Only in long-term studies that focus on an exploration of the biophysical limitations on production, such as the one presented in Chapter 6, can such issues be neglected.

Treatment of aggregation issues

Even though the SOLUS application discussed in this chapter considers the land use system, sub-regional and regional levels, it largely ignores the farm level. The regional level refers to the AZ. The sub-regions are 12 zones distinguished on the basis of transport costs (see Section 6.2.4). The level of the land use system consists of crop, pasture and forest land use systems, animal production systems and supplementary feeding systems, as generated by the LUCTOR and PASTOR models (Chapter 5).

In the case of the regional land use modeling of the AZ, it was necessary to abstract from the farm level, because the location in each sub-region of each prospective farm type could not be established. This fact made it impossible to determine the areas of each soil type per farm type in each sub-region. As a consequence, this source of aggregation bias had to be neglected. However, it can be shown (Bell *et al.*, 1982; Schipper, 1996; Jansen and Stoorvogel, 1998) that in cases when farm types only differ in relative resource availability (in this case land and labor) but have access to the same technological options, face the same price vector and share the same objective, while labor can be exchanged between farm types or otherwise hired without limits, the aggregation bias is not important in quantitative terms. That is to say, the objective function value will not differ much between a model containing farm types and a model without farm types, while the overall land use pattern will be very similar. The conditions mentioned apply to the model developed in the present chapter. An additional reason for not distinguishing between farm types involves considerations of model size, as many of the variables and constraints in the model would be indexed by the number of farm types. However, a logical result of not distinguishing between farm types in the model is that its results do not provide information about the distribution of income and land use over farm types, although such information is often important for policy analysis. A methodology for land use analysis on the regional level that explicitly incorporates the farm level is presented in Chapter 8.

The other aggregation issues are treated in a different way than in the earlier sub-regional models of settlements (Schipper *et al.*, 1995) and counties (Jansen and Stoorvogel, 1998). In the present model, the first aggregation issue - decision-making on both the farm as well as on the regional or national level — is approached in conjunction with the third aggregation issue — exogenous variables becoming endogenous. The purpose of the model developed in this chapter is to demonstrate how land use decisions by farm households are influenced by the biophysical and socio-economic environment. The latter includes the agricultural policies pursued by regional and national authorities. The model assumes competitive markets for agricultural products, where producers try to maximize their producer surplus and consumers maximize their consumer surplus. Because of the shape and direction of the demand and (implicit) supply functions, the intersection of the supply and demand functions of each product determines the equilibrium price that equates supply and demand quantity. At that point the sum of the — collective — producer and consumer surpluses is at its maximum (Section 6.2.5). Thus, by maximizing the sum of the producer and consumer surplus in each market, the model is able to find the equilibrium prices and quantities.

In this way, product prices become endogenous variables in the regional model. However, endogenization of product prices is only necessary if the region supplies a significant part of the relevant market, either domestic or export. That is, the supply originating from the region must be able to influence the price of the commodity. On the other hand, decision making on the farm level is incorporated by assuming that each individual farm maximizes its producer surplus. This results in an aggregate supply function for each commodity, representing the (marginal) cost of production. In the linear programming model, the supply function of each commodity is implicit, but can be traced by running the model with increasing prices, starting at a price of zero until the point at which the supply no longer increases. The result will be an incremental increasing of the supply function.

6.1.4 Sustainability

The SOLUS framework uses the definition of sustainable development put forward by Pearce and Turner (1990) as a starting point for sustainable land use: *Sustainable development involves maximizing the net benefits of economic development, subject to maintaining the services and quality of natural resources over time.* Maintaining the services and quality of the stock of resources over time implies the acceptance of the following three directives (Pearce and Turner, 1990):

a) Utilize renewable resources at rates less than or equal to the natural rate at which they regenerate.
b) Keep waste flows to the environment at or below the assimilative capacity of the environment.
c) Optimize the efficiency with which non-renewable resources are used, subject to substitutability between resources and technical progress.

Sustainable development, as defined above, can easily be translated into sustainable land use in order to have a simpler starting point for analysis. The given guidelines for resource use, such as the use of land, can then be applied to the particular circumstances in a specific area. Parameters can be designed to measure the quantity and quality of resources and the state of the environment. In developing SOLUS for the AZ of Costa Rica, a distinction in sustainability indicators was made between parameters that reflect the status of the natural resource "land" (rule a), operationalized in the soil nutrient balance, and environmental effect indicators (rule b), operationalized in waste flows and emissions of nutrients, biocides and greenhouse gases (Bouman *et al.*, 1998c, 1999a). These indicators were based on relevant sustainability issues in the AZ (Chapter 2). Moreover, they can be thought of as the relevant sustainability indicators of the "environmental (utilization) space" (Opschoor and Reijnders, 1991; Opschoor, 1992; Wetering and Opschoor, 1994) in which land in the research region is used.

Linear programming optimizes resource use given a certain objective. In other words, it strives for an optimum efficiency (rule c) for sustainable development.

Efficiency is optimized subject to the substitutability of resources and technological progress. The effects of substitution can be traced by observing the shadow prices in constraints and sensitivity analysis. The notion of "technological progress" (leading to a more productive use of resources) is part of the model: each land utilization type is specified according to technology and combined with a land unit (similar to the operations performed in the Technical Coefficient Generators discussed in Chapter 5). In the optimal solution the most efficient technologies, and thus resource use, are chosen after taking all options and constraints into account. For a more extensive discussion of sustainability issues, see Section 10.5.

Maintaining an objective based on economic behavior, *e.g.,* maximizing economic surplus, implies that biophysical sustainability criteria should be accounted for in terms of constraints. This is also the approach taken in a number of other economic models that consider agricultural land conservation and environmental improvement (Heady and Vocke, 1992). In such a set-up, each activity causes a positive or negative impact on a sustainability constraint, expressed in "technical" coefficients. The total impact of all activities can be restricted by the "Right Hand Side" coefficient, which should be an indication of the (renewable) resource availability and/or its regeneration rate, or, in case of pollutants, of the assimilative capacity of the environment.

6.2 Application of SOLUS to the AZ

6.2.1 The objective of the case study

The objective of this case study is *i*) to analyze land use on a regional basis, while simultaneously taking account of economic and environmental objectives and restrictions, and *ii*) to evaluate the influence of economic and agrarian policies on the agricultural sector. Land use is analyzed on three levels: the land use system, sub-regional and regional levels. The first aggregation issue — decision-making on different scale levels is approached in conjunction with the third aggregation issue — exogenous variables becoming endogenous (Section 6.1.3). Competitive markets are assumed for agricultural products where producers try to maximize their producer surplus and consumers maximize their consumer surplus. The labor market is also explicitly addressed. The regional distribution of resources, land and labor, and geographic variance in factor prices due to distance to markets and road infrastructure, is calculated and mapped using the GIS.

6.2.2 Structure of SOLUS

The linear programming model developed is called REALM (Regional Economic and Agricultural Land use Model). REALM selects, per sub-region, the optimal combination of land use systems, herds and feed supplements by maximizing the regional economic surplus generated by the agricultural sector in the AZ. The economic surplus is defined as the sum of the consumer and producer surplus (see also below).

Actual and alternative land use systems were generated using PASTOR and LUCTOR (Chapter 5) for 13 land utilization types: eight crops (banana, black beans, cassava, maize, palm heart, pineapple, plantain and natural forest for sustainable timber extraction) and five pastures (three fertilized improved grasslands, a grass-legume mixture, and a mixture of natural(ized) grasses). These land utilization types were combined with the three major land units identified in the northern AZ: Soil Fertile Well drained (SFW), Soil Infertile Well drained (SIW) and Soil Fertile Poorly drained (SFP), each of these major units being subdivided into mechanizable and non-mechanizable sub-units (see Chapters 2 and 5). Actual land use systems are quantified based on data derived from descriptions of the current best farm practices in the AZ, while alternative systems were generated using the target-oriented approach, with the restriction that soil nutrient balances must be kept in equilibrium (no changes in soil nutrient stocks for N, P and K). For alternative systems, different technology levels were generated by combining levels of fertilizer use and crop protection, varying between manual weeding and herbicide use, and, for pastures, changing the stocking rate. A total of 1352 crop land use systems and 1756 pasture land use systems were generated. Two herd types were distinguished (cattle breeding and cattle fattening), each were further sub-divided into four animal growth rates. Finally, five feed supplements were defined.

Technical coefficients were generated on a "per ha" basis and include such diverse aspects as labor requirements, costs of inputs, yield, change (Δ) in soil nutrient stock for nitrogen (N), phosphorus (P) and potassium (K), N-denitrification losses, N-leaching losses, N-volatilization losses, biocide use (BIOA), and a biocide index (BIOI). Technical coefficients are either averages per year (labor use, changes in soil nutrient stocks, N losses, biocide use and biocide index) or annuities of the present value over the life-span of the land use systems (yield, input costs, and labor use).

The relevant part of the simplified REALM model is given in mathematical notation in Appendix 6.1. Details of the complete formulation of the linear programming models in GAMS (Brooke *et al.*, 1992), as well as necessary assignments to (re-)calculate many of the matrix coefficients, can be found on the CD-ROM that accompanies this book.

A detailed biophysical and socio-economic description of the study area in the northern part of the AZ is given in Chapter 2, and detailed information about the generated land use systems and the technical coefficients involved is provided in Chapter 5.

6.2.3 One year model: use of annuities

The use of annuities in the regional land use model developed in this chapter needs a brief explanation. A number of land use systems represent perennials, while animal production systems also have a duration exceeding one year. Thus, such systems occupy a given piece of land for a number of years. In early years, costs usually exceed benefits, while the reverse is true in the later years. Since REALM is a one-year model (consisting of two sub-periods: a wet and a somewhat dryer season), values in different years must somehow be included. However, values that occur in an earlier year are worth more at present than those that occur later. A discount rate[2] is used to value future cost and benefit streams in today's terms in order to calculate their present values. In the cost-benefit analysis literature, discounting future values to the present is typically done in terms of monetary values, obtained by multiplying price by quantity. Nonetheless, assuming constant prices over time,[3] discounting can also occur in quantitative terms:

[2] As the level of the appropriate discount rate is not a subject of the present study, a discount rate of 7% was used which is considered a reasonable approximation of the opportunity cost of capital under the conditions that existed in Costa Rica in the late 1990s. In one of the scenarios, the sensitivity of the model for different discount rates is studied.

applied, for example, to production (in kg or other physical units) or labor use (in hours or days). After calculating such a "present quantity", the latter can be multiplied by its price to obtain the present value.

Different perennial land use systems occupy land for a different number of years, depending, among other factors, on the land unit and the technology involved (for example, two years for pineapple and 15 years for palm heart). Their present values must be made commensurable. Furthermore, the model also contains annual land use systems (*e.g.*, maize). Present values based on time periods of different length, including lengths of one year or less, can be made commensurable by converting them to an annuity. Based on a present value, an annuity is calculated through the capital recovery factor.[4]

6.2.4 Sub-zonation

Prices of outputs (*i.e.*, products) depend on geographical location within the AZ due to variation in the distance to markets and the quality of the roads. This geographical variation in product prices was addressed by dividing the AZ into sub-regions, each with its own specific transport costs to the most relevant market (depending on the type of product and final destination). These transport costs were calculated on the basis of a regression model estimated by Hoekstra (1996). The sub-regions are the result of a GIS-overlay of three zonification maps based on equal transport costs (Bouman *et al.*, 1998b). The first map concerns the transport costs of agricultural products to the road in the southwest of the AZ, where the products leave the AZ en route to the domestic market. The second map concerns livestock products shipped to the same destination. The third map concerns the transport of export products to Limón harbor in the southeast of the AZ. To keep the size of the model within limits, while still distinguishing meaningful transport zones, 12 "iso-transport cost" sub-regions were delineated (Figure 6.3). In the linear programming model, farm-gate prices were calculated per sub-region by subtracting the transport costs from the product prices in the respective market outlets of the AZ. Prices for agricultural inputs (*e.g.*, seed, fertilizer and biocides) were the same in all sub-regions, since they were found to show little variation in shops across the AZ.

For each sub-region, soil and labor endowments were calculated by map overlaying in GIS. Soil data were available from Stoorvogel and Eppink (1995) and Nieuwenhuyse (1996). The agricultural labor force in 1996 was estimated for each district on the basis of the proportion of the total population working in agriculture in 1984 (DGEC, 1987b). Based on population growth between 1984 and 1996 (calculated as the difference between annual registration of births and deaths; DGEC, 1997a) and the estimated migration to each county (based on migration rate data for the period 1979-1984) the agricultural labor force in each district was estimated (see also Table 2.1 in Chapter 2). The outcomes were compared with more recent survey information about the AZ as a

---

[3] Constant prices normally are a basic assumption of the linear programming models used in land use analysis. However, in models where we can not assume an infinitely elastic demand for a product (*i.e.*, in which downward sloping demand functions are introduced, implying prices are endogenous variables, as in the present regional model (Section 6.2.5 and 6.2.6)), we implicitly have to assume constant prices over time, even though they are determined endogenously by the model.

[4] In financial terms, the capital recovery rate can be described as the level of payment (A) to be made at the end of each of n periods to recover the present amount (P) at the end of the nth period at the discount rate of i (Gittinger, 1982: 433). In mathematical notation:

$$A = P \frac{i\,(1+i)^n}{(1+i)^n - 1}$$

whole (DGEC, 1997b) and deemed to be reasonably accurate. The labor force estimates per district were distributed over the 12 sub-regions on the bases of population density estimations derived from Stoorvogel and Eppink (1995). Finally, labor mobility costs between sub-regions were calculated on the basis of bus fares between the geographical centers of the sub-regions.

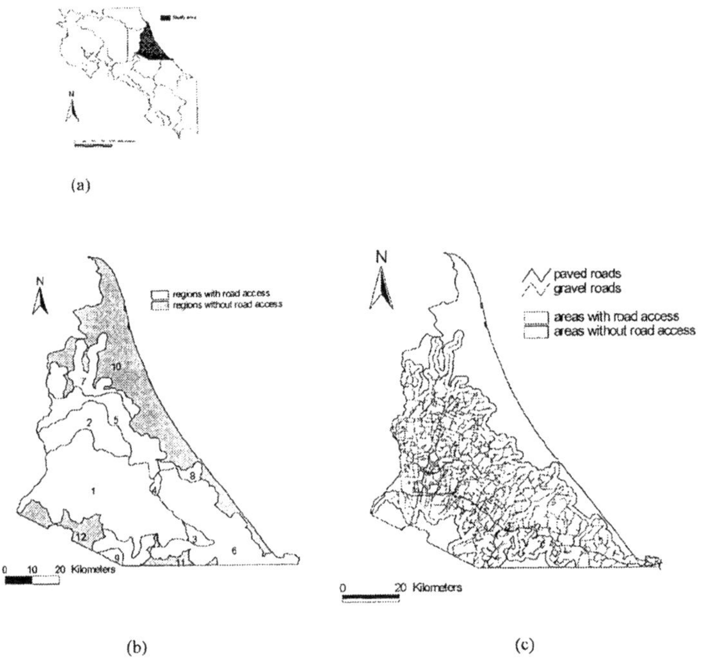

Figure 6.3. Map of northern Atlantic Zone in Costa Rica (a) with 12 sub-regions (b) and road infrastructure (c)

6.2.5 Downward sloping demand functions

For a number of products (*i.e.,* banana, palm heart, and plantain), the production of the AZ constitutes a large part of the national supply (see also Table 8.1 in Chapter 8), and a considerable part of the world supply (Table 8.1). Therefore, prices of these products are likely to be influenced by the supply originating from the AZ, *i.e.*, become endogenous. Based on research in Brazil (Kutcher and Scandizzo, 1981), Mexico (Duloy and Norton, 1973, 1983) and elsewhere, Hazell and Norton (1986) present a method of incorporating variable prices in linear programming models. Downward-sloping demand functions, based on econometrically estimated price-demand elasticities, are linearized on the basis of an observed base quantity and price. The relation between product prices and supply from the AZ is incorporated in REALM by estimating demand functions for a number of relevant products. For these products, the price P_j is a function of quantity Q_j. Even with a simple linear inverse demand function, $P_j = a_j - b_j * Q_j$, the linear programming model would become a quadratic model. Although models with quadratic elements in the objective function can be run with modern software packages, they require (much) more running time than linear models, particularly in the case of large-size models such as REALM. Therefore, we opted for a linear approximation method.

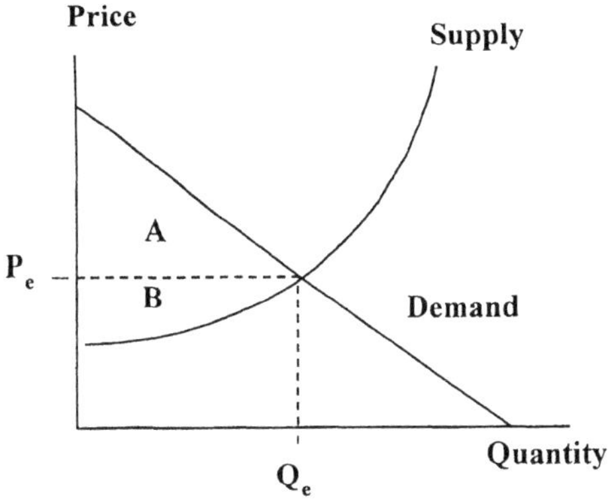

Figure 6.4. Relationships between price, demand and supply. Producer and consumer surplus is defined as area A plus B.

In models where prices are exogenous, the value of production (p_j * Q_j) is part of the objective function, as are the costs of production in terms of current inputs and labor costs. By maximizing the difference between the value of production and the costs of production, producer surplus is maximized (Schipper, 1996). In REALM, with endogenous product prices the area below the demand function of each product is calculated at different prices. These areas, minus the costs of production (input costs and labor costs, as well as transport costs, although strictly speaking they do not belong to production costs), represent the sum of producer and consumer surplus at different price-quantity combinations of a product. REALM selects those price-quantity combinations for all products that, taken together, maximize the sum of producer and consumer surplus (Figure 6.4). To do so, a number of assumptions were made. The inverse demand functions were assumed to be linear of the following form:

$$P_j = a_j - b_jQ_j \qquad (1)$$

in which P_j is the price of commodity j and Q_j is the quantity demanded, while a_j (intercept on P_j axis) and b_j (slope) are coefficients. Dropping the subscript j, each demand function has a price elasticity η at point (P^0,Q^0). Knowledge of the price elasticity at a certain point, for example in a base year, allows for the calculation of the coefficients a and b.

$$\mathrm{b} = \frac{-P^0}{\eta Q^0} \qquad (2)$$

$$a = P^0 + bQ^0 \qquad (3)$$

Having calculated the coefficients a and b, one can divide the demand function into a number of segments. Associated with each segment-limit d will be a dimensionless variable D^d, which is forced to take on values between 0 and 1. In REALM, the demand functions are divided into 100 segments over a length from Q^0/k to $k{*}Q^0$, as suggested by Norton (1995). For each of the segment-limits Q^d, one can calculate the price P^d, producer revenue R^d, and the area below the demand function W^d, as follows:

$$P^d = a - b\,Q^d \tag{4}$$

$$R^d = P^d\,Q^d \tag{5}$$

$$W^d = a\,Q^d - 0.5\,b\,(Qd)^2 \tag{6}$$

Quantity Q^d, producer revenue R^d and the area below the demand function W^d are coefficients to be associated with the segment-limit variables D^d.

The above equations apply to a country as a whole. For the AZ, the situation is more complicated, as one should take into account not only the supply originating from the AZ, but also supply from other regions in Costa Rica, as well as, in case of export products, the supply from other countries. Under these conditions, the demand function facing the producers in the AZ is different from the national demand function. It can be shown (Hazell and Norton, 1986) that the regional demand elasticity η_r can be expressed as follows:

$$\eta_r = \eta\,\frac{1}{K} - \sigma_{nr}\,\frac{1-K}{K} \tag{7}$$

In this equation η represents the national demand elasticity, K is the AZ share of the national production and σ_{nr} is the supply elasticity from other regions.

The necessary calculations in REALM accord with the suggestions made by Norton (1995). The parameters used for each product are a base price and quantity, a price demand elasticity, the share that the region contributes to the national supply (in the base year situation) and a price supply elasticity for the remaining regions (*i.e.*, regions outside the model) or countries in case of export products. The price demand elasticities used are taken from Geurts *et al.* (1997) and Van der Valk (1999). Base price and quantities, including the share of the region in the national supply, are based on 1996 data. In the current version of REALM, price supply elasticities are not estimated, but assumed to be 0.7. Other studies suggest that supply elasticities between 0.4 and 1.0 are not unreasonable (Mamingi, 1997; Sadoulet and De Janvry, 1995). Norton (1995), on the basis of Henneberry (1986), suggests long-run supply elasticities of 1.0. However, using high elasticities might have the effect of "driving other regions out of the market", because (much) lower prices are still economically attractive for producers in the AZ (at least according to the programming model).

6.2.6 Upward sloping labor supply function

Labor available for agriculture, and the remuneration of labor, has a considerable influence on production possibilities. In REALM, it is assumed that in each sub-region there is a certain amount of labor working in agriculture at a fixed wage (called the agricultural labor force). This labor can also work in the other sub-regions, in which case transaction costs are taken into account (Section 6.2.3). Apart from the agricultural labor force present in the AZ, it is assumed that the agricultural sector can attract labor from outside this sector and/or the AZ, depending on the wage the sector is willing to pay. In this case transaction costs are taken into account as well. These transaction costs are, in general, higher than the transaction costs that apply to labor already working in the agricultural sector of the AZ. However, the quantity of outside labor supplied depends not only on transaction costs, but is also a function of the wage. In the current version of REALM, it is assumed that the total labor supply function (*viz.* the sum of the agricultural labor force and the non-agricultural labor force) is a non-smooth curve with a vertex. For a well-defined supply (as a first assumption for the currently available agricultural labor force, whether employed or unemployed), the wage is fixed at the present market wage, while further demand for labor drives the cost of labor along an upward sloping curve (Figure 6.5).

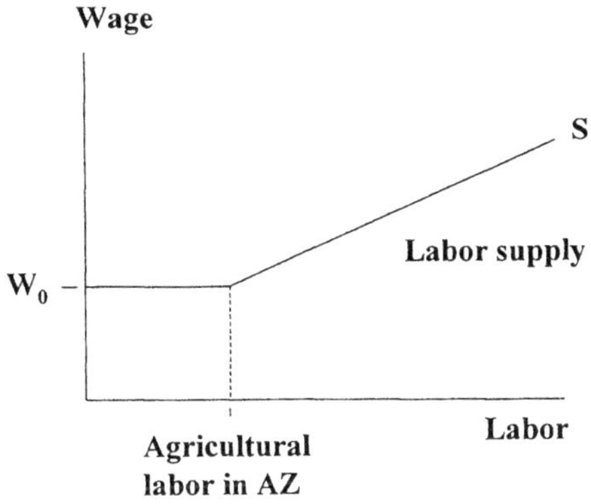

Figure 6.5. Upward sloping labor supply function (see text for explanation)

Similar to agricultural products, the market for agricultural labor in the AZ is only part of the national labor market. Therefore, the national labor supply elasticity has to be adjusted before it can be applied. Apart from the share of the AZ agricultural labor market in the national labor market (about 5%), the effect on labor demand in sectors/regions other than the AZ agricultural sector caused by an increased labor supply to the AZ agricultural sector and leading to increased wages has to be taken into account. In analogy to the situation for product markets (see above), the following relationship can be shown to exist (assuming no obstacles to labor mobility exist other than the previously mentioned transaction costs):

$$\varepsilon_r = \varepsilon \frac{1}{M} - \theta_{nr} \frac{1 - M}{M} \tag{8}$$

where ε_r is the labor supply elasticity for sector/region r, ε is the national labor supply elasticity, θ_{nr} is the labor demand elasticity in the remainder of the economy, and M is the share of the labor in sector/region r in the national labor market. In the current version of REALM, a labor supply elasticity of 0.2 has been assumed to exist at the national level, an estimate that is in line with other studies (Bosworth *et al.*, 1996). With regard to the labor demand elasticity in the remainder of the economy, -0.5 would be a plausible approximation (Bosworth *et al.*, 1996). Using equation 8 and in view of a labor share of 0.05, an ε_r of 13.5 for the labor supply elasticity in the AZ agricultural sector would thus be a reasonable approximation. Such a high elasticity implies a very gently upward sloping labor supply function, in case more labor would be required than presently available (*i.e.*, after the horizontal part of the labor supply function).

The effect of modeling the labor market in the way sketched above is that the agricultural sector can use more labor than its own labor force would permit, albeit at (slowly) increasing wages. In this way, the agricultural sector in the AZ competes for labor with other sectors and regions. Furthermore, the fixed wage (horizontal line) at the lower tail of the labor supply function incorporates the institutional feature of the labor market that no labor is supplied at wages below the current wage level. That is, wages either stay constant or increase, *i.e.*, they are "downward sticky".

6.3 Land use scenarios

The Costa Rican government is soliciting research that explicitly analyzes the trade-offs between socio-economic and environmental goals, for a range of policy options (SEPSA, 1997; see also Chapter 2). In this context, the capabilities of the SOLUS methodology in general, and that of the REALM model in particular, are demonstrated by using REALM to evaluate seven scenarios related to the projected macro-economic developments and to the policy and sustainability issues discussed in Chapter 2:

1. Technological change
2. Zero soil nutrient depletion
3. Quantitative limits on biocide use
4. Taxation of biocides
5. Conservation of natural forests
6. Lower interest rates
7. Increases in real wages

The effects of each scenario are studied by comparing its results with the results without such a policy or development, *i.e.*, the base run. Scenarios 1, 6 and 7 have to do with expected autonomous macro-economic developments in Costa Rica, scenario 2 with maintaining the resource base, and scenarios 3-5 with protection of the environment. Other sustainability and policy issues relevant to the AZ, namely the emission of greenhouse gases and the potential pollution by nutrients (Chapter 2), have been studied with a similar SOLUS set-up and are reported by Bouman *et al.* (1998c, 1999a) and Plant and Bouman (1999).

Results of the above scenarios are presented below. Monetary units are in US$, at the average 1994-1996 exchange rate of $ 1 = ¢ 181 (Costa Rican currency).

6.3.1 Technological change

The effect of technological progress was assessed by comparing the results of a model run that includes only actual land use options with those of the base run that includes both actual and alternative land use systems. Technological progress, essentially producing more with the same or less resources, has an important effect on economic surplus, employment, land use, and environmental indicators (Table 6.1).

Table 6.1. Impact of technological change: current technology versus technology used in base run.

		Scenarios	
	Units[1]	*Base*	*Current Technology*
Economic surplus	$ 10^6	268.4	220.5
Labor use	d 10^6	8.7	9.0
BIOA[2]	kg 10^6	1.9	2.9
BIOI[3]	10^6	84.1	100.0
Soil N depletion	kg 10^3 y^{-1}	7368.9	12775.8
Soil P depletion	kg 10^3 y^{-1}	-137.8	864.6
Soil K depletion	kg 10^3 y^{-1}	5337.0	28306.7
Crops	ha 10^3	61.2	76.7
Pastures	ha 10^3	189.9	174.4
Natural	ha 10^3	150.6	174.4
Grass-legume	ha 10^3	39.2	0
Animals, breeding	AU 10^3	253.5	197.7
Animals, fattening	AU 10^3	138.3	115.5
Area of important crops:			
Banana	ha 10^3	31.6	42.7
Pineapple	ha 10^3	2.3	2.8
Palm heart	ha 10^3	7.7	11.2
Plantain	ha 10^3	1.9	1.9
Cassava	ha 10^3	17.1	17.7
Production of important crops:			
Banana	t 10^3	2064.2	1851.7
Pineapple	t 10^3	194.3	196.8
Palm heart	t 10^3	82.2	60.1
Plantain	t 10^3	35.3	37.1
Cassava	t 10^3	87.2	90.2

[1] AU Animal Unit (live weight of 400 kg)
[2] BIOA Amount (kg) of active ingredient in all biocides
[3] BIOI Biocide Index (index of environmental effects of all biocides together)

Economic surplus increases by 21.4% when current technology is replaced by improved technology. Overall land productivity increases by the same percentage (the same area is used), while labor productivity increases 26.7% (higher surplus, using less labor). As a consequence, employment decreases by 4.2%. The use of biocides in the *base* run compares favorably with that in the *current technology* run: the amount of active ingredients applied (BIOA) is 33% lower and the biocide index (BIOI) is 16% lower. Thus, improved technology results in a win-win situation, *i.e.*, higher economic surplus and less environmental contamination from biocides.

In the *current technology* run, changes in soil nutrient stock are negative for all three nutrients nitrogen (N), phosphorus (P) and potassium (K). The soil mining of N and K is especially severe. In the *base* run, the situation improves considerably: soil N mining decreases by 42%, soil K mining decreases by 81% and P even accumulates slightly instead of being mined. The accumulation of P in the *base* run is caused by the relatively high use of fertilizer P (compared with the *current technology* run) and the (volcanic) soil's capacity for P fixation (Chapter 2). The negative changes in soil nutrient stock for N and K, however, indicate that the nutrient resource base is not maintained over time. As a consequence, soil productivity will decline. The selected production systems are, therefore, unsustainable. In the next section, the effects that the requirement of sustainability has on the soil nutrient resource base are investigated.

6.3.2 Zero soil nutrient depletion

Following the rule requiring farmers to "utilize renewable resources at rates less than or equal to the natural rate at which they regenerate" (Section 6.1.4; Pearce and Turner, 1990), a restriction of "zero change in soil nutrient stocks" was imposed in order to optimize the zero nutrient depletion scenario. This restriction was imposed on the levels of the land unit and of the sub-region. Results are presented in Table 6.2.

The economic surplus is 11% lower than in the base scenario, while employment more or less stays the same. The area with crops increases from 61 200 ha to 78 000 ha, while the area with pastures decreases from 189 800 ha to 30 700 ha. A large part of the land originally under pasture in the base scenario becomes forest (107 300 ha) in the zero nutrient depletion scenario, while 35 000 ha are not used at all. Crop areas do not change much, except that considerably less cassava is cultivated in the zero nutrient depletion scenario than in the base scenario; on the other hand, significantly more maize is cultivated (26 600 ha instead of 100 ha). These changes in land use are caused by a change from the nutrient-depleting actual technologies to non-depleting alternative technologies.

Any transition to land use systems that do not deplete the soil nutrient stock comes at a cost of 11% of the maximum achievable economic surplus. However, productivity levels in land use systems that deplete soil nutrients will decline over time. Bouman *et al.* (1999b) studied, for land under pastures, a future situation in which soil nutrient reserves have been depleted to a low equilibrium situation. In such a situation, land use systems with zero nutrient depletion would lead to a 3% higher economic surplus compared to the surplus obtained with nutrient depleting systems.

Table 6.2. Impact of zero soil nutrient depletion and limited biocide use compared to the *base* run.

		Scenario		
	Units	*Base*	*Zero soil nutrient depletion*	*Limits on biocide use*
Economic surplus	$ 10^6	268	239	267
Labor use	d 10^3	8661	8761	8638
BIOA[1]	kg 10^3	1935	1890	1924
BIOI[2]	10^6	84	48	17
Soil N depletion	kg 10^3 y^{-1}	7369	0	6977
Soil P depletion	kg 10^3 y^{-1}	-138	-125	-201
Soil K depletion	kg 10^3 y^{-1}	5337	0	5216
Crops	ha 10^3	61	78	55
Pastures	ha 10^3	190	31	196
Natural	ha 10^3	151	23	151
Grass-legume	ha 10^3	39	8	45
Forests	ha 10^3	0	107	0

[1] BIOA Amount (kg) of active ingredient in all biocides
[2] BIOI Biocide Index (index of environmental effects of all biocides together)

6.3.3 Quantitative limits on biocide use

Any attempt to reduce the impact of biocides on the environment, by imposing an upper limit on the biocide index (BIOI), complies with the rule stating that "waste flows to the environment must be kept at or below the assimilative capacity of the environment" (Section 6.1.4). It is assumed that the higher the biocide index, the higher the environmental damage. In the limited biocide use scenario, the biocide index is constrained relative to its value in the base run. Limiting the biocide index of the entire AZ to 20% of its value in the base scenario hardly changes the economic surplus nor the employment (Table 6.2). The stability in employment is the result of two countervailing changes. In order to reduce the biocide index, less herbicides (in particular paraquat) are used, particularly in banana and cassava production, requiring more weeding to be performed manually. On the other hand, the area of cassava is reduced from 17 100 ha in the base run to 11 200 ha in the proposed scenario. The difference of 5900 ha is mostly used for extending the pasture area with grass-legume mixtures in which no herbicides are used and which are less labor-intensive than cassava. The limited biocide use scenario shows that it is possible to substantially reduce the environmental impact of biocides without affecting the economic surplus nor the labor employment. The question of how to implement such a scenario, (*i.e.*, how to induce producers to use less biocides) is addressed in the next section, which evaluates the effects of different ways of taxing biocides.

6.3.4 Taxing biocides

Regulation and control of agricultural input use has been identified as an important policy option in a campaign to reduce certain negative externalities of agricultural production (SEPSA, 1997). Structural changes in the Costa Rican agricultural sector over the past decade have led to an increase in biocide use (Jansen *et al.*, 1998). In Costa Rica, biocide policies have traditionally consisted in legislative measures and the potential effects of economic instruments have been generally overlooked (Agne, 1996). There is little question that the taxing of an input that is currently not taxed, as is the case of biocides in Costa Rica, will lead to less use of this input. Such a tax can be implemented in a variety of ways. In this paper, we differentiate between a flat tax and a progressive tax. The latter is linked to the environmental damage caused by a specific biocide, which is approximated by the biocide index (BIOI). Taxing all biocides at a uniform rate of 100% leads to a reduction in the use of all the active ingredients in biocides (BIOA) by 13% below the level of the base scenario, while the BIOI is reduced by only 4% (Table 6.3). However, economic surplus is reduced by nearly 19%. Thus, a relatively modest environmental gain is obtained at a high economic cost. In contrast, a progressive tax regime, where different tax rates are applied to three categories of biocides depending on their degree of toxicity (*i.e.*, slightly, medium and very toxic), results in a much larger reduction of the BIOI, while at the same time preserving more of the economic surplus. For example, applying taxes of 20%, 50% and 200% (Tax System A) to the categories of slightly, medium and very toxic biocides, respectively, leads to a 4% reduction in the economic surplus, while reducing the BIOI by over 80%. When tax rates are reduced to the levels of 10%, 30% and 150% (Tax System C) respectively for the three categories of biocides, economic surplus decreases by just 2% while the same environmental improvement still occurs. However, Tax System B, in which the 200% tax on the very toxic biocide in Tax System A is lowered to 100%, does not lead to a substantial reduction of the BIOI.

Table 6.3. Effects of alternative ways of taxing biocide use

Type of biocide		*Base*	*Flat tax*	*Tax A*	*Tax B*	*Tax C*
				Progressive tax regimes		
Slightly toxic		0%	100%	20%	20%	10%
Medium toxic		0%	100%	50%	50%	30%
Very toxic		0%	100%	200%	100%	150%
Indicators	*Units*	*Value*	*% change*	*% change*	*% change*	*% change*
Economic surplus	$ 10^6	267.6	-18.7	-4.3	-4.1	-2.2
BIOA[1]	kg 10^6	1.9	-13.1	-3.9	-3.2	-3.8
BIOI[2]	10^6	84.1	-4.0	-81.9	-1.5	-81.9

[1] BIOA Total amount (kg) of active ingredient in biocides
[2] BIOI Biocide Index (index of environmental effects of biocides)

It can be concluded that, even though a flat tax results in the highest reduction in the absolute quantities of biocides applied, such a tax is relatively ineffective when it comes to protecting the environment and, at the same time, leads to a large decrease in economic surplus. In contrast, a biocide tax differentiated in relation to the degree of toxicity of each biocide can indeed reduce environmental damage, as measured by the BIOI, to a substantial degree at a relatively low economic cost. A comparable conclusion was reported by Jansen *et al.* (1997a) for Guácimo county in the AZ.

6.3.5 Conservation of natural forests

Agricultural policy in Costa Rica puts increasing emphasis on environmental protection (SEPSA, 1997). Recently, the government has introduced a policy to induce landowners to keep part of their property under natural forest. In return for not cutting down trees, as of 1997 a landowner can obtain a subsidy of \$ 43 ha^{-1} per year, initially for a period of five years. Taking into account an obligatory first year cost of \$ 13 for an officially approved forest management plan, this means an annuity of \$ 40 ha^{-1}.[5] The subsidy was created as a result of the international discussion about global warming. Maintaining or creating forest is seen as a means to sequestrate carbon dioxide. In the future, countries could sell carbon bonds if they maintain or create forests. The Costa Rican forest subsidy anticipates such a situation.

To analyze the effect of a forest subsidization policy on regional land use, premiums were allocated to the land use type *natural forest*. Natural forest can be exploited in a sustainable way, yielding about 0.6 m^3 of wood ha^{-1} y^{-1}, which means an annual return of about \$ 16 ha^{-1}. REALM was run with all the available land in the AZ (340 000 ha) potentially suitable for agricultural use, thus including the protected and semi-protected areas. In the base year of the model (1996), a subsidy of \$ 111 ha^{-1} y^{-1} (¢ 20 000 at the average 1994-96 exchange rate) is not sufficient (to induce landowners) to maintain their natural forests (Table 6.4). On the other hand, a subsidy of \$ 122 would lead to a forest area of about 120 000 ha and a subsidy of \$ 133 ha^{-1} would produce 200 000 ha of natural forest. These figures compare to the present 84 000 ha of primary or secondary forest in the area suitable for agriculture. As a result of such subsidies, which are considerably larger than the present subsidy, a large part of the land used for pastures in the base scenario is converted to natural forest, while the cropped area remains virtually unchanged.

Even though a subsidy of \$ 111 ha^{-1} y^{-1} would raise the annual return from natural forest to \$ 127 ha^{-1}, this is still lower than the shadow price of land in all sub-regions and for each land unit. In case of a subsidy of \$ 122 ha^{-1} y^{-1} however, returns of natural forest exceed the shadow prices of land belonging to the fertile poorly-drained (SFP) and infertile well-drained (SIW) land units in most sub-regions. On the other hand, the fertile well-drained (SFW) land unit has shadow prices between \$ 188 and \$ 204 ha^{-1} y^{-1} (depending on the sub-region) and a subsidy would have to exceed \$ 172 ha^{-1} y^{-1} for natural forest to become an economically attractive option. Land units SFP and SIW are mostly used for pasture, while SFW is mostly used for crops. These findings lend support to the hypothesis that pasture and natural forest are competing land use types for marginal land areas.

[5] A subsidy of ¢ 10 000 per ha for five years minus the ¢ 3000 costs for a forest management plan, converted at the average 1997 exchange rate of \$ 1 = ¢ 232.

The optimal level of a forest subsidy can be further defined by analyzing marginal land use from a different perspective, in which we suppose that a subsidy for natural forests does not exist. In the base scenario, only land outside the protected and semi-protected areas is considered (278 900 ha). Extending the availability of land by including the semi-protected areas (48 900 ha), or both the semi-protected and the protected areas (61 200 ha) provides an indication of the incremental increase in economic surplus that would occur if these areas could be used for agricultural purposes. Table 6.5 indicates that the increments in economic surplus are not substantial. Extending the area in the base run by including the semi-protected areas (17.5% more land) leads to a rather marginal (2.5%) increase in economic surplus. Average yearly returns ha^{-1} decrease from $ 960 to $ 837. The incremental economic surplus of the semi-protected areas is only $ 134 ha^{-1} y^{-1}. Similarly, extending the base and semi-protected area case with the land of the protected areas (another 19% more land) results in a further 0.6% increase in economic surplus. Average returns ha^{-1} drop to $ 811 per year, while incremental returns of the protected areas are $ 131 ha^{-1} y^{-1}. A comparison of land use patterns in each of the three above cases reveals that all extra land is used for pasture, because there is a limited demand for crop products at sufficiently remunerative prices.

6.3.6 Reductions in the interest rate

Costa Rica is currently trying to reduce the internal (public sector) debt, mainly through the (intended) sale of a number of public enterprises to private investors. A decrease in the size of the internal debt would imply lower debt servicing payments and a corresponding decrease in the demand for capital by the government, thus diminishing the crowding out of private investments by the public sector (Hausmann, 1998). As a result, interest rates can be expected to decrease. To simulate the effect of a reduction in the real interest rate on land use in the AZ, REALM was run with a discount rate of 3% (instead of 7% in the base scenario). The effect of a such a lower discount rate on land use turned out to be rather small (Table 6.6). The area under crops increases slightly, at the expense of pasture area. Within the area under pastures, there occurs a shift away from natural(ized) species towards grass-legume mixtures. Since grass-legume mixture require a substantial initial investment of about $ 400 ha^{-1} (Jansen *et al.*, 1997b), such an investment becomes more profitable at lower discount rates because of lower capital costs. Not surprisingly, economic surplus increases as well, mainly because most benefits occur later than costs and consequently suffer less from discounting. An explanation for the modest changes in land use that result from a more than 50% decrease in the discount rate, may be that REALM compares the relative profitability of each alternative land use system, thereby taking into account all constraints, including the market constraints implicitly imposed by the downward sloping demand functions. Since lowering the discount rate from 7% to 3% changes the relative profitability of each land use system only marginally, while the market constraints remain the same, land use is only slightly affected.

Table 6.4. Consequences of subsidizing natural forest in the Atlantic Zone.

Land unit	*Units*	*Available land*	*No non-timber value to forest*	*$ 111 ha⁻¹ y⁻¹ subsidy*	*$ 122 ha⁻¹ y⁻¹ subsidy*	*$ 133 ha⁻¹ y⁻¹ subsidy*
			Area with natural forest			
SFW	ha 10^3	118.4	0	0	0	0
SFP	ha 10^3	136.0	0	0	62.6	122.4
SIW	ha 10^3	85.7	0	0	56.4	77.1
Total	ha 10^3	340.1	0	0	118.9	199.5
Objective	$ 10^6		275.8	275.8	276.1	277.9

SFW Soil Fertile Well-drained
SFP Soil Fertile Poorly-drained
SIW Soil Infertile Well-drained

Table 6.5. Consequences of transforming semi-protected and protected areas in the Atlantic Zone into agricultural land on the regional economic surplus (objective) and on land use

Scenario	*Objective* $ 10^6	*Increase of objective* $ 10^6	*% change objective* %	*Area* ha 10^3	*Increase of area* ha 10^3	*% change area* %	*Average returns* $ ha⁻¹ y⁻¹	*Incremental returns* $ ha⁻¹ y⁻¹
Base	267.4	-	-	278.9	-	-	960	-
Base and semi-protected area	274.0	6.6	2.5	327.6	48.7	17.5	837	134
All land	275.8	1.7	0.6	340.1	12.5	3.8	811	131

Table 6.6. Effect of different discount rates on economic surplus, employment and land use in the Atlantic Zone.

	Units[1]	*Discount rate 3%*	*Discount rate 7% (base)*
Economic surplus	$ 10^6	281.4	267.6
Labor	d 10^6	8.6	8.7
Crops:	ha 10^3	62.7	61.2
Pineapple	ha 10^3	2.9	2.8
Palm heart	ha 10^3	8.0	7.7
Banana	ha 10^3	31.9	31.6
Plantain	ha 10^3	2.0	1.9
Cassava	ha 10^3	17.8	17.1
Pastures:	ha 10^3	188.3	189.8
Natural pasture	ha 10^3	148.7	150.6
Grass-legume	ha 10^3	39.6	39.2
Animals:			
Breeding	AU 10^3	250.7	252.5
Fattening	AU10^3	137.5	138.3

[1] AU Animal Unit (live weight of 400 kg)

6.3.7 Increasing real wages

International development banks expect a GDP growth in Cost Rica of 4.5-5% y^{-1} for the next five years. Given an expected population growth of 2% y^{-1}, this translates into a 2.5-3% annual increase of per capita GDP, similar to the average per capita GNP growth of 2.8% realized between 1985 and 1995 (World Bank, 1997). Assuming a continuation of such a per capita GDP growth in the future, it is likely that real wages will increase concurrently. A 2.8% increase per year during 20 years means a total wage increase of 74%. To explore the potential impact of real wage increases on land use, three scenarios were evaluated with total wage increases of 50%, 75% and 100%, respectively. A 75% aggregate wage increase can be expected on the basis of a continuation of current and past trends; the remaining two scenarios were evaluated to explore the sensitivity of the model to wage increases. A real wage increase of 75% results in less crop land and more land under pasture (Table 6.7). Wage increases of 50 and 100% result in similar land use changes. The principal reason for these results is that crops use relatively more labor than pastures, a fact also indicated by a decrease in the number of animal units per ha; the latter drops from 2.06 to 1.87, making animal husbandry less labor intensive.

Table 6.7. Effect of increasing real wages on economic surplus, employment and land use in the Atlantic Zone.

Scenario	*Economic surplus*	*Labor use*	*Labor income*	*Crops*	*Pasture*	*Animals: breeding*	*Animals: fattening*
Units	*$ 10^6*	*d 10^6*	*$ 10^6*	*ha 10^3*	*ha 10^3*	*AU 10^3*	*AU 10^3*
Base [1]	267.6	8.7	76.6	61.2	189.8	252.5	138.4
Wage+50%	232.2	7.6	100.7	54.9	196.1	233.0	128.2
Wage+75%	215.8	7.3	112.3	43.7	207.3	250.3	137.1
Wage+100%	200.2	7.0	124.1	42.0	209.0	252.7	138.4

[1] Wage in base run is $ 8.84 per day.

Not surprisingly, labor use in the agricultural sector of the AZ diminishes with increasing real wage rates. In a growing economy, labor can be expected to be increasingly employed in the non-agricultural sectors. Economic surplus also decreases with increasing real wages, as wages constitute a cost component in the model. On the other hand, wages also represent income to laborers. Thus, the sum of the economic surplus and total wage income (number of labor days times wage) better reflects the economic gains resulting from land use in the AZ on the Costa Rican society as a whole. In the base scenario this sum is $ 344 million, while at a 75% wage increase this sum is $ 333 million. Thus, as the economic surplus created in the agricultural sector in the AZ decreases by 13.2% as a result of a 75% increase in the real wage rate, the sum of economic surplus and labor income decreases by only 3.3%.

6.4 Conclusions

Policies aimed at influencing land use ideally should be based on a simultaneous evaluation of the socio-economic and environmental implications of both actual and alternative technological options for land use. The SOLUS methodology presented in this chapter provides a useful tool for such evaluations as shown here for the case study of the northern Atlantic Zone of Costa Rica. Important economic and agricultural policy issues relevant to the AZ are related to the impact of technological change; the desire to avoid land degradation by halting soil nutrient depletion; the desire to reduce externalities by minimizing biocide-related pollution; the policy objective intended to maintain as much natural forest as possible in order to restore ecological balance, avoid global warming, and promote tourism; and the strive to reduce interest rates in order to stimulate overall investment and economic growth. Another important question is what kind of effects on land use can be expected from continuous real wage increases, as a consequence of general economic development. The presented SOLUS framework has shown to be a useful instrument for the assessment of each of these issues, and as such is able to better inform policy makers.

Based on the policy simulations performed in this chapter, the following conclusions can be drawn:

1. Development of new options for land use through technological change can bring simultaneous economic as well as environmental gains, thus providing a "win-win" situation.
2. Many current land use systems deplete the soil resource base and are therefore unsustainable in the long run. Imposing long-term sustainability by foregoing maximum economic surplus in the short run through the exclusive implementation of sustainable, non soil nutrient depleting land use systems, comes at a cost of 11% of the maximum achievable economic surplus.
3. Limiting biocide use through a progressive biocide tax related to toxicity and persistence in the environment proved to be more efficient than a flat tax.
4. Subsidizing natural forest management - implying that society at large recognizes that the value of natural forests goes beyond potential timber revenues - could help to maintain existing natural forest areas or even create new ones, albeit that such subsidies have to be rather substantial to be effective (*i.e.,* higher than the current subsidy level based on carbon bonds). At the same time, extending the agricultural area (crops and pastures) into existing (semi-)protected areas only brings marginal economic benefits, mainly because these protected areas would be converted into pastures.
5. Lowering the interest rate increases regional economic surplus but hardly affects land use distribution because of product market constraints.
6. Higher wages as a consequence of overall economic development would lead to a reduction in the area of crops and an expansion of pastures. Not surprisingly, labor intensive activities are substituted by labor extensive ones. If real wage increases in Costa Rica would exceed wage increases in competing countries, Costa Rica may gradually lose its competitive advantage for crops like banana, plantain, palm heart, cassava and pineapple.

Appendix 6.1 Mathematical formulation of REALM

Table A6.1. Relevant part of simplified AZ model

Objective function: benefits less costs; area below domestic demand functions plus value of exports, less product market transaction costs, less value of current input and labor costs (wage sum, transaction costs and area below labor supply function) (¢ y^{-1})

$$\text{Max } Z = \sum_{j \in J1} p_{j \in J1} D_{j \in J1} + \sum_{j \in J2} p_{j \in J2} E_{j \in J2} + \sum_{j \in J3} \sum_{d} \omega^{d}_{j \in J3,d} D_{j \in J3,d} +$$

$$\sum_{j \in J4} \sum_{d} \rho^{x}_{j \in J4,d} E^{s}_{j \in J4,d} - \sum_{z} \sum_{j} t^{p}_{jz} S_{jz} - p\, C - \sum_{z} \sum_{\zeta} t^{l}_{\zeta z} L_{\zeta z} - \sum_{z} t^{0}_{z} O_{z} - \sum_{o} \bar{\omega}^{0}_{0} O_{0} \qquad \textbf{(1)}$$

Subject to:

- balances of product annuity per product per sub-region (t y^{-1})

$$\sum_{s} \sum_{l} \sum_{t} - y_{jslt} X_{zslt} + \sum_{h} - y_{jh} A_{zh} + S_{jz} \leq 0 \qquad \text{all } j, z \qquad \textbf{(2)}$$

- balances of product annuity per product for whole AZ (t y^{-1})

$$S_{j} - \sum_{z} S_{jz} \leq 0 \qquad \text{all } j \qquad \textbf{(3)}$$

- balance per product: domestic demand + export < production (t y^{-1}) (excluding imports not relevant for the products concerned)

$$D_{j} + E_{j} - S_{j} \leq 0 \qquad \text{all } j \qquad \textbf{(4)}$$

- segmentation of domestic demand per product (t y^{-1})

$$-D_{j} + \sum_{d} q^{d}_{jd} D^{s}_{jd} \leq 0 \qquad \text{all } j \in J3 \qquad \textbf{(5)}$$

- convex combination constraint for domestic demand segment-limit variables

$$\sum_{d} D^{s}_{jd} \leq 1 \qquad \text{all } j \in J3 \qquad \textbf{(6)}$$

- segmentation of export demand per product (t y^{-1})

$$-E_{j} + \sum_{d} q^{x}_{jd} E^{s}_{jd} \leq 0 \qquad \text{all } j \in J4 \qquad \textbf{(7)}$$

- convex combination constraint for export demand segment-limit variables

$$\sum_{d} E^{s}_{jd} \leq 1 \qquad \text{all } j \in J4 \qquad \textbf{(8)}$$

- annuity of input (sum of current input costs) balances per sub-region (¢ 10^3 y^{-1})

$$\sum_{s} \sum_{l} \sum_{t} c_{slt} X_{zslt} + \sum_{h} c_{h} A_{zh} + \sum_{s} \sum_{p} \sum_{r} c_{spr} P_{zspr} + \sum_{f} c_{f} F_{zfm} - C_{z} \leq 0 \qquad \text{all } j, z \qquad \textbf{(9)}$$

- annuity of input (sum of current input costs) balances for whole AZ (¢ 10^3 y^{-1})

$$-C + \sum_{z} C_{z} \leq 0 \qquad \textbf{(10)}$$

- feed balance per nutrition type per period per sub-region (Mcal y^{-1}; kg y^{-1})

$$\sum_{p} \sum_{s} \sum_{r} n_{sprmn} P_{zspr} + \sum_{f} n_{fn} F_{zfm} - \sum_{h} n_{hnm} A_{zh} \geq 0 \qquad \text{all } z, n, m \qquad \textbf{(11)}$$

- animal stock balance for pastures and herds per sub-region (AU y^{-1})

$$\sum_p \sum_s \sum_r s_{spr} P_{zspr} - \sum_h h_h A_{zh} = 0 \qquad \text{all } z \qquad (12)$$

- balance of calves (t y^{-1})

$$\sum_z y_{j=calves,\ h=breeding} A_{z,h=breeding} - \sum_z \sum_h v_{h=fattening,\ double\ purpose} A_{z,h=fattening,\ double\ purpose} \geq 0 \qquad (13)$$

- use of land units per sub-region per farm type by land use system (ha y^{-1})

$$\sum_l \sum_t X_{zfslt} + \sum_p \sum_r P_{zspr} \leq b_{zs} \qquad \text{all } z, s \qquad (14)$$

- annuity of labor use balanced by labor supply per sub-region (d y^{-1})

$$\sum_s \sum_l \sum_t l_{slt} X_{zslt} + \sum_h l_h A_{zh} + \sum_s \sum_p \sum_r l_{spr} P_{zspr} + \sum_f l_f F_{zfm} - \sum_\zeta L_{z\zeta} - O_z \leq 0 \qquad \text{all } z \qquad (15)$$

- agricultural labor force availability per sub-region (d y^{-1})

$$\sum_z L_{z\zeta} \leq a_\zeta \qquad \text{all } \zeta \qquad (16)$$

- calculation of agricultural labor force used in AZ (d y^{-1})

$$\sum_z \sum_\zeta L_{z\zeta} - L \leq 0 \qquad (17)$$

- segmentation of labor supply function (d y^{-1})

$$L + \sum_z O_z - \sum_o o_o O_o \leq 0 \qquad (18)$$

- convex combination constraint for labor

$$\sum_o O_o \leq 1 \qquad (19)$$

- calculation of environmental indicators per land unit per sub-region per indicator (kg y^{-1}; index y^{-1})

$$\sum_l \sum_t \delta_{slte} X_{zslt} + \sum_p \sum_r \delta_{spre} P_{zspr} - A_{sze} = 0 \qquad \text{all } s, z, e \qquad (20)$$

- limit to environmental indicator per sub-region per land unit per indicator (kg y^{-1}; index y^{-1})

$$A_{sze} \leq d_{sze} \qquad \text{all } s, z, e \qquad (21)$$

- limit to environmental indicator per sub-region per indicator (kg y^{-1}; index y^{-1})

$$\sum_s A_{sze} \leq d_{ze} \qquad \text{all } z, e \qquad (22)$$

- limit to environmental indicator per land unit per indicator (kg y^{-1}; index y^{-1})

$$\sum_z A_{sze} \leq d_{se} \qquad \text{all } s, e \qquad (23)$$

- limit to environmental indicator per indicator for whole AZ (kg y^{-1}; index y^{-1})

$$\sum_s \sum_z A_{sze} \leq d_e \qquad \text{all } e \qquad (24)$$

Table A6.2. Selected indices of the AZ model

Indices	*Description*	*Elements*
j	Products	Depends on selection of land use systems; the set *J* with elements *j* has four sub-sets: *J1*, products for domestic market without a market limitation; *J2*, products for export market without a market limitation; *J3*, products for domestic market with a downward-sloping demand function; and *J4*, products for export market with a downward-sloping demand function In the GAMS formulation this index is a combination between index C(PA) for crops (land use types) and index Q (product type/quality), or index HP(PA) for livestock product and the same index Q
d	segment-limits	1 to 100 in case of demand for products
z	sub-region	1 to 12, for sub-regions Rxxx; there is also index ζ as an alias for *z*
s	land units	SFP, SFW, SIW
l	land use types	depends on selection of land use systems, at present: pineapple, palm heart, melina, banana, plantain, cassava, (black) beans, teak, maize (corn), maize (cobs)
t	technology	depends on level (high/low) of fertilizer, biocides, herbicides and mechanization and on length of crop cycle (01, 02, 03, 10 or 15 years)
h	herd type	herds of 50 animals, either breeding, fattening or double purpose
p	pasture	depends on pasture (*Cynodon nlemfuensis*, *Brachiaria brizantha*, *Brachiaria radicans*, *B.brizantha-A.pintoi*, *B.humidicola-A.pintoi*, and "natural" which represents a mixture of the naturalized and native grasses *Ischaemum ciliare*, *Axonopus compressus* and *Paspalum* spp.), weeding type (only herbicides, manual/herbicides, only manual) and fertilization level (low to high)
r	stocking rate	low to high, at present: 1 to 5 animal units per ha
f	feed types	molasses of sugar cane, rejected bananas, chicken dung, P_2O phosphorus
n	nutrition types	metabolizable energy, crude protein, phosphorus
m	period	season 1 (dryer): January to March, season 2 (wetter): April to December
o	segment limit	1 to 100 in case of labor supply
e	environmental indicator	N balance, P balance, K balance, N denitrification, N leaching, N valorization, biocide active ingredients, biocide indicator

Table A6.3. Selected variables of the AZ model

Variables[1]	*Description*	*Unit of measurement*
Z	value of objective function	$¢\ y^{-1}$
S_j	annuity production per product	$t\ y^{-1}$
S_{jz}	annuity production per product per sub-region	$t\ y^{-1}$
D_j	domestic demand per product	$t\ y^{-1}$
D_{jd}^{d}	domestic demand segment-limit variable per product	
E_j	export demand per product	$t\ y^{-1}$
E_{jd}^{x}	export demand segment-limit variable per product	
C	annuity of current input use (materials and services)	$¢\ 10^3\ y^{-1}$
C_z	annuity current input use per sub-region	$¢\ 10^3\ y^{-1}$
X_{zslt}	land use system per sub-region per soil per land use type per technology	$ha\ y^{-1}$
P_{zspr}	pasture system per sub-region per land unit per pasture type per stocking rate	$ha\ y^{-1}$
F_{zfm}	feed acquisition system per sub-region per supplementary feed type per period	$Mcal\ y^{-1}$; $kg\ y^{-1}$
A_{zh}	animal production system per sub-region per herd type	$herds\ y^{-1}$
$L_{z\zeta}$	annuity of use of labor from the "agricultural labor force" per sub-region z, originating from sub-region ζ	$d\ y^{-1}$
L	annuity of use of labor from the "agricultural labor force" for whole AZ	$d\ y^{-1}$
O_z	annuity of use of labor not belonging to the "agricultural labor force" per sub-region	$d\ y^{-1}$
O_o	segment limit variable for total labor supply	
A_{sze}	environmental indicator variable per land unit per sub-region per indicator type	$kg\ y^{-1}$; $index\ y^{-1}$

[1] All variables in the model are continuous and larger than, or equal to, zero, except for Z and A_{sze}, which are "free" variables (larger than minus infinity). Furthermore, ¢ = *Colón*, the currency unit of Costa Rica.

Table A6.4. Selected coefficients of the AZ model

Coefficients	*Description*	*Units of measurement*
P_j	product price per product (OBJ)[1]	$¢\ t^{-1}$
P	price of current inputs (OBJ)	$¢¢^{-1}10^{-3}$
t_{jz}	product market transaction costs per product per sub-region (OBJ)	$¢\ t^{-1}$
ω_{jd}^{d}	areas below domestic demand function related to each segment limit of domestic demand functions (OBJ)	$¢\ y^{-1}$
ρ_{jd}^{x}	producer revenue related to each segment limit of export demand functions (OBJ)	$¢\ y^{-1}$

Table A6.4. Continued

Coefficients	*Description*	*Units of measurement*
q_{jd}^{d}	quantity of domestic demand related to each segment limit of domestic demand functions	$t\ y^{-1}$
q_{jd}^{x}	quantity of export demand related to each segment limit of export demand functions	$t\ y^{-1}$
$t_{\zeta z}^{l}$	agricultural labor force transaction costs from sub-region ζ to sub-region z (OBJ)	¢ d^{-1}
w	wage of agricultural labor force (OBJ)	¢ d^{-1}
t_{z}^{o}	transaction costs for labor from outside the agricultural sector (OBJ)	¢ d^{-1}
ϖ_{o}^{o}	area under total labor supply function sector (OBJ)	¢ d^{-1}
y_{jslt}	annuity yield of a land use system per product	$t\ ha^{-1}$
y_{jh}	annuity yield of an animal production system per product	$t\ herd^{-1}$
c_{slt}	annuity use of current inputs by a land use system	¢ $10^{3}\ ha^{-1}$
c_{h}	annuity use of current inputs by an animal production system	¢ $10^{3}\ herd^{-1}$
c_{spr}	annuity use of current inputs by a pasture system	¢ $10^{3}\ ha^{-1}$
v_{h}	calves as input for APST in case of fattening or double purpose	$t\ herd^{-1}$
c_{f}	annuity use of current inputs by a feed acquisition system	¢ $10^{3}\ kg^{-1}$
n_{sprmn}	feed elements yielded by a pasture system per land unit per pasture type per stocking rate per period per nutrition type	$Mcal\ y^{-1}$; $kg\ y^{-1}$
n_{fn}	feed elements yielded by a feed acquisition system per supplementary feed type per nutrition type	$Mcal\ y^{-1}$; $kg\ y^{-1}$
n_{hnm}	feed elements required by an animal production system per herd type per nutrition type per period	$Mcal\ y^{-1}$; $kg\ y^{-1}$
s_{spr}	stocking rate per land unit per pasture type per stocking rate	$AU\ ha^{-1}$
h_{h}	herd size per herd type	$AU\ herd^{-1}$
b_{zs}	land availability per farm type per land unit (RHS)[2]	$ha\ y^{-1}$
l_{slt}	annuity use of labor by a land use system	$d\ ha^{-1}$
l_{h}	annuity use of labor by an animal production system	$d\ ha^{-1}$
l_{spr}	annuity use of labor by a pasture system	$d\ ha^{-1}$
l_{f}	annuity use of labor by a feed acquisition system	$d\ ha^{-1}$
o_{o}	annuity use of labor related to each segment limit of the total labor supply function	$d\ y^{-1}$
a_{ζ}	(annuity of) agricultural labor force availability (RHS)	$d\ y^{-1}$
δ_{slte}	sustainability indicator of land use system per indicator type	$kg\ ha^{-1}$; $index\ ha^{-1}$
δ_{spre}	sustainability indicator of pasture system per indicator type	$kg\ ha^{-1}$; $index\ ha^{-1}$
d_{zse}	limit to sustainability indicator per sub-region per land unit per indicator type (RHS)	$kg\ y^{-1}$; $index\ y^{-1}$
d_{ze}	limit to sustainability indicator per sub-region per indicator type (RHS)	$kg\ y^{-1}$; $index\ y^{-1}$
d_{se}	limit to sustainability indicator per land unit per indicator type (RHS)	$kg\ y^{-1}$; $index\ y^{-1}$
d_{e}	limit to sustainability indicator per indicator type for whole AZ (RHS)	$kg\ y^{-1}$; $index\ y^{-1}$

[1] OBJ: objective function coefficient
[2] RHS: right hand side coefficient

7 Exploring future land use options: combining biophysical opportunities and societal objectives

JANETTE BESSEMBINDER, MARTIN K. VAN ITTERSUM,
ROBERT A. SCHIPPER, BAS A.M. BOUMAN,
HUIB HENGSDIJK, and ANDRÉ NIEUWENHUYSE

Abstract

Explorative land use studies that combine biophysical opportunities and societal objectives make stakeholders more aware of land use options. In this chapter, the methodology of such explorative studies is presented for a case study of the northern Atlantic Zone of Costa Rica. The SOLUS framework (explained in Chapter 6) for land use analysis is used: *i*) a GIS database, ii) the Technical Coefficient Generators PASTOR (Pasture and Animal System Technical coefficient generatOR) and LUCTOR (Land Use Crop Technical coefficient generatOR) to generate alternative and innovative, target oriented land use systems, and iii) a multiple goal linear programming model called GOAL-AZ (General Optimal Allocation of Land use for the northern Atlantic Zone of Costa Rica). GOAL-AZ optimizes for economic surplus, employment, biocide use and N losses by selection of combinations of land use systems, herds and feed supplements. Biophysical constraints of the model relate to the availability of land, and to requirements and availability of feed for cattle. Results indicate that trade-offs exist between economic and environmental objectives and the associated optimum use of resources. In the last part of this chapter, results of GOAL-AZ are compared with those of REALM (Regional Economic and Agriculture Land use Model; see Chapter 6), both being explorative land use models implemented within the SOLUS framework. While the former is a multiple goal model, the latter is a single objective (economic surplus) model, including market mechanisms for products and labor. This chapter will show how economic constraints affect future land use options and how GOAL-AZ sheds light on possible land uses beyond projected economic constraints. The complementarity of both approaches in supporting land use analysis is discussed.

7.1 Introduction

Explorative land use studies are meant to provide stakeholders with alternatives to current land use that meet a set of well-defined objectives. Different types of explorative land use studies exist. Depending on the aims of the study, emphasis may be on the biophysical possibilities or on the economic constraints, perspectives that affect the basic assumptions made, methodology used, and the interpretation of the results of the study. In studies emphasizing the ways in which biophysical features restrict various societal desires, information on socio-economic aspects and constraints is less important than in studies that analyze opportunities resulting from prevailing and expected market conditions (in Chapter 6 an example of the latter type of study is presented). Explorative studies emphasizing biophysical elements intend to clarify policy debates by enhancing understanding about the possibilities and limitations of agricultural development, as defined by biophysical processes, technical options and societal desires. This chapter presents an example of such an explorative study of the northern Atlantic Zone of Costa Rica using the SOLUS framework presented in Chapter 6. In addition, results of this study are compared with the results of the approach analyzing future options while taking socio-economic constraints into account (Chapter 6). Emphasis is put on the complementarity of both approaches in assessing future land use options.

7.2 Concepts and methodology of exploring biophysical land use options

Explorative studies focusing on the biophysical aspects of land use reveal future land use options by confronting biophysical and technical possibilities and constraints with the societal objectives and priorities of stakeholders. Differences in the priorities of each group of stakeholders are driven by their perceived needs and prospective risks as participants in both the socio-economic and the eco-systems (WRR, 1995). Exploration of future agricultural land use options is possible and useful, since potentials and limitations of agricultural production can be quantified using knowledge of the natural resource base. The aim of these explorations has three clear consequences for the methodology, the required technical information, and the interpretation of the results.

First, science-based biophysical and agro-technical information is segregated from information related to stakeholder priorities (normative or value-driven information). Biophysical and technical options are explored without projecting current socio-economic and institutional constraints into the future. This enables us to distinguish between technical and behavioral or economic factors, all of which strongly influence actual policy development (Spharim *et al.,* 1992). Using a scenario approach, different stakeholder priorities can be analyzed and their consequences for land use revealed.

Second, only technically or ecologically efficient production techniques are considered. Knowledge of the natural resource base and underlying biophysical processes, *e.g.*, photosynthesis and effects of growth factors, is used to quantify such alternative or innovative production techniques. Land use systems are defined using the target-oriented approach: target production levels are predefined and the optimal combinations of inputs required to realize these targets are subsequently quantified using knowledge of the processes involved (De Wit *et al.*, 1988; De Koning *et al.,* 1995; Chapter 5 of this book). This enables us to determine the technically most efficient combinations of inputs to realize particular production levels, according to current levels of knowledge and available techniques (Van Ittersum and Rabbinge, 1997). The input-output combinations are defined in such a way that they can be repeated in time without changing input requirements. This means that, for instance, no depletion of soil nutrients is allowed.

Third, the time horizon is related to the period within which adoption of the new land use systems is technically feasible. A distinct definition of the time horizon for this type of exploration is not possible and even not necessary, because there is no pretension to predict the future. Figure 7.1 schematically shows the concept of explorative land use studies, illustrating the different elements of the methodology for explorations combining biophysical and technical factors with societal priorities (see also Figure 6.2). Multiple Goal Linear Programming (MGLP) is used as an integrative tool. With this optimization procedure, field-level information on possible land use systems is confronted with regional-level constraints and objective functions. The result is that several possible combinations of regional land use systems are generated.

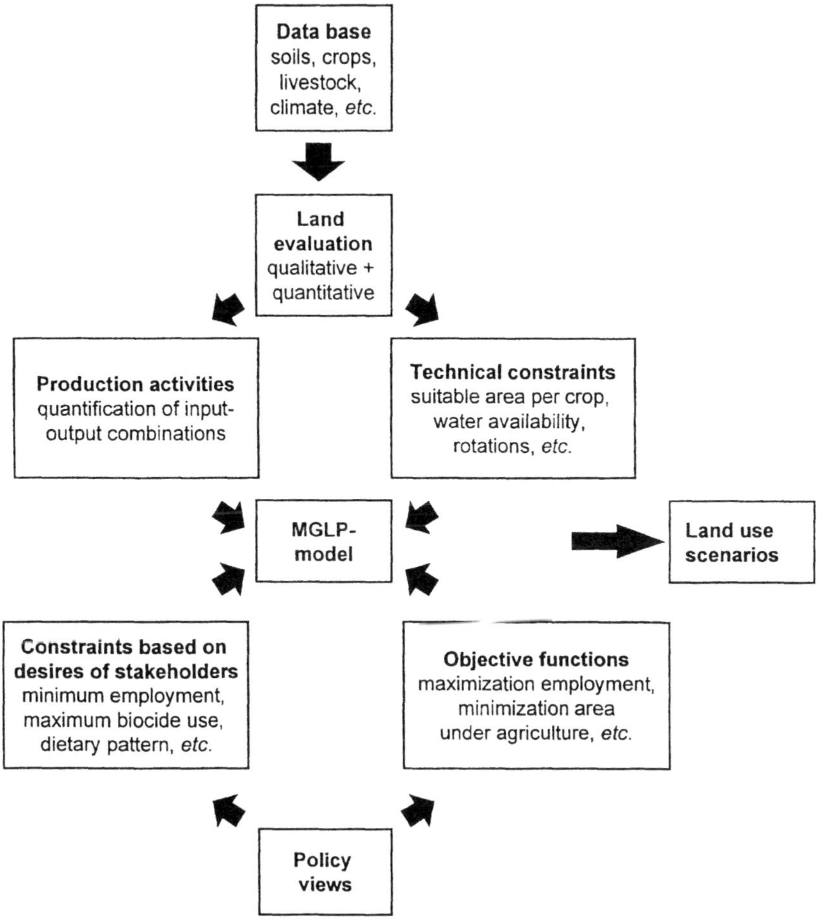

Figure 7.1. Diagram of the methodology for explorative land use studies (based on Bessembinder, 1997).

The consecutive steps in the diagram displayed as Figure 7.1 are as follows. First, an inventory is made of policy objectives relevant to the study area. Based on this inventory and the aim of the entire study, the system is delimited, *i.e.*, the region is defined, the boundaries of the agricultural sector are set and the interactions with other economic sectors are identified. Then, the various feasible forms of land-bound agricultural production (crop, livestock, forestry) in the region are identified. For these forms of land use, the potentially suitable physical production environments are provided in a qualitative land evaluation. Physical production environments are characterized by soil, terrain and climate. Next, in a quantitative land evaluation, potential and attainable yields are calculated or estimated for all suitable production environments, using knowledge of underlying biophysical processes. Crop growth simulation models, empirical information and expert knowledge may all be used in this step to explicitly address the effects of different climate and soil conditions on agricultural production (WRR, 1992; Rabbinge, 1993; Bouman *et al.,* 1996, 1999a).

Land use systems (production activities in Figure 7.1) are quantified by their input-output combinations with the help of Technical Coefficient Generators. These inputs and outputs differ by crop, production orientation, production technique and physical production environment (Rabbinge *et al.,* 1994; Van Ittersum and Rabbinge, 1997; Chapter 5 of this book). In the present application of the Technical Coefficient Generators, only alternative land use systems are identified, *i.e.*, those land use systems that are characterized by a zero change in the soil nutrient stock.

Input-output combinations are quantified using a target-oriented approach. In this approach, the production target (*i.e.*, a well-defined yield or emission level) is set, and then the set of inputs to realize the target is defined. No more nutrients, biocides or labor to realize a particular production level are used than strictly necessary from a technical point of view, *i.e.*, inputs are used with the highest technical efficiency according to the available knowledge and techniques. No substitution is allowed among inputs such as water and nutrients, which are taken up by the plant to fulfil specific complementary roles. Other inputs, such as biocides, labor, and mechanization, can substitute each other up to a certain degree. A detailed description of the target-oriented way of quantifying land use systems (and herds) is given in Chapter 5 of this book.

The choice of production techniques depends on the aim of the land use system or the production orientation: high soil productivity, low emissions of nutrients or biocides, *etc.* This in turn determines the target yield. Subsequently, the target yield determines the combination of inputs. For instance, in an activity geared towards high land productivity, and accordingly, with a high target yield, control of diseases and pests takes place in such a way that high productivity is achieved with the efficient use of biocides. In an environment-oriented activity, biocides are excluded as much as possible, and lower target yields per unit of land are consequently accepted (WRR, 1992).

Constraints in MGLP models can be divided into two groups. Biophysical constraints are determined by the biophysical and technical features. An example is the suitability of a certain land unit for a particular crop or for mechanized production as determined by climate and terrain conditions. The second type of constraint reflects the desires of the stakeholders or of society in general, *e.g.*, a certain dietary pattern, unemployment rate, or set of socio-economic conditions. Such constraints have a normative nature, as the strength of them often varies with different groups of stakeholders, all of whom have different priorities concerning the objectives, *i.e.*, different policy views (De Wit *et al.,* 1988). Policies involved in land use problems focus on such issues as self-sufficiency in food, income maximization, nature conservation, or environmental problems. These views are operationalized with one or more objective functions, such as the maximization of agricultural production, the maximization of gross regional income, or the minimization of the area used for agriculture. Not all policy views can be quantitatively expressed in objective functions. For instance, nature development and conservation have strong spatial components that are generally hard to catch in linear programming models (WRR, 1992), even though the use of the GIS within the SOLUS methodology enables delineation of designated geographic areas and modeling of some spatially explicit relationships (Bouman *et al.*, 1998b; Stoorvogel, 1993, 1995).

Finally, land use options are explored with a MGLP model. Objective functions that are not optimized in specific runs can be used as the parameters with which target values for these objectives are set (De Wit *et al.*, 1988). First, the range of land use options is determined by optimizing each of the objective functions in separate model runs and by putting no, or hardly any, restrictions on the other objective functions. In this way, the initial freedom of choice (*i.e.*, the worst and the best values) for each objective is made explicit to the stakeholders. Next, the optimal land use scenario for a group of stakeholders is obtained as described by Spronk and Veeneklaas (1983) and De Wit *et al.* (1988). In an interactive way, stakeholders have to select the objective with the worst value that they consider most unacceptable. A tighter, more acceptable, limit to that objective is then formulated. Subsequently, stakeholders are confronted with the results of a new round of optimization runs. Then again, they have to select an objective with a value that is unacceptable to them. This procedure continues until the stakeholders are satisfied with the level of attainment of each of their objectives. The procedure can be repeated with different groups of stakeholders, resulting in different objective function values and land use allocations, *i.e.*, land use scenarios. Comparison of different land use scenarios reveals the possibilities to realize objectives and resolving conflicts between objectives.

In addition to scenario analysis, trade-offs can be revealed using MGLP, while alleviating or tightening the target values of objectives. Because the aim of this chapter is to describe and illustrate the methodology, rather than to operationalize the methodology in an interactive setting, this chapter deals with such trade-offs rather than with the interactive definition of scenarios in order to support different policy views. However, in anticipation of future interactive stakeholder analysis, Wilhelmus (1998) interviewed a number of stakeholders in the AZ in order to gain better insight into their goals.

7.3 The methodology applied to the northern Atlantic Zone

The methodology presented in the previous section is implemented using the SOLUS framework (Chapter 6) with the following specifications: *i*) a MGLP model (called GOAL-AZ) was developed, and *ii*) generated land use systems were oriented especially towards the biophysical profiles described above, *i.e.*, only alternative systems with balanced soil nutrient pools were considered. Current biophysical and socio-economic conditions in the northern AZ are discussed in Chapter 2. In GOAL-AZ, no direct economic or physical interaction with other regions is included, although the immigration of laborers and the import and export of agricultural products is allowed. The same sub-zonation into 12 sub-regions presented in Section 6.2.4 is used. This sub-zonation is based on the distribution and the quality of the current road infrastructure and associated transport and labor transaction costs. The current road infrastructure, however, does not necessarily represent an optimal situation; the possible effect of an improved road infrastructure is therefore analyzed in additional model runs (Section 7.4.2). The derivation of land and labor resources by sub-region was already described in Section 6.2.4.

7.3.1 Land use systems

Land use systems and their input-output combinations are generated with PASTOR and LUCTOR as described in Chapter 5. Only *alternative* land use systems based on the target-oriented approach were used. Soil nutrient balances were kept in equilibrium. Consequently, the generated land use systems can be practiced theoretically for an infinite number of years without changes in input-output combinations. This means that alternative or innovative land use systems are included that, in general, are technically more efficient than the ones currently practiced. Crop systems are defined by a combination of a land unit and a land utilization type (*e.g.*, crop type, variety, and production technique).[1] Three major land units were identified, fertile well drained (SFW), infertile well drained (SIW) and fertile poorly drained (SFP), and each of these major units were further subdivided into mechanizable and non-mechanizable sub-units (Chapter 5). The crops included are: banana, black beans, cassava, maize, palm heart, pineapple and plantain. Table 7.1 shows the feasible crop-land unit combinations and the relevant crop-technique combinations. The production technique involves a specific combination of fertilizer level, pesticide level, herbicide level and mechanization level. For all crops, five fertilizer levels are included, corresponding to respective target production levels of 20%, 40%, 60%, 80% and 100% of the maximum attainable production.

Livestock activities are described as a combination of pasture production systems, herd types, and feed supplements. The pasture production systems consist of a combination of a grass species, land unit, fertilizer level, and stocking rate. Pasture systems were generated for three grass species, all improved and fertilized: *Cynodon nlemfuensis*, *Brachiaria brizantha* and *Brachiaria radicans*. These pastures can be produced on one or more of the three major land units distinguished. Ten fertilizer application levels were used, ranging from 10% to 100% of the fertilizer amount needed for the maximum attainable production. Stocking rates ranged from 1.0 to 6.5 animal units (AU; 1 AU = 400 kg live weight) per hectare, in steps of 0.5 AU per hectare. Separate breeding and fattening herds were generated with target animal growth rates ranging from 0.6 to 1.0 kg animal^{-1} d^{-1} in steps of 0.1 kg animal^{-1} d^{-1}. Green rejected bananas, sugar cane molasses, two types of chicken dung- based concentrates, and a P mineral salt were included in the feed supplements of livestock.

The generated technical coefficients for the land use systems include: labor requirements; costs of inputs; yield; changes in soil nutrient stock for nitrogen (N), phosphorus (P) and potassium (K) (set at zero in alternative land use systems); N losses

[1] See definition of land use system in the Appendix in this book.

to the environment (via leaching, volatilization and (de)nitrification); and biocide use calculated both as the amount of active ingredients (a.i.) applied (BIOA), and by using the biocide index (BIOI) (see Chapter 5).

Table 7.1. Crop-land use systems included in the study of the biophysical options for the northern Atlantic Zone.

	Land units[a]			*Production techniques*[b]							
Crop:	SFW	SIW	SFP	HHH	HHL	HLH	HLL	LHH	LHL	LLH	LLL
Banana	+[c]	-	+	+	+	+	+	-	-	-	-
Beans	+	-	-	+	+	+	+	+	+	+	+
Cassava	+	+	-	+	+	+	+	+	+	+	+
Maize	+	-	-	+	+	+	+	+	+	+	+
Palm heart	+	+	-	-	-	-	-	+	+	+	+
Pineapple	+	+	-	+	+	+	+	+	+	+	+
Plantain	+	-	+	+	+	+	+	-	-	-	-

[a] Land units: SFW = fertile well drained, SFP = fertile poorly drained, and SIW = infertile well drained;
[b] Production techniques: first letter indicates low (L) or high (H) pesticide use level, second letter indicates low (L) or high (H) herbicide use level, third letter indicates low (L) or high (H) mechanization level.
[c] + = included combination, - = excluded combination.

The crop land use systems presented in Table 7.1 cover three types of production practices, whose priorities were: 1) to improve yields, 2) to protect the environment, or 3) to provide employment. Yield-oriented production focuses on high production per unit area and refers to land use systems with close to maximum attainable productions (*e.g.*, the production techniques HHH and HHL in Table 7.1). The second orientation tries to lower the environmental burden. In contrast with yield-oriented production, yield reduction is acceptable if more environment-friendly production is possible due to a lower biocide use (separated into pesticides and herbicides) and/or lower N losses per unit area. For palm heart and pasture, no land use systems with high pesticide use are formulated, because pesticide use is always low. The third production orientation is intended to create high employment and is represented by land use systems with a low level of mechanization. Some yield reduction in crop-land use systems due to a lower level of mechanization is acceptable. Finally, mechanization is not an option in pasture production systems.

7.3.2 The Multiple Goal Linear Programming model GOAL-AZ

GOAL-AZ (General Optimal Allocation of Land use for the AZ[2]) is a MGLP model that considers not only which land use systems are biophysically and technically possible, but also what are the societal objectives for the northern Atlantic Zone. The mathematical description of the model is given in Appendix 7.1. See also the CDROM that accompanies this volume and on which the complete GAMS (Brooke *et al.*, 1992)

[2] The name of the model, *i.e.*, GOAL, was derived from WRR (1992).

model, as well as the necessary assignments to (re-)calculate matrix coefficients, can be found. Four diverging objective functions were implemented in GOAL-AZ, derived from an analysis of the policy issues (Chapter 2) and the stakeholder concerns in the AZ (Bessembinder, 1997; Wilhelmus, 1998):

1. Maximization of economic surplus generated by the agricultural sector
2. Maximization of employment in the agricultural sector
3. Minimization of total biocide use in pasture and crop production
4. Minimization of total N losses in pasture and crop production

In each run of GOAL-AZ, one objective is optimized while the other objectives serve as constraints, with maximum values set to comply with the requests of stakeholders or the desires of researchers wanting to analyze the trade-offs among objectives. Other constraints on GOAL-AZ relate to the availability of land, and to the requirements and availability of cattle feed. Both product prices and wages are assumed to be fixed and unaffected by supply-demand mechanisms, though transaction costs are taken into account for both products and labor. Below a short description is given of the objective functions, constraints and balances in the GOAL-AZ model.

Economic surplus

Economic surplus is defined as the value of production minus labor costs and input costs. Since prices are kept exogenous in the model, the economic surplus here equals the accumulated producer surplus. The value of production is calculated as the quantities of product multiplied by their respective farm-gate prices. To determine farm-gate prices, transport costs are deducted from prices at market destinations. Transport costs are differentiated per sub-region on the basis of the distance to markets and the quality of roads as explained in Section 6.2.4. Alternatively, GOAL-AZ was also run with zero and uniform transport costs throughout the region (see Section 7.4.2). Labor costs are calculated by multiplying the required labor by a fixed wage, and adding labor transaction costs. The calculation of the labor transaction costs per sub-region is explained in detail in Section 6.2.4. The transaction costs of agricultural workers within the region are lower than those for labor from outside the agricultural sector or from outside the region. The costs of inputs include those of fertilizers, biocides, machinery, feed supplements, corrals, and calves needed as input for fattening systems. The value and costs of production are expressed as an annuity to account for the investments in materials with a life span longer than one year and the production patterns of perennial crops and livestock systems over their lifetime. A discount factor of 7% was used (Chapter 5 and 6). Economic surplus and costs are presented in US$, at the average 1994-1996 exchange rate of US$ 1 = ¢ 181 (Costa Rican currency).

Employment

Total employment in the agricultural sector is calculated as the labor needed per hectare multiplied by the area under crops and pastures, plus the labor needed for herd management and application of feed supplements. In the northern AZ, a limited pool of agricultural labor is available, but there is no restriction on the use of labor from other sectors or on the immigration of labor from outside the region. It is assumed that the estimated existing pool of agricultural labor in the northern AZ in 1996 (see Section 6.2.4) will not increase in the future, despite population growth, since employment in non-agricultural sectors is expected to rise.

Biocide use and N losses

Total biocide use is calculated as the amount of active ingredients used per hectare times the area under crops and pastures. Similarly a biocide index for the northern AZ is calculated. N losses are defined as the losses through volatilization, leaching and denitrification. Total N losses are calculated as the product of the N losses per hectare and the area under crops and pastures.

Land availability and suitability

The area used for crops and pastures per land unit per sub-region can not exceed the available area per land unit per sub-region. Only feasible combinations of land units and crops or pastures are allowed (see Table 7.1 and Chapter 5). This means, *e.g.*, that palm heart production is not allowed on poorly drained soils, because this crop is susceptible to water logging. The area used for mechanized production of crops can not exceed the available area for mechanized production (slope <25% and stoniness <1.5%) per land unit per sub-region.

Feed requirements and supply

The metabolizable energy, crude proteins and phosphorus required by the selected herds in each sub-region have to be supplied by the selected pastures and feed supplements. Balancing simultaneously supply and demand for the three nutritional elements is difficult, therefore the overproduction of some elements is allowed. Pasture land use systems and herd production systems are described separately. GOAL-AZ combines pastures and feed supplements with herds. An animal number balance assures that the area under pasture times the stocking rate of the selected pastures per sub-region is equal to the number of selected animal units per sub-region (Bouman and Nieuwenhuyse, 1999; Bouman *et al.*, 1999b).

7.4 Results

7.4.1 Land use optimization

Maximum values for each objective function are determined in a first optimization round, *i.e.*, the zero-round (Table 7.2). In the zero-round, no limits are set on objectives that are not optimized, *e.g.* when maximizing employment, there are no constraints on economic surplus. As a result, the ultimate consequence of minimizing N losses is no agricultural land use at all, so the other objectives are zero as well. When minimizing biocide use, some economic surplus can be generated without the use of biocides, because certain land use systems for palm heart can be selected that do not require biocides. Maximum economic surplus and employment are obtained when all available land is used for agriculture. Pineapple and plantain are the most profitable crops and cover 99% of the area when maximizing economic surplus. Pineapple and banana production with a low mechanization rate and low herbicide use are the most labor intensive land use options, and are selected on 99% of the available area when maximizing employment.

Table 7.2. Maximum values of the objective functions of GOAL-AZ (discount rate 7%, differentiated transport and labor transaction costs)

	Optimized objectives			
	Economic surplus	*Employment*	*Biocide use*	*N losses*
Economic surplus ($ 10^6 y^{-1})	**1632**[a]	*0*[a]	321[b]	*0*
Employment (d 10^6 y^{-1})[c]	30 826	**45 988**	*4452*	*0*
Biocide use (t a.i. y^{-1})	6249	*10 000*	**0**	0
N losses (t y^{-1})	96 071	*115 949*	36 231	**0**

[a] Bold figures indicate optimal values; italic figures indicate worst values.
[b] Maximum value for biocide use, with the highest economic surplus.
[c] Employment is expressed in days of 8 working hours.

To put the results of the zero-round in perspective, they are compared with recent data for the (northern) AZ and for Costa Rica as a whole. In 1996 in Costa Rica, 19% of the Gross Domestic Product (GDP) was obtained from agriculture. While in 1995 the northern AZ accounted for about 36% of the total Costa Rican agricultural value of production, it generated about 10% of the value added of this sector (producer surplus plus labor income). The producer surplus in 1995 of the agricultural sector of the northern AZ is estimated at about $ 64 10^6 (based on DGEC, 1997b and Chapter 4). This means that the maximum producer surplus obtained in the zero-round ($ 1632 10^6) of the present model is 25 times higher than the currently obtained economic surplus in the northern AZ. In 1996, the available work force in the agricultural sector of the study area in 1996 is estimated at about 40 000 man-years, representing 12 10^6 d y^{-1} (see Section 6.2.4). The maximum employment obtained in the zero-round is 3.8 times higher. In 1989, the average biocide use in Costa Rica was 6 kg active ingredient (a.i.) per hectare arable land per year (Wesseling *et al.*, 1993). The average value in the northern AZ is probably higher, because biocide use in banana production is very high, and about 95 % of all Costa Rican banana plantations are in the northern AZ. Data on the total N losses in the AZ are not available. However, the high amounts of intensive rainfall combined with highly permeable soils and large amounts of N applied to crops such as banana and pineapple in the AZ suggests that total N losses are probably considerable (Bouman and Nieuwenhuyse, 1999; Nieuwenhuyse, 1996).

The highest levels of biocide use and N losses are obtained when optimizing for economic surplus and employment. Figure 7.2 presents the trade-offs among economic surplus, biocide use and N losses over the entire range of their maximum values. The data were obtained by removing all constraints on biocide use (increased in steps of 5 10^5 kg a.i. y^{-1}) and N losses (increased in steps of 10^7 kg N y^{-1}) in each new run intended to maximize economic surplus. Figure 7.3 shows the associated use of land for different crops and pasture. Examining trade-offs in this way reveals the relative performance of various land use systems with respect to the different objectives.

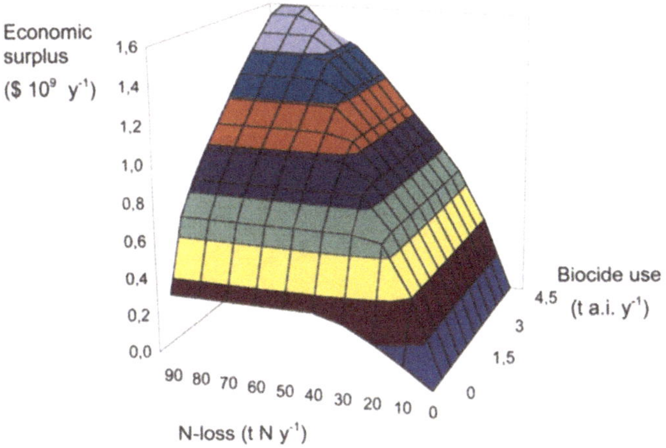

Figure 7.2. Trade-offs among economic surplus (maximized), biocide use and N losses. A discount rate of 7% and differentiated transport and labor transaction costs are used.

When no biocide use is allowed at all, still an economic surplus of \$ 321 10^6 y^{-1} can be obtained, because of the inclusion of palm heart land use systems that do not use any biocides at all but still generate a considerable profit (Figure 7.3d). If a higher economic surplus is desired, more profitable crops and techniques have to be selected, using at least some biocides. Minimum biocide use increases from 0 kg a.i. ha^{-1} at an economic surplus of \$ 321 10^6 y^{-1} to 19.8 kg a.i. ha^{-1} at an economic surplus of \$ 1381 10^6 y^{-1}, without restrictions on N losses. In this "trajectory", first cassava production and pasture land replace palm heart. Plantain and pineapple are the most profitable crops (Figure 7.3b and c), but in the cultivation of both crops large amounts of biocides are used. The maximum biocide use per hectare is 56 kg a.i. y^{-1} in banana land use systems.

Agricultural land use always results in some N losses to the environment. With increasing economic surplus, the area used for agriculture increases, with corresponding increases in total N losses (Figures 7.2 and 7.3f). Land use systems with higher production levels are selected, and the average N losses increase from 74.6 kg ha^{-1} at an economic surplus of \$ 276 10^6 y^{-1} to 325.7 kg ha^{-1} at an economic surplus of \$ 1381 10^6 y^{-1}, without restrictions on biocide use. Cassava, with a production level of 20 % of the maximum attainable production, has the lowest N losses per hectare, but when economic targets increase, less productive cassava gets increasingly replaced by more productive cassava production technologies, as well as by pineapple and plantain (Figure 7.3). The highest N losses per ha are 563 kg y^{-1} in plantain, reached at the maximum attainable production.

For the options presented in Figure 7.2 and involving low biocide use and low N losses, only the fertile well-drained land units (SFW) with the highest productivity are used in the sub-regions with the lowest transport costs. This is because production on these land units results in the highest economic surplus. When higher biocide use and higher N losses are allowed, the area used for agriculture is extended to other sub-regions and to the infertile well-drained land units (SIW). Finally, also the poorly drained soils (SFP) in the sub-regions with the highest transport costs are taken into production.

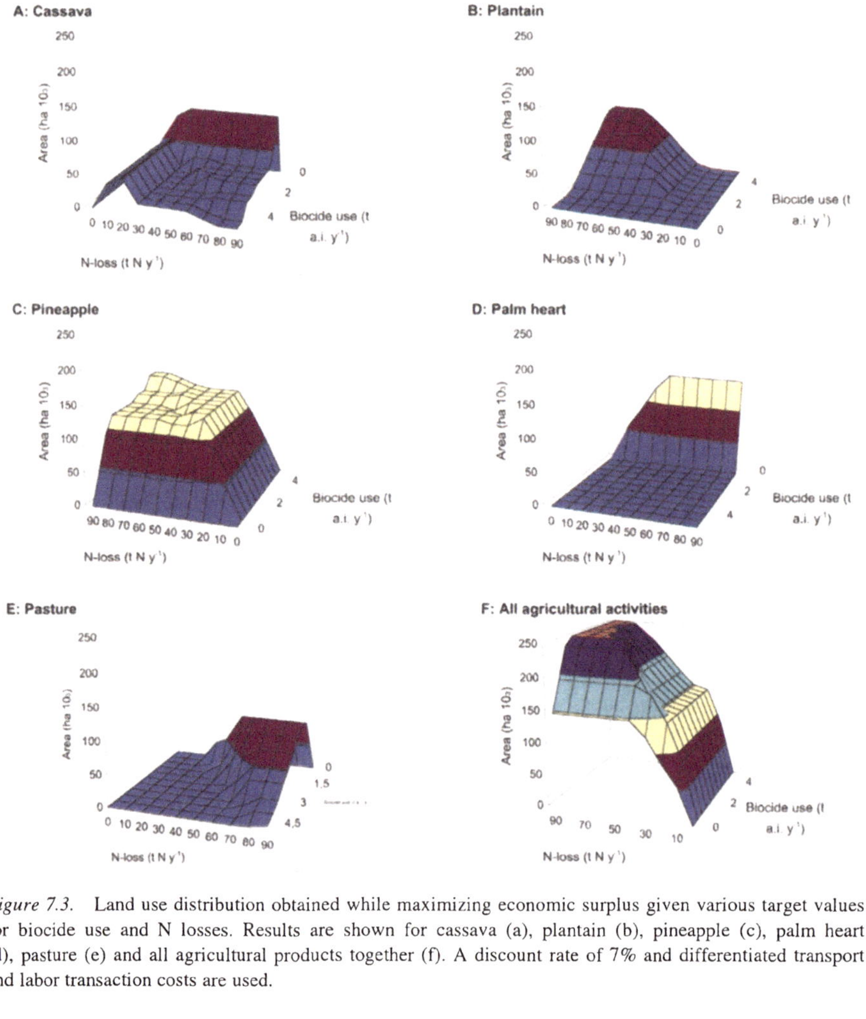

Figure 7.3. Land use distribution obtained while maximizing economic surplus given various target values for biocide use and N losses. Results are shown for cassava (a), plantain (b), pineapple (c), palm heart (d), pasture (e) and all agricultural products together (f). A discount rate of 7% and differentiated transport and labor transaction costs are used.

7.4.2 The effects of production techniques, discount rate and road infrastructure

The trade-offs presented in the previous section show the sensitivity of GOAL-AZ to changes in the constraints on farming practices. In general, it is important to do a sensitivity analysis concerning other aspects as well, *e.g.*, the specification of objective functions, the uncertainty in the technical coefficients and prices, and the inclusion or exclusion of particular constraints and land use systems (Bessembinder, 1997). However, in view of the objectives presented in Section 7.1, only the effects of different production techniques, discount rates and road infrastructure are assessed in this section.

Production techniques

The runs presented in Figures 7.2 and 7.3 involve, almost exclusively, production techniques with a high level of biocide use, with the exception of the land use systems for growing palm heart. In additional runs, the effect of excluding land use systems with high biocide use levels was studied. In these cases, the maximum economic surplus is only \$ 455 10^6 y^{-1}, or 28% of the maximum level obtained in scenarios incorporating production techniques with high biocide use. This reduction is caused by the lower yields of land use systems with low biocide use, and by the exclusion of plantain, for which no land use systems with low biocide use were included.

Discount rate

The discount rate is used in the calculation of annuities and affects costs and benefits of land use systems. In perennials, costs and benefits are calculated as annualized net present values using the capital recovery rate (Chapter 5 and 6). Therefore, a sensitivity analysis was performed on the discount rate. Figure 7.4 shows that a discount rate of 0% (instead of 7% as used in the previous runs) hardly affects the objective values in this study. This is because land use patterns hardly change,

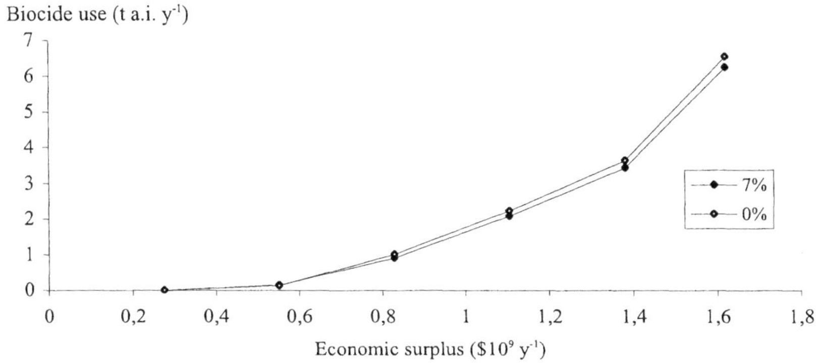

Figure 7.4. Effect of the discount rate on the trade-off between economic surplus and biocide use without any restrictions on N losses (differentiated transport and transaction costs).

i.e., mostly the same crops are selected, only the area per crop differs somewhat (data not shown). For instance, the area under plantain, a perennial crop, is larger with a discount rate of 0%. The relatively small effect of the discount rate may be explained by the fact that the perennial crops included in this study (banana, plantain, palm heart) and pastures, all produce their first yield shortly after establishment and maintain a constant yield level during their life span.

Road infrastructure

Current product transportation costs and labor transaction costs were the basis for dividing the northern AZ into 12 sub-regions (Chapter 6). These costs were used in the calculations presented in Section 7.4.1. Here, the sensitivity of the GOAL-AZ model to transport and transaction costs is investigated by assuming hypothetical road improvements according to the following four scenarios:

1. Base run (current costs differentiated over the 12 sub-regions).
2. Uniform product transport costs and labor transaction costs, based on current costs.
3. Zero product transport costs and labor transaction costs.
4. Current differentiated product transport costs and zero labor transaction costs.

Results are presented in Figure 7.5. The optimal cases involving either scenario 1 or 2 do not differ. Mostly, the objective function values and the area per selected land use system are hardly affected; only the distribution of land use systems over the region is in some cases slightly different. A similar result is obtained when plotting N losses versus economic surplus. Apparently, in studies exploring the biophysical potential of a relatively small region such as the northern AZ, there is no response to discriminate between sub-regions with different transport and labor transaction costs. When there are no transport and transaction costs at all (scenario 3), economic surplus increases

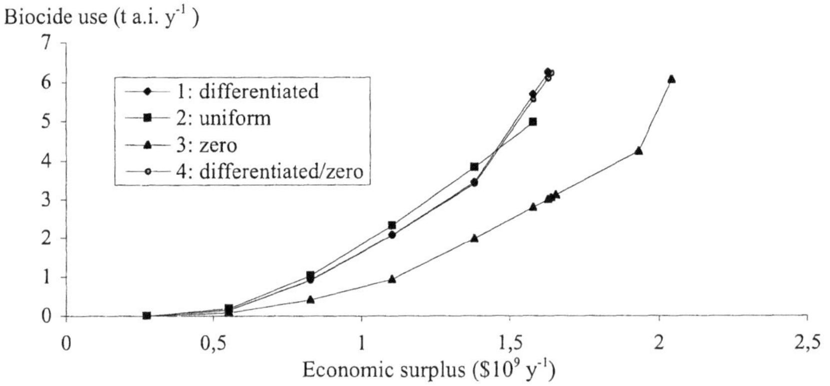

Figure 7.5 Effect of the costs of transporting products and transacting labor on the trade-off between economic surplus and biocide use without any restrictions on N loss. ♦: Base run (differentiated product transport and labor transaction costs), ■: Uniform product transport and labor transaction costs, Δ: Zero product transport and labor transaction costs, O: Differentiated product transport and zero labor transaction costs. Discount rate is 7% in all runs.

significantly. This increase is mainly caused by the zero product transport costs, since zero labor transaction costs alone (scenario 4) do not results in large differences from the base situation (scenario 1). This fact suggests that the maximum economic surplus can be increased considerably if current transport costs can be lowered, *e.g.*, by improvements in the road infrastructure. However, such increase has to be weighed against the costs of upgrading the infrastructure.[3]

7.5 Incorporating economic constraints: implications for land use options

7.5.1 Differences between GOAL-AZ and REALM

Both GOAL-AZ presented in this chapter and the REALM model presented in Chapter 6 use the SOLUS framework along with the same data about the resource base and sub-zonation. Therefore, differences in results between both models, when using the same objective functions, can be attributed to differences in *i*) the inclusion of economic constraints and *ii*) the land use systems that are considered. Table 7.3 summarizes the differences in the SOLUS set up as used in this chapter (using GOAL-AZ) and in Chapter 6 (using REALM).

Table 7.3. Main differences in the SOLUS set-up using GOAL-AZ (this chapter) and using REALM (Chapter 6).

Goal: Biophysical exploration with little consideration given to socio-economic constraints.	Goal: Integrated biophysical and socio-economic exploration.
Linear programming model GOAL-AZ:	Linear programming model REALM:
• Four objective functions: *i*) maximizing producer surplus, *ii*) maximizing employment, *iii*) minimizing biocide use, and *iv*) minimizing N losses.	• One objective function: maximizing producer and consumer surplus.
• No imposed limitations in market demand and labor supply.	• Limited market demand and labor supply.
• Fixed prices and wage rate.	• Price and wage formation in response to supply-demand mechanisms.
• Free import of calves from outside AZ.	• No import of calves from outside AZ (optional).
Generated land use systems:	Generated land use systems:
• Only alternative (target oriented) systems.	• Actual and alternative (target oriented) systems.

[3] At present there are 300 km of paved roads and 2500 km of gravel roads in the study area. Paving, *e.g.*, the 2500 km of gravel roads at a modest cost of \$ 50 000 per km would require an investment of \$ 125 10^6. Annualized investment costs (20 years, 7% discount rate) would thus be about \$ 12 10^6. Under perhumid conditions like those in the AZ, annual maintenance costs are likely to be substantial.

A direct comparison between results of the linear programming models GOAL-AZ and REALM is presented in Table 7.4. Both models were run with differentiated transport and transaction costs and a discount rate of 7%. REALM was run with the same crops, pastures, herds and feed supplements as GOAL-AZ in the present chapter, but in addition to the target-oriented alternative land use systems for annual and perennial crops, actual systems were included.[4]

Land use options generated with GOAL-AZ show more extreme results than those generated with REALM: fewer land use systems are selected each on relatively large areas, resulting in quite extreme objective function values (compare the *base* column of GOAL-AZ in Table 7.4 with its *minimize biocide use* column). Differences in the results of the land use scenarios generated with GOAL-AZ may be relatively large because the conflicting objective functions used are widely different. Many land use results obtained with GOAL-AZ would require clear breaks with current land use practices and their drivers. These scenarios reveal the extreme consequences of different objectives and biophysical conditions on long-term land use.

The differences in results among the various land use scenarios performed with REALM are relatively small (compare its *base* column in Table 7.4 with its *zero nutrient depletion* column). Also, the results are much closer to the currently prevailing situation in the AZ, since REALM includes supply and demand relationships that, apparently, determine actual land use to a significant extent. If large amounts of the same products are produced, the prices of these products decrease and other crops become more attractive, and consequently land use becomes more diverse. Supply-demand relationships make prices endogenous, resulting in a much lower economic surplus (defined as the sum of the producer and the consumer surplus; Chapter 6) in the REALM scenarios as compared to the economic surplus (defined as the producer surplus only) obtained with the GOAL-AZ model.

Table 7.4. Results of the GOAL-AZ and REALM models, as part of the SOLUS methodology (with a 7% discount rate and with differentiated transport and labor transaction costs).

		GOAL-AZ		*REALM*	
	Unit	*Base*[a]	*Minimize biocide use*[b]	*Base*[a]	*Zero nutrient depletion*[c]
Economic surplus[d]	$\$ 10^6\ y^{-1}$	1632	230	233	230
Employment[e]	$d\ 10^3\ y^{-1}$	30 826	4135	8351	8018
Biocide use	$t\ a.i.\ y^{-1}$	6249	0	2031	1996
N losses	$t\ y^{-1}$	96 071	25 797	25 435	26 661
Pineapple	ha	100 238	0	3470	3205
Palm heart	ha	0	146 657	9303	9307
Banana	ha	12 250	0	33 456	33 456
Plantain	ha	136 786	0	2333	910
Cassava	ha	0	0	23 986	11 406

[4] Compared to Chapter 6, natural pastures and grass-legume mixtures were not included when running REALM. Since fertilized improved grasslands have low economic returns given the current price structure (Bouman and Nieuwenhuyse, 1999; Bouman *et al.*, 1999b), very few pastures are selected by REALM as shown in Table 7.4 (10 522 ha) compared to the base run results shown in Table 6.1 (where natural pastures and grass-legumes were selected on 190 000 ha), and the regional economic surplus in the base run is 15% lower than in Chapter 6.

Table 7.4. Continued

	Unit	GOAL-AZ		REALM	
		Base[a]	*Minimize biocide use*[b]	*Base*[a]	*Zero nutrient depletion*[c]
Maize	ha	0	0	126	126
Pasture	ha	1736	0	10 522	14 370
Total agricultural area	ha	251 010	146 657	83 196	72 780
Δ soil N stock	kg ha^{-1} y^{-1}	0	15[f]	-11	2[f]
Δ soil P stock	kg ha^{-1} y^{-1}	0	1	3	7
Δ soil K stock	kg ha^{-1} y^{-1}	0	14	-30	0

[a] The base scenarios with GOAL-AZ and REALM maximize economic surplus and do not impose any limits on biocide use or N losses.
[b] The objective function minimizes biocide use under the constraint that economic surplus is at least $ 230 10^6.
[c] Economic surplus is maximized under the restriction that the changes in soil nutrient balances are ≥ zero.
[d] In REALM, the economic surplus is the combined producer and consumer surplus; in GOAL-AZ, the economic surplus is the producer surplus. Therefore, both economic surpluses can not directly be compared.
[e] Employment is expressed in days of 8 working hours.
[f] After a growing period of 15 years, palm heart plantations or other perennial crops are renewed. The new crop does not require all the released nutrients from the plant material of the previous crop. As a consequence, positive nutrient balances are obtained.

Inclusion of the actual land use systems for growing annual and perennial crops in the REALM base run results in changes in soil nutrient stocks that are negative for N and K for the entire AZ. This fact implies that agriculturally non-sustainable options are generated. Bouman and Nieuwenhuyse (1999) further analyze the issue of nutrient depletion, the trade-off with economic surplus, and the economic feasibility of introducing land use systems with zero nutrient depletion. The scenarios performed with GOAL-AZ only included alternative land use systems that do not result in nutrient depletion.

Another point where the results of GOAL-AZ and REALM can be compared concerns the effect of decreasing costs of transport by improving road quality. In this chapter, it was suggested that the maximum economic surplus would increase considerably if the road infrastructure was improved (Section 7.4.2), in particular because of the resulting savings in transport costs. However, when a road improvement scenario was run with REALM (Schipper *et al.*, 1998), the effects on economic surplus and land use distribution were much smaller than the ones presented in this chapter. This finding can be explained by the fact that the most profitable current land use systems are located in land units with relatively good access to better quality roads, and they already optimally exploit market possibilities. With improved roads, these land use systems can theoretically move into areas that were previously farther away from these roads, but at the moment there is no economic incentive to do so (since the land area devoted to these forms of land use already optimally satisfies demand for the crops produced by these land use systems). This example illustrates a case where socio-economic conditions (in this case, supply and demand relationships) constrain biophysical options.

7.5.2 Effects of introducing economic constraints in explorative land use studies

To determine which socio-economic constraints affect model results most, the economic constraints of REALM are introduced one at a time into GOAL-AZ. In Table 7.5, the results of such inclusions and of including actual land use systems in the analysis are presented. Inclusion of actual land use systems in the otherwise unmodified GOAL-AZ has hardly any effect on optimization results. The alternative land use systems are more profitable than most actual ones because higher resource use efficiency is assumed (see Chapter 5). Including the restriction that no calves can be imported from outside the northern AZ (and that calves for fattening systems thus have to be produced by breeding herds within the AZ) has practically no effect on the final results. The small amount of pasture selected in the base run (1736 ha) disappears because of the lower profitability of cattle breeding systems compared with that of cattle fattening systems. Next, including labor market considerations (see also Section 6.2.6) significantly affects scenario results: economic surplus decreases to 70% of its base run value, employment in the agricultural sector to 46%, biocide use to 68%, and N losses to 59%. However, the most dramatic effect is obtained when product markets are considered in the GOAL-AZ model (see Section 6.2.5): economic surplus decreases to only 17% of its base run value, employment to 34%, biocide use to 33%, N losses to 67%, and the land use pattern further diversifies. Including the supply-demand function for labor, in addition to the supply-demand functions for products, did not affect the land use options. In this case the labor demand remained lower than the available regional labor pool. Now adding the restriction that the number of calves used for cattle fattening should be produced by cattle breeding systems within the AZ to the modified GOAL-AZ model, further constrains scenario results: economic surplus is decreased by 15%, employment by 25%, biocide use by 4%, and N losses by 59% of their respective values in the product market restricted scenarios. The acreage of pasture decreases from 198 582 ha to only 14 370 ha. This decrease in livestock activities is explained by the fact that fattening systems are about twice as profitable in the AZ as breeding systems (Bouman and Nieuwenhuyse, 1999). In the scenario with unrestricted import of calves, only specialized (relatively profitable) beef fattening herds were selected, grazing on (relatively expensive) fertilized grass pastures. With the calve import restriction, fattening of cattle is only possible when linked to cattle breeding, which is relatively less profitable on the (expensive) fertilized grass pastures. Therefore, other land use systems, such as cassava and plantain, become more attractive than combined cattle breeding-fattening, and the herd size and the area under pasture decrease.

7.6 Discussion and conclusions

The case study for the northern AZ presented in this chapter, using the GOAL-AZ multiple goal linear programming model and alternative, target-oriented land use systems, enables us to optimize regional land use and to evaluate the relative performance of new land use systems in terms of a well-defined set of policy objectives. Trade-offs among objectives are revealed, as well as the consequences of releasing or tightening the restrictions placed on them. For the AZ in Costa Rica, two major conclusions can be drawn:

Table 7.5. Effect of transforming GOAL-AZ to REALM by the step-by-step inclusion of economic constraints. In all runs, economic surplus was maximized.

		GOAL-AZ	*Modified GOAL-AZ*						*REALM*
	Unit	*Base*[a]	*With actual land use systems*	*With calve import restriction*	*With labor market*	*With product market*	*With product and labor market restriction*	*With product and labor market and calve import restriction*	*Base*[a]
Economic surplus[b]	$ 10^6 y^{-1}	1632	1635	1630	1144	272	272	230	233
Employment[c]	d 10^3 y^{-1}	30826	30968	30817	14228	10562	10562	7965	8351
Biocide use	t a.i. y^{-1}	6249	6293	6248	4221	2091	2091	2007	2031
N loss	t y^{-1}	96071	95900	95678	56767	64452	64452	26487	25435
Pineapple	ha	100238	100238	100238	119638	3067	3067	3205	3470
Palm heart	ha	0	0	0	18735	8169	8169	9303	9303
Banana	ha	12250	12250	12250	0	32176	32176	33456	33456
Plantain	ha	136786	138522	136786	76460	860	860	910	2333
Cassava	ha	0	0	0	3787	8062	8062	8938	23986
Maize	ha	0	0	0	0	94	94	126	126
Pasture	ha	1736	0	0	0	198582	198582	14370	10522
Total agricultural area	ha	251010	251010	249274	218620	251010	251010	70308	83196
Δ soil N stock	kg ha^{-1} y^{-1}	0	-1	0	1[d]	1[d]	1[d]	3	-11
Δ soil P stock	kg ha^{-1} y^{-1}	0	0	0	0	15	15	8	3
Δ soil K stock	kg ha^{-1} y^{-1}	0	-2	0	2	1	1	2	-30

[a] The base runs with GOAL-AZ and REALM do not include any bounds on biocide use and N losses. In the REALM base run, both actual and alternative land use systems are included.

[b] In explorations with product markets (supply-demand functions), economic surplus is the combined producer and consumer surplus, in other scenarios, economic surplus is the producer surplus.

[c] Employment is expressed in days of 8 working hours.

[d] After a growing period of 15 years, palm heart plantations or other perennial crops are renewed. The new crop does not require all the released nutrients from the plant material of the previous crop. As a consequence, positive nutrient balances are obtained.

1. Combining different land units, crops and production techniques results in a large number of land use systems with a wide performance range in terms of agricultural productivity, economic surplus, biocide use, N losses, and labor requirements. For the regional level, this results in a wide range of long-term land use options, depending on the prioritization of objectives. This finding implies that an explicit consideration of objectives is very important in determining future land use.
2. The model results show that a considerable economic surplus can be obtained on a relatively small area in the AZ. This finding suggests that, in the AZ, no serious competition needs to exist for space between agriculture and nature conservation. Where they do not already exist, conflicts are much more likely to arise regarding the various external effects associated with agricultural production. In this context, such conflicts may center around the biocide or nutrient flows from the agricultural areas into the nature conservation areas.

In Section 7.5 differences in results between GOAL-AZ (this chapter) and REALM (Chapter 6), and the way they were used in the SOLUS framework for land use exploration, were presented and discussed. The land use explorations carried out with these two models attempt to make stakeholders aware of the alternatives to current land use and to gain insight into the trade-offs among economic, ecological and agricultural objectives. Exploratory studies stimulate the imagination of stakeholders and contribute to the perception that the future is not necessarily a continuation of the past and present. The two models GOAL-AZ and REALM have different aims and use different assumptions, affecting the results and their interpretation. For well-founded and balanced land use analysis in the short term as well as the long term, information is needed on current land use, on the problems and alternatives of land use under a wide range of constraints and for various policy views. Explorations that combine biophysical factors with societal priorities, such as the study present in this chapter, reveal the options for optimal use of resources and production techniques beyond economic constraints that drive current land use to a large extent. Studies taking economic constraints into account when analyzing future land use options, such as the ones presented in Chapter 6 and Section 7.5, show such opportunities may be curtailed by product and labor markets.

Emphasis in this chapter was on the scientific presentation of the methodology and the discussion of results. However, the presented approach can only be successful in supporting policy-making processes when used in interactive and iterative settings, with intensive participation by representatives of all the relevant stakeholders. Then, the modeling approach can be used to give immediate feedback on the consequences of societal preferences for various land use objectives and on the biophysical and technical conditions and limitations within the system. As such, the approach can be used to enhance the transparency of discussions and societal learning processes (Röling, 1994). In Chapter 10, some experience with the adaptation and application of the SOLUS framework to a study area in the Pacific region of Costa Rica with intensive participation by representatives of the Agricultural Research Department, the Extension Department, and the Planning Department of the Costa Rican Ministry of Agriculture and Livestock (MAG) is reported (see also Sáenz *et al.*, 1998).

Appendix 7.1 Mathematical description of the explorative MGLP-model GOAL-AZ

Objective functions

Maximization of economic surplus: Economic surplus or producer surplus is defined as the amount of products produced times a price, minus transport costs, input costs, labor costs, and labor transaction costs ($ y^{-1}).

$$\text{Max } Z = \sum_z \sum_j (p_j S_{jz}) - \sum_z \sum_j (t_{jz} S_{jz}) - C - \sum_z \sum_\zeta (wL_{z\zeta}) - \sum_z (wO_z) - \sum_z \sum_\zeta (t^l_{z\zeta} L_{z\zeta}) - \sum_z (t^o_z O_z)$$

Maximization of physical labor use: Total physical labor requirements (from the northern AZ and from outside) are calculated as the physical labor requirements multiplied by the area under crop and pasture land use systems, and the physical labor requirements by the number of livestock production systems and feed supplements (d y^{-1}).

$$\text{Max } PL = \sum_z \sum_s \sum_l \sum_t (pl_{slt} X_{zslt}) + \sum_z \sum_h (pl_h A_{zh}) + \sum_z \sum_s \sum_p \sum_r (pl_{spr} P_{zspr}) + \sum_z \sum_f \sum_m (pl_f F_{zfm})$$

Minimization of total biocide use: Total biocide use is defined as the biocide use per hectare multiplied by the area under crop and pasture production (kg a.i. y^{-1}).

$$\text{Min } B = \sum_z \sum_s \sum_l \sum_t (\delta_{slt,e=BIOA} X_{zslt}) + \sum_z \sum_s \sum_p \sum_r (\delta_{spr,e=BIOA} P_{zspr})$$

Minimization of total N losses: Total N losses are calculated as the denitrification, N leaching and N volatilization per area multiplied by the area under crop and pasture production (kg N y^{-1}).

$$\begin{aligned}\text{Min } N = & \sum_z \sum_s \sum_l \sum_t (\delta_{slt,e=denitrification} X_{zslt}) + \sum_z \sum_s \sum_p \sum_r (\delta_{spr,e=denitrification} P_{zspr}) \\ & + \sum_z \sum_s \sum_l \sum_t (\delta_{slt,e=N\text{-}leaching} X_{zslt}) + \sum_z \sum_s \sum_p \sum_r (\delta_{slt,e=N\text{-}leaching} P_{zspr}) \\ & + \sum_z \sum_s \sum_l \sum_t (\delta_{slt,e=N\text{-}volatilization} X_{zslt}) + \sum_z \sum_s \sum_p \sum_r (\delta_{slt,e=N\text{-}volatilization} P_{zspr})\end{aligned}$$

Product balances

Balance of product annuity per product per sub-region (t y^{-1}).

$$\sum_s \sum_l \sum_t (-y_{jslt} X_{zslt}) + \sum_h (-y_{jh} A_{zh}) + S_{jz} \leq 0 \qquad \text{all } z, j$$

Labor

Total annuity of labor requirements per sub-region (from the northern AZ and outside) is calculated as the annuity of labor requirements multiplied by the area under crop and pasture land use systems, and the annuity of labor required multiplied by the number of livestock production systems and feed supplements per sub-region (d y^{-1}).

$$\sum_s \sum_l \sum_t (l_{slt} X_{zslt}) + \sum_h (l_h A_{zh}) + \sum_s \sum_p \sum_r (l_{spr} P_{zspr}) + \sum_m \sum_f (l_f F_{zfm}) - \sum_\zeta L_{z\zeta} - O_z \leq 0 \qquad \text{all } z$$

Labor use from the agricultural labor force in the sub-regions can not exceed the available labor in the sub-regions (d y^{-1}).

$$\sum_z L_{z\zeta} \leq a_\zeta \qquad \text{all } \zeta$$

Inputs

Annuity input costs are defined as the annuity input costs multiplied by the area under crop production and livestock systems, the annuity input costs by the number of livestock systems and feed supplements, and the annuity costs of calves as input for fattening systems (\$ y^{-1}).

$$\sum_z \sum_s \sum_l \sum_t (c_{slt} X_{zslt}) + \sum_z \sum_h (c_h A_{zh}) + \sum_z \sum_h (p_j v_h A_{zh}) + \sum_z \sum_s \sum_p \sum_r (c_{spr} P_{zspr}) + \sum_z \sum_f \sum_m (c_f F_{fmh}) - C \leq 0$$

Land availability

The sum of the area with selected pastures and crops per soil type per sub-region is less than or equal to the available area per soil type per sub-region.

$$\sum_l \sum_t X_{zslt} + \sum_p \sum_r P_{zspr} \leq b_{zs} \qquad \text{all } z, s$$

The sum of selected mechanized crop production per soil type per sub-region is less than or equal to the available area for mechanized production per soil type per sub-region.

$$\sum_l \sum_t P_{zslt} \leq m_{zs} \qquad \text{all } z, s, \textit{for } t\textit{=mechanized}$$

Animal balances

The nutritional value of pasture production and the feed supplements used per sub-region per season is larger than or equal to the required nutritional value of selected livestock production systems in each sub-region per season.

$$\sum_s \sum_p \sum_r (n_{sprnm} P_{zspr}) + \sum_f (n_{fn} F_{zfm}) - \sum_h (n_{hnm} A_{zh}) \geq 0 \qquad \text{all } z, m, n$$

The product of the area under pasture and the stocking rate per sub-region is equal to the product of the herd size and the selected livestock production systems per sub-region.

$$\sum_s \sum_p \sum_r (s_{spr} P_{zspr}) + \sum_h (h_h A_{zh}) = 0 \qquad \text{all } z$$

Environmental aspects

The environmental criteria for the whole northern AZ are calculated as the environmental criterion per hectare multiplied by the area under crop and pasture land use systems.

$$\sum_z \sum_s \sum_l \sum_t (\delta_{slte} X_{zslt}) + \sum_z \sum_s \sum_p \sum_r (\delta_{spre} P_{zspr}) - A_e = 0 \qquad \text{all } e$$

Maximum on total N losses and biocide use

$B \leq biomx$
$N \leq nlosmx$

Table A7.1. Indices in the GOAL-AZ model

Index	*Description*	*Elements*
e	Environmental aspects	N balance, P balance, K balance, denitrification, N leaching, N volatilization, biocide active ingredients use (BIOA), biocide index (BIOI)
f	Feed types	Sugar cane molasses, rejected bananas, chicken dung, P_2O
h	Herd type	Herds of 50 animals for fattening or breeding; target growth rates of 0.6 to 1.0 kg animal^{-1} day^{-1}, in steps of 0.1 kg animal^{-1} day^{-1}
j	Products	Banana, beans, cassava, maize corn, maize cobs, palm heart, pineapple, plantain, meat, calves; three qualities: export, domestic, refuse
l	Land use types	Banana, beans, cassava, maize, palm heart, pineapple, plantain
m	Season	Dryer season: January to March, wetter season: April to December
n	Nutritional elements	Metabolizable energy, crude protein, phosphorus
p	Pasture types	*Brachiaria brizantha, Cynodon nlemfluensis, Brachiaria radicans*; fertilizer levels of 10% to 100% for maximum yield, in steps of 10%
r	Stocking rate	1.0 to 6.5 animal units per hectare, in steps of 0.5 animal unit
s	Soil types	Fertile poorly drained, fertile well-drained, infertile well-drained
t	Technology	High and low levels biocide use, herbicide use and mechanization; length of crop cycle; target growth rates of 20%, 40%, 60%, 80% and 100% of maximum yield
z	Sub-regions	1 to 12 with different infrastructure, ξ used as alias for sub-regions

Table A7.2. Variables in the explorative GOAL-AZ model

Variable	*Description*	*Unit*
A_{zh}	Livestock systems per sub-region	herds y^{-1}
A_e	Environmental situation for northern AZ	kg y^{-1}; y^{-1}
B	Total biocide use	kg a.i. y^{-1}
C	Total annuity costs of input use	\$ y^{-1}
F_{zfm}	Supplementary feed systems per sub-region per season	kg y^{-1}
$L_{z\zeta}$	Annuity labor use per sub-region z from sub-region ζ	d y^{-1}
N	Total N losses	kg N y^{-1}
O_z	Annuity of labor use per sub-region from outside the agricultural labor force of the northern AZ	d y^{-1}
P_{zspr}	Pasture land use systems per sub-region per soil type	ha y^{-1}
PL	Total physical employment	d y^{-1}
S_{jz}	Annuity production per product per sub-region	t y^{-1}
X_{zslt}	Land use systems per soil type per sub-region	ha y^{-1}
Z	Total economic surplus	\$ y^{-1}

Table A7.3. Coefficients in the GOAL-AZ model

Coefficient	*Description*	*Unit*
δ_{slte}	Environmental indicators for crop production systems	kg ha^{-1} y^{-1}; ha^{-1} y^{-1}
δ_{spre}	Environmental indicators for pasture land use systems	kg ha^{-1} y^{-1}; ha^{-1} y^{-1}
a_ζ	Available labor from the agricultural sector per sub-region	d y^{-1}
b_{zs}	Land availability per sub-region per soil type	ha y^{-1}
$biomx$	Maximum of total biocide use	kg a.i. y^{-1}
c_{slt}	Annuity costs of inputs for crop production systems	\$ ha^{-1}
c_h	Annuity costs of inputs for livestock systems	\$ $herd^{-1}$
c_{spr}	Annuity costs of inputs for pasture land use systems	\$ ha^{-1}
c_f	Annuity costs of inputs for supplementary feed systems	\$ kg^{-1}
h_h	Herd size per herd type	AU $herd^{-1}$
l_{slt}	Annuity labor use in crop production systems	d ha^{-1} y^{-1}
l_{spr}	Annuity labor use in pasture land use systems	d ha^{-1} y^{-1}
l_h	Annuity labor use in livestock systems	d $herd^{-1}$ y^{-1}
l_f	Annuity labor use for supplementary feed	d kg^{-1}
m_{zs}	Land availability for mechanized production per sub-region per soil type	ha y^{-1}
n_{hnm}	Feed elements required by livestock system per season	Mcal.$herd^{-1}$ $season^{-1}$; kg $herd^{-1}$ $season^{-1}$

Table A7.3. Continued

Coefficient	*Description*	*Unit*
n_{fn}	Feed elements yielded by supplementary feed system	Mcal kg^{-1}; kg kg^{-1}
n_{sprmn}	Feed elements yielded per pasture land use system per season	Mcal ha^{-1} $season^{-1}$; kg ha^{-1} $season^{-1}$
nlosmx	Maximum of total N losses	kg N y^{-1}
p_j	Price per product	\$ t^{-1}
pl_{slt}	Physical labor use in crop production systems	d ha^{-1} y^{-1}
pl_{spr}	Physical labor use in pasture land use systems	d ha^{-1} y^{-1}
pl_h	Physical labor use in livestock systems	d $herd^{-1}$ y^{-1}
pl_f	Physical labor use for supplementary feed	d kg^{-1}
s_{spr}	Stocking rate of pasture land use system	AU ha^{-1}
t_z^o	Transaction costs for labor from outside the agricultural sector	\$ d^{-1}
$t_{z\zeta}^l$	Transaction costs for labor from sub-region ξ to sub-region z	\$ d^{-1}
t_{jz}	Transport costs per product per sub-region	\$ t^{-1}
v_h	Calves as input for livestock systems for fattening	t $herd^{-1}$
w	Wage for agricultural labor	\$ d^{-1}
y_{jslt}	Annuity yield of crop production system	t ha^{-1}
y_{jh}	Annuity yield of livestock system	t $herd^{-1}$

8 Farm modeling for policy analysis on the farm and regional level

PETER C. ROEBELING, HANS G.P. JANSEN, ROBERT A. SCHIPPER, FERNANDO S·ENZ, EDMUNDO CASTRO, RUERD RUBEN, HUIB HENGSDIJK, and BAS A.M. BOUMAN

Abstract

The Atlantic Zone of Costa Rica accounts for nearly half of the total agricultural income in Costa Rica, and therefore is an important region for agricultural policy. Policy analysis can be performed on the regional level, using regional agricultural sector models that maximize regional welfare, as well as on the farm level, where different farm types are characterized by their specific objectives, production possibilities and resource endowments. While the former approach fails to model farm type-specific characteristics, the simple aggregation to the regional level of the representative farm type results ignores the interaction between the farm types that occurs in product and factor markets. The present study presents a regional equilibrium modeling approach that incorporates farm type characteristics as well as the equilibrium equations for product markets. Compared to a simple aggregation of representative partial results with exogenous output prices, the results produced by the regional equilibrium model indicate that the endogenization of product prices leads to lower levels of specialization in agricultural production, as well as lower incomes, profits and labor use. As such, the results obtained in a model with endogenous prices better reflect reality. A number of policy simulations are performed, including a 20% decrease in transaction costs, a 40% tax on biocide prices, and a 20% increase in credit availability. The former as well as the latter lead to increased cash crop production and corresponding increases in agricultural income, while taxing biocides leads to less biocide intensive cropping systems. The effectiveness of policy measures is, however, overestimated when product prices are assumed exogenous, since endogenously determined product prices limit specialization in the most profitable crops or in crops that require relatively low levels of biocides.

8.1 Introduction

Research efforts aimed at identifying and evaluating efficient policy instruments designed for inducing certain desired changes in land use, often make use of farm household modeling approaches along the lines originally proposed by Singh *et al.* (1986). While most of this work has paid ample attention to adequate simulation of farmer behavior and objectives (typically in the context of a utility maximizing framework), adequate modeling of the production side has received comparatively less attention. The UNA-DLV research project on "Agrarian Policies for Sustainable Land Use and Food Security in Costa Rica" has attempted to overcome this deficiency by developing an integrated bio-economic modeling approach that combines econometric farm (household) models with linear programming techniques and agronomy-based models.[1] The UNA-DLV methodology is a decision-support system for policy makers, which offers insight into the likely reactions of farm types to different kinds of economic incentives. Based on a detailed understanding of farmer priorities and their resource allocation decisions, the methodology simulates the impact of policy instruments on patterns of land use, labor allocation, market production, and farm and regional income levels.

In the UNA-DLV methodology, linear programming models are employed to optimize land use, simultaneously taking into account farmer objectives and resource constraints, as well as the relevant policy environment. The methodology has been applied in the northern Atlantic Zone of Costa Rica (Kruseman *et al.,* 1995, 1997; Ruben *et al.*, 1994; Roebeling *et al.*, 2000). A number of models were constructed, each of which considers an individual representative farm type with its own specific objectives regarding production and consumption, as well as with its own resource endowments. All these factors are widely considered as important determinants of the farm-level decision-making process (Singh *et al.*, 1986). The UNA-DLV methodology exhibits close similarities to the SOLUS framework presented in Chapter 6. The agronomic models applied in the UNA-DLV methodology (*i.e.*, the Technical Coefficient Generators LUCTOR and PASTOR; see Chapter 5) were developed in close cooperation with the REPOSA project, and a common biophysical and socio-economic database was developed.

To determine the aggregate effect of certain policy measures, model simulation results for each individual farm model can be "scaled-up" to the regional (*i.e.*, sector) level through weighted aggregation based on the number of farms per representative farm type. The obvious shortcoming of this approach is that it considers each individual farm type in isolation, without taking into account the aggregate market effects of decisions made on all farms together. Aggregate demand and supply influence regional product and factor price levels, and thus investment, production and consumption decisions on the farm level. Thus, ignoring aggregate supply and demand conditions not only causes aggregation bias (Chapter 6; Jansen and Stoorvogel, 1998), it may also lead to erroneous conclusions regarding the effectiveness of agricultural policies. Consequently, the main objective of this chapter is to evaluate the importance of incorporating the equilibrium conditions for product markets in the farm-level analysis of the effectiveness of different agrarian policies. The case study presented is one for the Atlantic Zone of Costa Rica.

The remainder of this chapter is structured as follows: the next section describes the construction of partial models for individual farm types and their implementation in the Atlantic Zone. The third section outlines the aggregation method used to arrive at a regional agricultural land use model that incorporates farmer behavior. In the fourth section, scenario runs are presented that support policy design, and results are discussed with emphasis on the comparison between the aggregate model and the partial farm models. The final section offers concluding remarks and observations.

[1] The approach has been developed by the Wageningen Agricultural University research group on Sustainable Land Use and Food Security (Dutch abbreviation: DLV; *Duurzaam Landgebruik en Voedselzekerheid*) in close cooperation with the International Center for Economic Policies for Sustainable Development (CINPE; *Centro Internacional de Política Económica*) of the Automous National University (*Universidad Nacional Autónoma*) in Heredia, Costa Rica.

8.2 Stakeholders and the policy priorities for regional development

The Atlantic Zone of Costa Rica plays an important role in the national agricultural production in general and export production in particular (Roebeling *et al.*, 1999), while it historically provided the space for the establishment of new rural settlements (Jones, 1990). This has resulted in a rather dynamic and heterogeneous agrarian structure, where banana plantations and extensive livestock ranches coexist with mixed cropping systems on small and medium-size peasant farms.

During the last decade, structural adjustment policies intend to reinforce the competitiveness of the agricultural sector within a more liberal market environment. Since further expansion on the agrarian frontier is restricted, selected incentives are given to enhance diversification into non-traditional activities and to intensify production in the traditional sectors in order to improve factor productivity. The second aspect of regional policy in the Atlantic Zone involves natural resource management. Sustainable practices for cropping and pasture activities should rely on efficient nutrient management and reduced use of biocides. From the viewpoint of nature conservation (and for maintaining bio-diversity), the externalities of agricultural production have to be reduced and deforestation has to be controlled (Estado de la Nación, 1997; Kruseman *et al.*, 1994).

In brief, regional development in the Atlantic Zone attempts to combine agricultural growth and sustainable land use and to contribute simultaneously to farmer wealth while maintaining the natural resource base. Agricultural modernization and intensification could reduce the existing pressure on natural resources, but market failures and institutional constraints tend to inhibit the responsiveness of local farmers to the need of making adjustments in their land use practices.

Different policy instruments could be used to improve the effectiveness of economic incentives meant to influence farmer behavior. The current policy debate in Costa Rica requires guidelines that could provide assistance to the rural transition process (Estado de la Nación, 1997). Comparisons should be made between price and (infra-) structure policies, which both tend to favor commercial agricultural production but might have quite opposite effects on regional land use. Moreover, market policies could be selective for certain inputs (*e.g.*, fertilizer, biocides) or could offer a general incentive on the output side. Distributive effects of price and fiscal policies have to be considered as well. Finally, from a public choice viewpoint, different policy instruments have to be compared with respect to their budgetary implications.

8.3 Methodology and specification of partial models for individual farm types in the Atlantic Zone

Contrary to the other studies presented in this book, this study concerns both the Northern and the Southern parts of the Atlantic Zone of Costa Rica (Figure 8.1). In total, the study area covers some 918 000 ha, of which nearly 40% is suitable for agriculture. Since information on the Southern part of the AZ is mostly lacking, it is assumed that the Southern part is similar to the Northern part in terms of both biophysical and socio-economic characteristics.

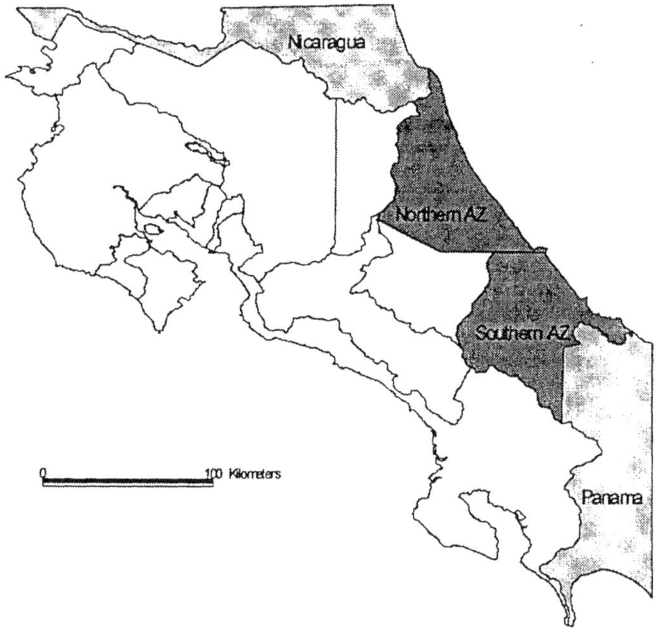

Figure 8.1. Study area: Atlantic Zone of Costa Rica.

For a biophysical and socio-economic description of the study area, the reader is referred to Chapter 2. Current land use is dominated by natural forests (48%), cattle ranching (39%) and banana plantations (10%), with secondary crops including plantain, palm heart, root and tuber crops, maize, pineapple and ornamental plants (3%). In monetary terms, however, the relative importance of these crops is substantial. For example, whereas as in 1984 (the year of the latest agricultural census) some 37% of the agricultural area in the AZ was dedicated to pasture-livestock activities, the livestock sector accounted for less than 4% of the value added of the agricultural sector in the AZ (DGEC, 1987a). In terms of its importance to the total national agricultural production, the AZ is the main producer of banana, plantain and palm heart, as well as a significant supplier of livestock products and, to a lesser extent, pineapple (Table 8.1).

8.3.1 Farm household stratification and characterization

Farmer decisions about land use and technology choice are guided by their objectives and subject to available resources, production possibilities and external economic and bio-physical constraints, all of which may vary substantially between individual farms. Therefore, in farm modeling, a proper farm stratification is important to develop individual models for each representative farm type. Within the AZ different farm types can be identified according to dominant land use systems and related farming systems, resource availability, main production activities and perceived objectives. The 1984 agrarian census is the latest source of available information upon which it is possible to base a proper farm stratification (Table 8.2). This resulted in the identification of the following farm types: small size farm households, medium size farm households, banana plantations and extensive beef cattle farms (*haciendas*) (Kruseman *et al.*, 1994). This stratification covers 97% of all farms and 70% of the total agricultural area in the region.

Table 8.1. Production and percentage share of the Atlantic Zone in national and world production (1996)

	Unit	*Production:*			*Share of AZ production in:*	
		World	*Costa Rica*	*AZ*	*World*	*Costa Rica*
Maize	10^6 kg	-	24.3	0.3	-	1.3
Beans	10^6 kg	-	21.4	0.1	-	0.3
Cassava[1]	10^6 kg	-	137.8	0.1	-	0.1
Local pineapple	10^3 units (1.2 kg)	-	613.7	2.0	-	0.3
Export pineapple	10^6 units (1.5 kg)	931.8	205.0	20.5	2.2	10.0
Banana	10^6 boxes (18.1 kg)	483.9	106.5	100.9	20.9	94.8
Plantain	10^6 boxes (25 kg)	42.0	9.2	8.7	20.7	93.9
Palm heart	10^6 units (1.3 kg)	56.1	12.4	6.2	11.0	50.0
Beef	10^6 kg	-	49.1	6.3	-	12.9

[1] Data based on PIMA figures, representing product quantities for the domestic market and therefore not reflecting total cassava production.

Source: World production data are obtained from the FAO statistical database. National and regional production data for maize, beans, palm heart, export pineapple and beef are obtained from the National Production Council (CNP), for banana from the National Banana Corporation (CORBANA) and for cassava, local pineapple and plantain from the National Center for Supply and Distribution of Food Products (CENADA).

Table 8.2. Farm stratification in the Atlantic Zone (1984)

	Unit	*Farm type*				*Total stratification*	*Total Atlantic Zone*
		Small 0-20 ha	*Medium 20-50 ha*	*Hacienda >50 ha*	*Banana >100 ha*		
Objective(s)		Utility and full income		Quasi-rent	Profit		
Farming system		Mixed	Mixed	Livestock	Bananas		
Number of farms	#	6480	1690	803	83	9056	9316
Average farm area	ha	7.1	28.7	108.3	226.3	22.1	30.6
Total agricultural area	ha	46255	48472	86991	18780	200498	285315
Protected areas	ha	-	-	-	-	-	558097
Agrarian frontier area[1]	ha	-	-	-	-	-	74903
Total area	ha	-	-	-	-	-	918315

[1] Areas that border protected areas but that are not (yet) used for agriculture.

Source: DLV calculations based on DGEC (1987a). Protected areas include national parks, forest reserves and Indian reserves (Jones, 1990).

As of 1984, extensive *haciendas* and banana plantations covered about 67% of the total agricultural area, while representing only 12% of all farms in the AZ (Table 8.2). Economic decision-making on extensive *haciendas* is guided by a quasi-rent objective, taking into account not only returns obtained from cattle production (meat and milk) but also expected returns from investments in land (Roebeling *et al.*, 1998). On banana plantations, profit maximization is considered the main objective (Kruseman *et al.*, 1994). Small and medium size farm types represent 70% and 18%, respectively, of all farms in the AZ, while covering respectively 16 and 17% of the total agricultural area. Both farm types are characterized by mixed farming systems as well as by utility and income objectives (Kruseman *et al.*, 1994).[2] Agricultural activities of small farm types are mainly geared towards basic grains, cassava and plantain production, in combination with cash crops such as palm heart and pineapple. Production is mainly for the market, while basic grain, milk and cassava production is to a significant extent used for family consumption (Castro *et al.*, 1996). Sixty-six percent of livestock activities in the AZ are devoted to cattle breeding and fattening for beef production, while the remainder of the total herd in the AZ can be characterized as double purpose and dairy cattle (Van Loon, 1997).

Table 8.3. Farm characteristics

Farm type	*Farm area[1] (ha)*	*Labor (d)*	*Savings ($)*	*Cattle (AU)*	*Equipment (units)*	*Sprayers (units)*	*Vehicles (units)*	*Time discount rate (%)*	*Savings coefficient (%)*
Small[2]	8.9	491.9	281.8	4.1	0.9	1.1	0.2	7.5	25
Medium[3]	39.2	412.7	773.5	48.3	0.9	1.8	0.6	7.5	25
Hacienda[4]	170.4	570.0	5524.9	188.1	-	-	-	7.5	48
Banana	226.3	-	+∞	-	-	-	-	7.5	-

[1] The farm area differs from that presented in Table 8.2, since the calculations are based on more recent field work.
[2] DLV calculations based on Castro *et al.* (1996).
[3] DLV calculations based on Kuiper (1996).
[4] DLV calculations based on REPOSA field research data.

Table 8.3 shows the initial resource base per farm type. The average agricultural area of small and medium size farm types is 8.9 and 39.2 ha, respectively, while availability of family labor is 41 and 34 days per month, respectively. Labor can be hired in as well as hired out against the prevailing wage rate of $ 7.70 per day. Capital availability of both farm types is low, as reflected in the limited household savings and equipment ownership. Access to formal credit is restricted due to strong collateral requirements. While informal credit is more widely used, its costs may be very high (Quirós *et al.*, 1997). A representative extensive *hacienda* would have an initial cultivable area of 170.4 ha, with almost two permanent laborers, in addition to any hired labor. Personal savings of *hacienda* owners are assumed to amount to $ 5525, while credit availability is modeled as a proportion of the value of land and cattle stock. The latter consists of 188 animal units (AU; 1 animal unit = 400 kg

[2] Utility is defined as the capacity of a good or service to satisfy a necessity or desire, while income is defined as the difference between net revenues and the monetary value of nutrient losses (Kruseman *et al.*, 1996; Roebeling *et al.*, 2000).

live weight). Finally, the representative banana plantation's size is 226 ha, with operating capital assumed to be available in unlimited quantities. Equipment and labor can be hired without limitations at given prices. With the exception of banana plantations, most farmers in the Atlantic Zone face considerable marketing problems, including (Jansen *et al.*, 1996): *i*) limited sales outlets, *ii*) limited market and countervailing power, *iii*) high transport costs, and *iv*) shortage of transport facilities contributing to low farm-gate prices.

8.3.2 Small and medium size farm households

The methodology used for the small and medium size farm types is a farm household modeling approach (Singh *et al.*, 1986) that links an econometrically specified behavioral expenditure module with a linear programming optimization procedure of the production structure (Ruben *et al.*, 1994; Roebeling *et al.*, 2000). The methodology allows us to appraise the impact of policy measures on farm profits and utility, factor allocation, and land use in the short term (up to 5 years). To combine these two approaches, two linear programming models are linked: a production and a consumption model.

The production model permits the analysis of the ways in which farm households allocate savings and credit to cover their operation costs and investments, while the consumption model permits the analysis of farm household consumption preferences (based on an econometrically derived non-linear utility functions) and the way in which these preferences determine crop choice. Given the production and consumption model, the selection of production activities takes place in a stepwise optimization procedure (Roebeling *et al.*, 2000). First, production decisions are made on the basis of expected prices in order to maximize income, subject to cash and credit constraints, production technology, minimum consumption requirements, and initial farm characteristics (Table 8.3). Secondly, consumption decisions are taken on the basis of actual prices in order to maximize utility, subject to net farm income, a time constraint, and the farm resource availability adjusted in the first step. A goal weight generator is used to attach weights to the different objectives, thus solving the non-separable model and allowing for non-recursive relations between the production and consumption parts of the farm household model (Kruseman *et al.*, 1997).

Farm household decisions take place in specific contexts governed by certain objectives, production options and resource constraints, within a specific policy environment. Farm household resource constraints are defined by the household's initial resource endowments, while also its savings behavior (expressed in a savings coefficient) and relative valuation of future and present income (expressed in a time discount rate) determine the optimal solution. Resources available to the household include land, labor, savings, cattle and fixed capital inputs (Table 8.3). Based on real interest calculations by Zuñiga (1996), the farm household's time discount rate is set at 7.5% per year. Savings available for the financing of operation costs and investments are a proportion (set at 25%) of net farm income obtained in the previous year, consumption being the remainder (Bade *et al.*, 1997).[3] While formal credit is limited to a fraction of the farm household's land value (25%) and value of the cattle stock (20%), informal credit is assumed to be limited to a fraction of the expected marketed crop production value (10%). Annual real interest rates for the use of formal and informal credit capital are 12% (Zuñiga, 1996) and 47% (Quirós *et al.*, 1997), respectively.

[3] The amount of income that is set aside for savings depends on (1) the time discount rate as it values future and present consumption, and (2) the degree to which savings can generate future income (Kruseman *et al.*, 1997).

Farm household options include both on- and off-farm activities. Off-farm activities refer to external employment possibilities for family labor. On-farm production activities are determined by specific cropping and livestock systems, the latter of which are sub-divided into pasture land use systems, herds and feed supplements. Technical coefficients are generated for cropping systems using the Technical Coefficient Generator LUCTOR, and for pasture land use systems, herds and feed supplements using PASTOR (see Chapter 5).

Optimization takes place for utility and income objectives. Utility is obtained through the consumption of on-farm production as well as leisure (Q_j^{cons}), and purchased products (Q_j^{buy}). Utility is maximized subject to net farm income *NFI*, which is defined as returns from marketed production (Q_j^{sold}) and off-farm employment (O^{off}), net of the costs related to the use of fixed and variable inputs (I_i), capital resources (C_b), and consumption. Thus, net farm income *NFI* is defined as follows:

$$NFI = \{\sum_j p_j Q_j^{sold} + w^{off} O^{off}\} - \{\sum_i p_i I_i + \sum_b p_b C_b + \sum_j p_j (Q_j^{cons} + O_j^{buy})\} \tag{1}$$

where p refers to the prices related to commodities j, fixed and variable inputs i and capital sources b, and w is the off-farm wage rate.

Direct utility functions are used applying the methodology proposed by Kruseman *et al.* (1997), which permits estimation of utility functions on the basis of Engel curves. Utility functions are derived from the latest National Household Income and Expenditure Survey (DGEC 1988, 1990), using a negative exponential utility function for basic food crops (characterized by decreasing marginal utility which asymptotically reaches a maximum) and an exponential utility function for other food products, non-food products and leisure (characterized by decreasing marginal utility but without maximum). Since utility is assumed to be additive and separable, the objective utility function Z_1 is given by:

$$Z_1 = UTIL = \sum_j U_j^{max} (1 - e^{-\rho_j (Q_j^{cons} + O_j^{buy})}) + \sum_j \rho_j (Q_j^{cons} + O_j^{buy})^{\sigma_j} \tag{2}$$

where U_j^{max} denotes the maximum attainable utility with commodity j, ρ_j is the conversion factor of consumption to utility, and σ_j is the exponent of consumption commodity j.

The income objective is defined as net farm income corrected for the expected monetary value of nutrient losses (Van der Pol, 1993), thereby using the farm household's nutrient reservation price for the valuation of current nutrient gains or losses as reflected by the change in soil nutrient stock (B_n). The income objective Z_2 is given by:

$$Z_2 = INCOME = NFI - \sum_n p_n * B_n \tag{3}$$

where p_n represents the reservation price related to nutrient n.[4] The investment and consumption component of the model are calculated separately, and final model results are obtained by their weighted sum (Kruseman *et al.*, 1997). Following Romero (1993), the relative importance of each objective is determined through a comparison of

[4] Since changes in soil nutrient availability affect future production potentials, the producer attaches a value to changes in soil nutrient availability according to a time discount rate in combination with future production possibilities, resource availability and objectives. Farm household models were run sequentially, thereby (linearly) reducing future production potentials on the basis of nutrient losses. This allows the calculation of the cost per kilogram of lost nutrient, given the farmers' time discount rate, which represents the nutrient reservation price. The focus in this chapter will be on the changes in soil nitrogen stock, since it is the most mobile macro-nutrient while it also catalyses the use of other nutrients.

the base run production plan that is obtained using actual production activities and prices with the actual production plan as derived from farm survey data (Castro *et al.*, 1996). Calibration of the models resulted in relative goal weights for utility and income of 60% and 40% for the small size farm type, and respectively 40% and 60% for the medium size farm type (Roebeling *et al.*, 2000).

Expected as well as actual market prices are used for model optimization, while transaction costs are taken into account for the calculation of farm gate prices. The model makes use of expected prices for medium-term investment decisions, while actual market prices are used for short-term consumption decisions. Expected prices are based on a weighted average of market prices over the last three years (coefficients of expectation set at 0.5 for year (t-1), and 0.25 for years (t-2) and (t-3)). Transaction costs are defined as the margin between market and farm-gate prices, resulting from transport costs, marketing margins and imperfect market information.

8.3.3 Extensive *haciendas*

The dynamic linear programming model for extensive *haciendas* evaluates a large number of options for pasture based beef cattle production and its associated management schemes. These options are analyzed according to the long term quasi-rent objective and subject to resource and cash flow restrictions (Roebeling *et al.*, 1998). Dynamic properties are incorporated in the model through multi-period livestock activities and a savings and investment module, within the long-term planning horizon of the *hacienda* owner (set at 10 years). The time aspect in livestock activities allows for intertemporal consideration of growth, fertility, mortality and feed requirements over the years related to buying, marketing and feed purchasing strategies. Moreover, intertemporal savings and investment considerations allow for adjustments in the capital stock (consisting of cattle and land) as well as changes in future credit availability.

Restrictions are mainly defined by resource endowments, while the optimal solution also depends on the *hacienda* owner's attitude towards savings as well as the time discount rate. Available resources include land, livestock, own capital and credit, all of which are allowed to change over time within the *hacienda* owner's planning horizon. Contrary to the models for peasant farm households and banana plantations, land is considered a variable in the *hacienda* farm model. As a consequence, land that is assumed to be located on the agricultural frontier can be purchased at market prices.[5]

The multi-period optimization procedure requires the specification of intertemporal choices in farm household behavior. Operating capital is defined as a proportion of net returns obtained in the previous year, while the availability of formal credit is assumed to be limited to a proportion (25%) of the land and cattle value of the previous year. Both capital sources can be used to finance investments in land and cattle, as well as for the financing of operation costs. Relative valuation of present and future income is expressed in the time discount rate, which for the *hacienda* owner is relatively low (due to the long-term quasi-rent objective) and set equal to the opportunity cost of capital of 7.5% per year. The real credit interest rate is set at 10% per year. The extensive *hacienda*'s savings coefficient is determined using the bequest theory approach as proposed by Phimister (1993), which, given a time discount rate of 7.5%, results in a savings coefficient of 48%.

[5] Due to lack of data on land markets, competition for land among farm types is not included in the aggregate modeling framework.

Production options are determined by the on- and off-farm allocation of capital. Off-farm capital allocation refers to investments in the capital market, with an expected return equal to the opportunity cost of capital. On-farm production and investment possibilities include livestock activities on the one hand, and (related) investment opportunities in land and cattle on the other. Technical coefficients for the three components involved in livestock production (pastures, herds and feed supplements) are generated by PASTOR, in the same way that they are generated for small and medium size farm types (see also Chapter 5). For herds, six animal classes are differentiated on the basis of sex and age (starting age 1, 2 or 3 years). Animal classes are either owned (initial resource base and birth) or purchased, and remain productive until they are sold at a certain age corresponding to one of four identified marketing strategies (*i.e.*, animals can be sold at 3 to 6 years of age).

The decision making of *hacienda* owners is mainly guided by net returns and rent objectives (Crotty, 1980; Jarvis, 1986), the combination of which defines the long term quasi-rent objective for which model optimization takes place. Net returns refer to the present value of the difference between the value of cattle sold (Q_j) on the one hand, and expenditures on variable inputs (Q_i) related to pasture and cattle production, investments in cattle (I^{cattle}) and land (I^{land}), capital costs (C_b) and tax levies (τ_r) over the *hacienda*'s resource value (R_r), on the other hand. The net return objective (NR) over the 10 year (y) planning period is given by:

$$NR = \sum_{y=1}^{y=10} \frac{1}{(1+i)^y} * \{(\sum_j p_j Q_{jy}) - (\sum_i p_i Q_{iy} + \sum_b p_b C_{by} + \sum_j p_j I_{jy}^{cattle} + \sum_s p_{sy} I_{sy}^{land} + \sum_r \tau_r R_{ry})\} \quad (4)$$

where i is the time discount rate, and p represents prices related to animal classes j, variable input types i, capital sources b, and land units s. Tax levies (τ_r) are differentiated for the resources (r) land and cattle, both being determined at 0.5% of the resource value per year according to the actual land tax and assets tax, respectively. Minimum yearly net returns are set at $ 5525 per year, in line with the initial capital availability.

Rent refers to the present value of expected returns derived from the ownership of assets in the long run, obtained over and above the returns obtained from the productive use of these assets (Henderson and Quandt, 1980). In the case of the extensive *hacienda*, land is the major asset and considered to be the most important long-term investment (Van Hijfte, 1989). The rent objective is thus given by the expected present value of land assets at the end of the 10 year planning period, as follows:

$$\text{RENT} = \sum_{y=1}^{y=10} \sum_{s=1}^{s=3} \frac{1}{(1+i)^{y=10}} * (p_{s,y=10} I_{sy}^{land}) \quad (5)$$

The model makes use of expected market prices, thereby taking into account transaction costs. Cattle prices differ by animal class, while land prices are differentiated according to fertility characteristics. The real growth rate of the land price is assumed to be higher than the time discount rate, and set at 12.5% per year. Prices for labor and feed supplements are locally determined, while prices of inputs and beef are determined by their respective world markets (Jarvis, 1986).

8.3.4 Banana plantations

The banana plantation model is a linear programming model that analyzes the production side to determine technology choice. Banana production technologies are evaluated according to a profit objective, subject to availability of certain resources. While availability of land is restricted, capital availability is assumed to be unlimited. Labor and other inputs can be obtained at fixed market prices, assuming unlimited supply. Production possibilities are limited to banana cultivation, with technical coefficients for banana land use systems generated by LUCTOR (see Chapter 5).

Banana plantations are guided by a profit objective (Kruseman *et al.*, 1994). Profit is defined as the returns from banana export ($Q_{j=ban}$) sold at world market prices ($p^{wm}{}_{j=ban}$), net of the costs related to the use of fixed and variable inputs (I_I including labor) and invested capital ($C_{b=own}$). Mathematically, the profit objective can be expressed as follows:

$$PROFIT = \{\sum_{j=ban} p_j Q_j^{sold}\} - \{\sum_i p_i I_i + \sum_b p_b C_b\} \qquad \textbf{(6)}$$

where p_i the price of fixed and variable inputs i, and p_b the price related to the use of operating capital sources b. The latter is set equal to the opportunity cost of capital of 7.5% per year. Model optimization takes place using actual prices, while accounting for transaction costs in order to determine prices at the farm gate level.

8.3.5 Generation of technical coefficients

PASTOR and LUCTOR were used to generate technical coefficients for seven crop land utilization types (banana, black bean, cassava, maize, palm heart, pineapple and plantain) and five pastures (three fertilized improved grasslands, a grass-legume mixture, and a mixture of natural(ized) grasses). These land utilization types were combined with the three major land units identified in the northern Atlantic Zone (Chapter 2): Soil Fertile Well drained (SFW), Soil Infertile Well drained (SIW) and Soil Fertile Poorly drained (SFP). Actual land use systems are derived from descriptions by the current best farmers in the AZ, and alternative systems were generated using the target-oriented approach (Chapter 5). For crops, the alternative systems were generated with a zero soil nutrient loss restriction, while for pastures, seven nutrient mining levels were defined, ranging from 0 to -60 in steps of 10 kg ha^{-1} y^{-1}. For alternative land use systems, different technology levels were generated by combining levels of fertilizer use, crop protection, substitution between manual weeding and herbicide use, and the pasture stocking rate. For beef cattle herds, four production systems were generated based on target growth rates (Roebeling *et al.*, 1998).

Technical coefficients include labor requirements, costs, inputs, yield, change in soil nutrient stock for nitrogen (N), phosphorus (P) and potassium (K), N denitrification loss, N leaching loss, N volatilization loss, and biocide use in terms of both active ingredients (a.i.) applied (BIOA) and a biocide index (BIOI). Detailed information on the generated land use systems, their technical coefficients and calculation procedures are given in Chapter 5.

8.4 Partial and aggregate simulation methodology

8.4.1 Aggregation issues in linear programming models

In the various levels of analysis in Chapter 6, three aggregation issues noticed by Erenstein and Schipper (1993) were again raised:

1 The use of land is often considered without sufficient knowledge regarding the behavior of the farm households responsible for the actual use of land.
2 Aggregation bias resulting from the fact that individual farmers have resources at their disposal in different proportions from the aggregated resources of a region.
3 Variables that are exogenous on the micro level become endogenous on higher levels.

The first two issues were addressed in Section 8.3 which explains how the methodology presented here uses farm stratification and the derivation of farm type specific objective functions to deal with these two aggregation issues.

The third aggregation issue refers to the earlier-mentioned phenomenon of various farm types operating in the same region, supplying their products to the same product markets and operating in the same factor markets. The implication is that prices, while exogenous on the individual farm level, become endogenous on the regional level, insofar as the region of study accounts for a significant proportion of domestic supply or demand. In regional agricultural sector models, such as the one presented in Chapter 6, such endogenization is achieved by maximizing aggregate consumer and producer surplus under the assumption of competitive product markets. On the other hand, in modeling efforts that simulate the agricultural sector of a particular region by simultaneously considering a number of partial models for individual farm types, simple maximization of regional consumer and producer surplus is no longer possible, since each farm type has different objectives and therefore cannot be part of an aggregated model on the sector level. In this kind of sector analysis, aggregation and simulation of product markets has to occur outside the models for individual farm types, through an iterative procedure as explained below.

8.4.2 Operationalization of aggregation in farm modeling

Even though single farm decisions will not affect product and factor prices, on the regional level total product supply and factor demand of all farms in the region might affect subsequent equilibrium prices. Partial model results for the different farm types are used to determine the total regional product supply and factor demand through weighted aggregation, based on the number of farms per farm type (Table 8.2). In the presented methodology, aggregation takes place on the product side for all considered crops,[6] to determine market clearing prices on the regional or world market level. Regional product supply in combination with the respective product demand curves faced by producers in the region, determine market clearing equilibrium prices for products. In turn, these newly determined equilibrium prices form the input of subsequent partial model runs for each farm type. This procedure is repeated until product prices deviate less than 1% from corresponding prices determined in the previous iteration.

[6] Beef and milk areè not included in the aggregation, since beef prices are determined by world market prices, while milk prices are highly protected by government policies (Kaimowitz, 1996).

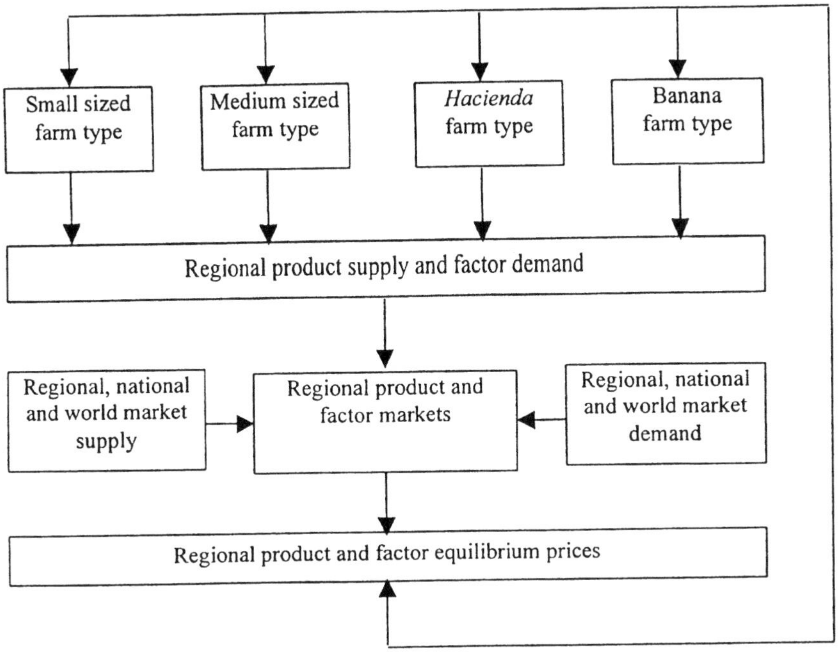

Figure 8.2. Multiple farm model structure.

In a regional model, demand functions faced by producers rarely coincide with national demand functions. Product demand may be met by products purchased in other regions, while demand for local products may also arise in other regions. Kutcher (1972, 1983) therefore argues that the demand function faced by the local producer is determined by the national demand function as well as by supply functions of producers in other regions of the country, and depends on the relative importance of local production (Q_{10}) in the total national production (Q_0). Given the national demand elasticity (η_N), the other regions' supply elasticity (σ_2), and the other regions' supply (Q_{20}), the demand elasticity faced by the local producer (η_l) can be defined as follows (Hazell and Norton, 1986; equivalent to equation 7 in Section 6.2.5):

$$\eta_l = \eta_N \frac{Q_0}{Q_{10}} - \sigma_2 \frac{Q_{20}}{Q_{10}} \qquad (7)$$

where total national production (Q_0) is the sum of local production (Q_{10}) and production by other regions in the country (Q_{20}). Equation (7) shows that the demand elasticity faced by the local producer approaches the national demand elasticity when local production approximates national production. On the other hand, the demand elasticity faced by the local producer approaches perfect elasticity when local production forms a relatively insignificant part of the total national production.

Table 8.4. Derivation of demand elasticities for the Atlantic Zone of Costa Rica

	Supply	*elasticities*[1]	*Demand elasticities*[2]		
	world	*Costa Rica*	*Actual world*	*Actual Costa Rica*	*Derived Atlantic Zone*
Maize	n.a.	0.6	n.a.	-0.90	-113.1
Beans	n.a.	0.2	n.a.	-0.90	-336.7
Cassava	n.a.	0.2	n.a.	-0.60	-1102.3
Local pineapple	n.a.	0.5	n.a.	-0.70	-367.7
Export pineapple	0.7	0.7	-1.1	-0.88	-81.1
Banana	0.7	1.0	-0.5	-0.91	-5.1
Plantain	0.7	0.5	-1.2	-0.80	-8.5
Palm heart	0.7	0.7	-1.8	-1.20	-22.0

[1] Supply elasticities for the world are assumed values. Supply elasticities for Costa Rica are based on Roebeling *et al.* (1999).

[2] Demand elasticities for the world are obtained from the REALM model (Chapter 6). Demand elasticities for Costa Rica and for the regional level are obtained from Roebeling *et al.* (1999).

Demand elasticities faced by producers in the Atlantic Zone of Costa Rica are shown in Table 8.4. The AZ is a major producer of banana, plantain and palm heart, while production of basic food crops and pineapple represents only a small part of the total national production. As a result, the derived demand for these latter crops is highly elastic, with prices that are hardly influenced by the production originating in the Atlantic Zone. On the basis of the derived demand elasticity (η_j), the initial price level (p_j^0), the initial production level (q_j^0) and the (model derived) production level (q_j^1), the new equilibrium price (p_j^1) per product j can be calculated using the following definition for the calculation of the demand elasticity:

$$p^1_j = p^0_j + \frac{dq_j}{\left(\dfrac{\eta_j}{p^0_j / q^0_j}\right)} \tag{8}$$

As a first approximation it is assumed that production originating from other regions in Costa Rica is not influenced by price changes.

8.5 Model implementation and results

8.5.1 Scenario definition

The methodology described in the previous sections was implemented by running policy scenarios for the Atlantic Zone. Aggregation bias that stems from not accounting for the fact that various farm types operate in the same region and compete on the same product markets is quantified. In this context, a distinction is made between partial and aggregate model results. Partial model results are obtained by running the model without taking into account the effect of regional product supply on regional equilibrium prices, while aggregate model results are obtained when product equilibrium prices are influenced by regional product supply. Model runs were made for a so-called base scenario and for three policy scenarios. For each individual farm type model, the base scenario maximizes the corresponding objective function subject to factor constraints, while neither normative restrictions nor limitations on biophysical sustainability indicators are included. The policy simulations are based on the current debate regarding agricultural policy priorities in Costa Rica. On the national level, the Costa Rican government is solliciting applied research that analyzes the effects of alternative policy measures on land use. Explicit attention has to be given to the question of how to achieve a more sustainable land use and thereby to the possible trade-offs between socio-economic and environmental goals (SEPSA, 1997). Insofar as the AZ is concerned, two major regional development objectives include improvement of the competitiveness of agricultural production under increasing trade liberalization; and improved natural resources management (Ruben *et al.*, 1994; Chapter 2 of this book). The policy simulations therefore investigate the effects of:

1. Increased infrastructure investments. The relatively poor state of the infrastructure in general, and of the road structure in particular, has been identified as a serious obstacle towards a more competitive position of the Costa Rican agricultural sector (Echandi, 1998; Hausmann, 1998). Investments in infrastructure in a wide sense (*i.e.*, better marketing facilities and market information, improved road network) would lead to a reduction in transaction costs, resulting in a decline in farm gate input prices and a rise in farm gate output prices.
2. A tax on biocides. In order to reduce the negative externalities of agricultural production, regulation and control of input use is considered an important policy option by the Costa Rican government (SEPSA, 1997). Structural adjustment policies implemented in Costa Rica since the 1980s have contributed to increased use of biocides in agricultural production. While the government designed some legislation aimed at reducing the use of biocides in agriculture, the potential effect of these policy measures remains largely unknown (Agne, 1996).
3. Improved credit availability. In December 1997, the *Productive Reconversion Program* was approved by the Costa Rican Congress. The program's main objective is to improve the integration of small and medium size farmers into international markets (SEPSA, 1997; La Gaceta, 1998). The Program enables the National Production Council (CNP) to issue certain guarantees in favor of organized farm groups, in this way providing the latter with access to loans from the public banking system. Such improved capital availability should widen production and investment options.

The effects of each policy are studied on three different levels. First, partial policy simulations are compared to the partial base run situation. Second, partial policy simulations are compared to the aggregate policy simulations. Finally, aggregate policy simulations are compared to the aggregate base run situation. Policy simulations include a 20% decline in transaction costs, a 40% tax on the price of biocides, and a 20% increase in credit availability.[7]

8.5.2 Base run results

Farm level

Indicators of the base run are given in Table 8.5 per farm type, for model results generated with exogenous (partial analysis) and endogenous (aggregate analysis) product prices. Base run results are obtained for each farm type given objectives for optimization under expected and actual market prices, initial resource availability, and actual (mostly unsustainable) as well as alternative (sustainable) options for crop, pasture and livestock land use systems. Indicators include the objective function value, production structure, and indicators for sustainability and factor use intensity. Sustainability of production systems is reflected in the level of soil nitrogen gain or loss as well as in the use of active ingredients (a.i.) in applied biocides (BIOA), while the intensity of production systems is reflected in labor and capital use per hectare per year.

Table 8.5. Partial and aggregate base run results on the farm level

Indicator	*Unit*	*Partial base run*				*Aggregate base run*			
		Small	*Medium*	*Hacienda*	*Banana*	*Small*	*Medium*	*Hacienda*	*Banana*
Economic gain[1]	$ 10^3 y^{-1}	11.2	89.0	945.3	1260.2	9.4	55.4	945.3	1192.3
Production structure:									
Maize	ha	0.1	0.0	-	-	0.0	0.0	-	-
Beans	ha	0.0	0.0	-	-	0.0	0.0	-	-
Pineapple	ha	1.2	5.2	-	-	0.6	4.9	-	-
Banana	ha	-	-	-	169.7	-	-	-	155.6
Plantain	ha	2.1	0.0	-	-	1.8	0.0	-	-
Palm heart	ha	0.0	0.0	-	-	1.3	0.0	-	-
Cassava	ha	0.1	0.0	-	-	0.0	0.0	-	-
Pasture	ha	0.9	20.3	190.3	-	0.9	20.8	190.3	-
Fallow	ha	4.5	13.7	-	56.6	4.3	13.5	-	70.7
Livestock	AU	1.2	58.0	304.3	-	1.1	58.7	304.3	-
Sustainability:									
Δ soil N stock	kg ha^{-1} y^{-1}	-104.8	-46.2	-50.2	0.0	-93.6	-45.6	-50.2	0.0
Biocide use (BIOA)	kg a.i. ha^{-1} y^{-1}	14.8	4.5	0.9	54.8	12.5	4.4	0.9	54.9
Resource use intensity:									
Labor intensity[2]	d ha^{-1} y^{-1}	95.1	29.3	5.3	203.5	80.0	27.7	5.3	224.0
Capital intensity[3]	$ ha^{-1} y^{-1}	1423.0	1640.2	185.3	4836.4	1171.2	1306.5	185.3	4836.3

[1] Economic gain refers to net farm income for small and medium size farm types, value of land and cattle stock for the *hacienda* farm type, and profit for the banana farm type.
[2] Labor intensity refers to the number of labor days (d).
[3] Capital intensity refers to the amount of variable inputs and labor spent.

[7] A 40% biocide tax is applied because no response reactions were obtained with the 20% biocide tax.

Partial results for small farms show a specialization in cash crop production (plantain and pineapple), while food crops (maize and cassava) as well as beef and milk production prove important for household consumption. The medium size farm type shows a complete specialization in pineapple production and in livestock fattening for beef production. The resulting high regional levels of production lead to product price decreases in the aggregate base run, particularly for pineapple (see Table 8.6). As a consequence, the small size farm type abolishes its food crop production meant for on-farm consumption and diminishes its pineapple and plantain production, in favor of palm heart production.

Even though responding less strongly than the small size farm type, the medium size farm type diminishes its pineapple production in favor of cattle raising since beef prices are determined on the world market and therefore considered exogenous. In both cases, endogenous pricing of products leads to a slight increase in total land use. Agricultural production on medium size farms is more capital intensive than on small size farms due to better access to capital resources, but it is also less labor intensive due to the larger importance of labor-extensive pasture production. Even though relatively unimportant in absolute terms, production of maize and cassava for farm household consumption completely disappears in the aggregate run for two main reasons. First, lower prices result in lower net margins, which in turn limit the use of formal and (especially) informal credit. Second, since off-farm employment as a means of generating work capital becomes more important, the availability of family members as a source of cheap labor in the cultivation of low yielding food crops for on-farm consumption decreases.

Haciendas are oriented towards production of the pastures needed for animal production, with an average stocking rate of about 1.6 AU ha^{-1} and a form of production characterized by low labor and capital intensity. Exogenous beef prices assure that regional beef production does not influence regional beef prices, and as a consequence no differences are observed between the partial and the aggregate model results. Finally, banana plantations do not use all of their land for banana production, since prevailing price levels do not permit profitable cultivation on inferior soils, while endogenization of product prices leads to slightly lower banana prices and correspondingly larger fallow areas. Not surprisingly, both labor and capital intensities in banana production are high.

The overall effect of aggregation using endogenous prices is a tendency towards lower price levels and less specialization, with a more diversified production of cash crops, while on-farm food production disappears. Consequently, objective values are overestimated for all farm types in the partial base run. In addition, less specialization in highly N-depleting and biocide-demanding cash crop production (mainly plantain and to a minor extent pineapple) leads to slightly lower nitrogen depletion and biocide use levels. It must be noted that these effects take place on the small and medium size farm types, which cover 45% of the cultivated crop area, though only 12% of the total cultivated area in the Atlantic Zone.

In both partial and aggregate model results, soil nitrogen losses are observed. This means that the soil nitrogen resource base is being depleted and that soil productivity will decline over time, and thus the selected land use systems can be qualified as unsustainable. For the time horizon of this study (*i.e.*, up to five years ahead), the consequences of declining productivity due to nitrogen losses will probably not be notable. However, it should be realized that in the long run, model results will probably be different from the ones obtained here. A way to explore sustainable land use options with a longer time horizon is to only use alternative land use systems that are not soil nutrient depleting (*e.g.*, as in Chapter 7).

Regional level

Table 8.6 shows partial and aggregate base run results in terms of total regional land use, total regional production, prices and total regional monthly labor use. While in the partial base run, maize and cassava production were important for household consumption only, both crops disappear in the aggregate analysis. Major shifts are observed in cash crop production, which becomes more diversified than the partial base run simulation and increases (in terms of cultivated area) in comparison with the actual cash crop production (Table 8.1). Consequently, prices of these cash crops decline in the aggregate model simulation.

Table 8.6. Partial and aggregate base run results on the regional level

	Partial base run				*Aggregate base run*			
	Land use[1]	*Production*[2]	*Prices*[3]	*Total*	*Land use*[1]	*Production*[2]	*Prices*[3]	*Total*
Production:								
Maize	251	1.7	0.26	-	0	0.0	0.26	-
Beans	0	0.0	0.64	-	0	0.0	0.64	-
Pineapple	16 337	892.5	0.33	-	12 300	716.2	0.25	-
Banana	37 509	116.1	8.21	-	34 384	108.1	8.13	-
Plantain	13 526	19.4	5.71	-	11 439	16.4	5.62	-
Palm heart	0	0.0	0.45	-	8121	80.4	0.22	-
Cassava	155	1.3	0.42	-	0	0.0	0.42	-
Cattle	193 215	20.0	1.77	-	193 727	20.0	1.77	-
Total	260 993	-	-	-	259 970	-	-	-
Income ($ 10^6 y^{-1}):								
Crops and pastures	-	-	-	1398.8	-	-	-	1205.7
Crops	-	-	-	1363.4	-	-	-	1170.3
Labor use (10^3 d mth^{-1})	-	-	-	966.7	-	-	-	884.6
Δ soil N stock (kg ha^{-1} y^{-1}):								
Crops and pastures	-	-	-	-48.1	-	-	-	-46.4
Crops	-	-	-	-42.1	-	-	-	-35.4
a.i. use (kg ha^{-1} y^{-1}):								
Crops and pastures	-	-	-	10.7	-	-	-	10.4
Crops	-	-	-	38.6	-	-	-	38.3

[1] Land use in hectares.

[2] Production in 10^6 kg of maize, beans and cassava; in 10^6 units of 1.4 kg of pineapple; in 10^6 boxes of banana (18 kg box^{-1}) and plantain (25 kg box^{-1}); in 10^6 units of 1.3 kg of palm heart; in 10^6 kg carcass weight of cattle (beef).

[3] Market prices in $ per kg of maize, beans and cassava; in $ per unit of 1.4 kg of pineapple; in $ per box of 18 kg and 25 kg of banana and plantain respectively; in $ per unit of 1.3 kg of palm heart.

Product price endogenization leads to a slight reduction in the cultivated crop area (-2.3%), while the pasture area shows an insignificant increase (+0.3%). As a result of the lower specialization, labor use declines (-8.5%), while labor requirements can be fully met by the locally available agricultural labor force of 1.0 10^6 labor days per month. Regional agricultural income is overestimated by nearly 14% when prices are assumed exogenous, agricultural income declining from \$ 1398.8 10^6 in the partial base run to \$ 1205.7 10^6 in the aggregate base run.[8] Nitrogen losses as well as (even though to a lesser extent) biocide use levels decrease. Total nitrogen losses in crop production are lower than those in pastures, as crops include sustainable banana activities, which cover 55% of the cultivated crop area, while natural pasture production depletes about 50 kg of N per hectare per year.

8.5.3 Improved infrastructure

Comparing partial base run with partial simulation

A 20% reduction in transaction costs leads to a larger focus on cash crop production in the partial policy simulation (Tables 8.7 and 8.8) as compared to the partial base run (Table 8.6). Small and medium size farms increase their cash crop production (especially pineapple) at the expense of pasture for beef production, since the difference between farm gate sales prices and farm gate purchase prices becomes smaller, while net margins of (especially high-input high-output) cash crops increase. Moreover, the smaller price band reduces the necessity of food production for on-farm consumption. The *hacienda* farm type increases its pasture area at the expense of the agricultural frontier area, due to the higher margins in beef fattening production systems. Banana production remains the same, since transaction costs for banana are already relatively low.

Both total crop area and pasture area increase (7.1% and 5.8% respectively), as a result of higher net margins. The stronger focus on labor-intensive cash crop production leads to an increase in total labor requirements of 1.6%, which is met by an increased use of hired labor as well as by lower off-farm employment of family members. Increased production of cash crops and beef lead to a rise in regional agricultural income from \$ 1398.8 10^6 to \$ 1470.4 10^6 (+5.1%). Neither nitrogen depletion nor biocide use levels are significantly affected by the reduction in transaction costs.

Comparing partial simulation with aggregate simulation

Even though relatively unimportant in terms of cultivated area, comparison of the partial and the aggregate policy simulations again shows that food crop production (important for on-farm consumption) disappears due to the earlier mentioned labor and capital constraints. In addition, significant shifts are observed in the production of both cash crops and (to a lesser extent) beef (Tables 8.7 and 8.8). Due to the relatively low levels of actual pineapple production (Table 8.1), increased pineapple production in the partial policy simulation results in a sharp drop in pineapple prices at the aggregate level of

[8] Agricultural income in the Atlantic Zone in 1996 equaled \$ 900 million, as calculated on the basis of Table 8.1 and the actual prices presented in Table 8.6 (partial base run). The difference in agricultural income in 1996 compared to the partial base run is determined by: (1) the defined production systems that are considerably more productive than actual production systems, (2) factors and limitations that are not incorporated into the model, and (3) the aggregation bias.

Table 8.7. Partial and aggregate policy simulation on the regional level: 20% decrease in transaction costs

	Partial policy simulation				*Aggregate policy simulation*			
	Land use[1]	*Production*[2]	*Prices*[3]	*Total*	*Land use*[1]	*Production*[2]	*Prices*[3]	*Total*
Production:								
Maize	158	1.2	0.26	-	0	0.0	0.26	-
Beans	0	0.0	0.64	-	0	0.0	0.64	-
Pineapple	21 155	1097.2	0.33	-	8734	579.0	0.31	-
Banana	37 509	116.1	8.21	-	37509	116.1	8.03	-
Plantain	13 614	19.5	5.71	-	20236	29.0	5.45	-
Palm heart	0	0.0	0.45	-	232	2.3	0.48	-
Cassava	161	1.4	0.42	-	0	0.0	0.42	-
Cattle	204 330	21.4	1.77	-	215 437	22.1	1.77	-
Total	276 926	-	-	-	282 149	-	-	-
Income ($ 10^6 y^{-1}):								
Crops and pastures	-	-	-	1470.4	-	-	-	1309.0
Crops	-	-	-	1432.4	-	-	-	1270.0
Labor use (10^3 d mth^{-1})	-	-	-	982.0	-	-	-	1012.0
Δ soil N stock (kg ha^{-1} y^{-1}):								
Crops and pastures	-	-	-	-48.2	-	-	-	-52.9
Crops	-	-	-	-42.7	-	-	-	-61.7
a.i. use (kg ha^{-1} y^{-1}):								
Crops and pastures	-	-	-	10.3	-	-	-	10.4
Crops	-	-	-	36.6	-	-	-	41.1

Notes: See Table 8.6.

Table 8.8. Response reactions comparing simulations on the partial and aggregate levels (% change).

	Transaction costs -20%			*Biocide tax +40%*			*Credit accessibility +20%*		
	PBR-PSI[1]	*PSI-ASI*[2]	*ABR-ASI*[3]	*PBR-PSI*[1]	*PSI-ASI*[2]	*ABR-ASI*[3]	*PBR-PSI*[1]	*PSI-ASI*[2]	*ABR-ASI*[3]
Land use:									
Maize	-36.9	-100.0	-	-21.2	-100.0	-	0.5	-100.0	-
Beans	-	-	-	-	-	-	-	-	-
Pineapple	29.5	-58.7	-29.0	4.4	-51.9	-33.3	-0.7	-51.6	-36.2
Banana	0.0	0.0	9.1	-33.3	0.0	-27.3	0.0	-8.3	0.0
Plantain	0.6	48.6	76.9	-70.4	349.4	57.4	17.2	-14.0	19.3
Palm heart	-	-	-97.1	-	-85.0	-70.7	-	-	32.1
Cassava	4.0	-100.0	-	34.6	-100.0	-	4.0	-100.0	-
Cattle	5.8	5.4	11.2	0.8	-0.3	0.3	0.0	4.9	4.6
Total	6.1	1.9	8.5	-1.5	-3.6	-4.7	0.9	2.3	3.5
Income:									
Crops and pastures	5.1	-11.0	8.6	-19.5	-6.5	-12.7	1.3	-14.8	0.1
Crops	5.1	-11.3	8.5	-20.0	-6.7	-13.1	1.3	-15.4	-0.1
Labor use	1.6	4.0	15.4	-29.3	12.4	-13.1	2.9	-13.0	-2.1
Δ soil N stock[4]:									
Crops and pastures	-0.3	-9.7	-13.9	13.6	-30.3	-16.7	-3.3	6.6	-0.0
Crops	-1.4	-44.5	-74.1	65.3	-369.9	-93.5	-14.8	27.5	1.1
a.i. use:									
Crops and pastures	-4.0	1.3	-0.2	-32.1	20.4	-16.1	1.3	-14.3	-10.9
Crops	-5.2	12.1	7.3	-29.8	37.1	-2.8	-0.9	-9.0	-9.0

[1] % difference between the partial base run (PBR) and the partial simulation (PSI).
[2] % difference between the partial simulation (PSI) and the aggregate simulation (ASI).
[3] % difference between the aggregate base run (ABR) and the aggregate simulation (ASI).
[4] A negative response reaction in Δ soil N stock refers to a higher level of N depletion.

analysis, conditional upon the demand elasticity for pineapple faced by producers in the Atlantic Zone. Consequently, on the aggregate level of analysis, pineapple production decreases in order to sustain pineapple prices, which in turn induces a 50% increase in the production of plantain. The latter becomes relatively more attractive in the aggregate run since output prices are hardly affected by the increase in production due to the relatively high level of actual plantain production. Exogenous beef prices assure non-diminishing returns from beef production, and therefore more resources are dedicated towards beef production given the lower net margins in crop production resulting from lower crop prices on the aggregate level of analysis. Although banana prices decrease in the aggregate policy simulation, this price effect is compensated by the reduction in transaction costs and therefore does not result in a decrease in the area devoted to banana cultivation.

Endogenously determined crop product prices lead to a small increase in the cultivated area (+1.9%), which is mainly due to a rise in the pasture area (+5.4%). The increased focus on labor intensive plantain production, in combination with an increase in pastures for beef production, leads to a 4.0% rise in the demand for agricultural labor, despite a 8.0% decline in crop area. Thus, total labor requirements in the aggregate policy simulation exceed the regional agricultural labor availability by about 2%, leading to the import of labor from outside the Atlantic Zone. An extension of the aggregate model could therefore consist of an endogenization of wages. Regional agricultural income is upwardly biased in the partial policy simulation, and decreases by 11.0% to \$ 1309.0 10^6 in the aggregate policy simulation. Finally, nitrogen depletion as well as biocide use are considerably higher in the aggregate policy simulation, and can be explained by the increased importance of plantain production.

Comparing aggregate base run with aggregate simulation

On the aggregate level of analysis, a 20% reduction in transaction costs leads to increased cash crop production, with the latter oriented towards plantain instead of pineapple. Plantain production is characterized by high input costs as well as high output values. The latter are also relatively stable at increasing production levels given the relatively low demand elasticity for plantain. Similarly, lower input prices in combination with a fixed output price, favors pasture-based beef production. Contrary to the partial simulation, banana production increases in the aggregate model simulation as the reduction in transaction costs is now sufficiently high, given the higher marginal production at lower levels of production.

Because of endogenous prices, net margins in crop production are less affected by a decrease in transaction costs when analyzed on the aggregate level than when analyzed on the partial level, as higher production levels result in lower output prices. Therefore, the area under crops hardly changes, while the pasture area increases by 11.2%. The orientation towards labor-intensive plantain and banana production results in a 15.4% rise in the demand for agricultural labor, and also explains the sharper rise in agricultural income in the aggregate simulation (8.6%) as compared to the partial policy simulation. Moreover, both biocide use and (especially) nitrogen depletion levels rise sharply.

8.5.4 Taxation of biocides

Comparing partial base run with partial simulation

A 40% tax on the price of biocides leads to a decline in crop area (-8.0%) in the partial policy simulation (Tables 8.9 and 8.8) compared to the partial base run (Table 8.6). On the other hand, the increase in pasture area is insignificant (0.8%), as an effect of lower net margins that limit the use of both hired labor and formal as well as informal credit. Production of biocide-intensive crops (banana and plantain) decreases, in favor of crops that are less biocide-intensive (local pineapple and palm heart) and pastures. Production of food crops remains about equal, in view of their relatively biocide-extensive production technologies. Similarly, pasture production is hardly affected by the biocide tax, since *i*) biocide use in pasture production is extremely low, and *ii*) there exist sufficient substitution options between crops, preventing a move of resources to pasture production.

Table 8.9. Partial and aggregate policy simulation at the regional level: 40% tax on biocide use

	Partial policy simulation				*Aggregate policy simulation*			
	Land use[1]	*Production*[2]	*Prices*[3]	*Total*	*Land use*[1]	*Production*[2]	*Prices*[3]	*Total*
Production:								
Maize	198	1.8	0.26	-	0	0	0.26	-
Beans	0	0.0	0.64	-	0	0	0.64	-
Pineapple	17 049	926.1	0.33	-	8202	544.7	0.31	-
Banana	25 060	82.5	8.21	-	25 006	82.5	8.44	-
Plantain	4005	5.7	5.71	-	18 001	25.8	5.55	-
Palm heart	15 903	157.5	0.45	-	2379	23.6	0.32	-
Cassava	208	1.4	0.42	-	0	0	0.42	-
Cattle	194 752	20.1	1.77	-	194 241	20.1	1.77	-
Total	257 121	-	-	-	247 829	-	-	-
Income ($ 10^6 y^{-1}):								
Crops and pastures	-	-	-	1126.6	-	-	-	1053.1
Crops	-	-	-	1091.1	-	-	-	1017.6
Labor use (10^3 d mth^{-1})	-	-	-	683.9	-	-	-	768.6
Δ soil N stock (kg ha^{-1} y^{-1}):								
Crops and pastures	-	-	-	-41.6	-	-	-	-54.2
Crops	-	-	-	-14.6	-	-	-	-68.6
a.i. use (kg ha^{-1} y^{-1}):								
Crops and pastures	-	-	-	7.3	-	-	-	8.7
Crops	-	-	-	27.1	-	-	-	37.2

Notes: See Table 8.6.

The shift from highly labor intensive (banana and plantain) to less labor intensive (pineapple and palm heart) cash crop production in combination with a decrease in total crop area, results in a decrease in labor use of nearly 30%. The decrease in total crop area in combination with a more diversified production plan, results in a decline in regional agricultural income of nearly 20%, to \$ 1126.6 10^6. The biocide tax efficiently reduces biocide use in crop production, as evidenced by a decrease in biocide use (in terms of a.i.) of about 30%. At the same time, nitrogen losses in crop production are reduced to about one-third of the nitrogen losses observed in the partial base run, since pineapple and palm heart production are characterized by relatively low biocide use and nitrogen depletion levels.

Comparing partial simulation with aggregate simulation

A comparison of the partial and the aggregate policy simulations shows a more than quadrupled plantain production at the expense of (biocide intensive) export pineapple and palm heart production. Low initial levels of actual pineapple and palm heart production (Table 8.1) do not allow for a large increase in production without a significant drop in their prices. Lower equilibrium prices in combination with the biocide tax lead to less favorable net margins for these crops than can be obtained from plantain production, which is also less biocide intensive than export pineapple production. Food crop production is once again abandoned as product prices are endogenized, largely because of labor and capital constraints. Since the projected banana production falls below the level of actual banana production, the price of banana increases in the aggregate as compared to the partial policy run. However, since the increase in the price of banana is not sufficient to compensate for higher input prices resulting from the 40% biocide tax, banana production remains unchanged.

Endogenous pricing of products leads to a 14% decline in the cultivated crop area, while hardly affecting the pasture area. The substantial increase in the plantain area leads to a more than 12% increase in labor demand which, however, does not exceed the regional agricultural labor availability. Endogenization of product prices in the aggregate policy run shows, once again, that the regional agricultural income is overestimated in the partial simulation. Agricultural income in the latter is upwardly biased by about 6.5%. In addition, the efficiency of a biocide tax in inducing lower biocide application levels is shown to be overestimated when product prices are assumed exogenous, while nutrient depletion levels are underestimated.

Comparing aggregate base run with aggregate simulation

The biocide tax policy simulation performed on the aggregate level of analysis leads to a shift towards plantain production at the expense of (biocide intensive) export pineapple and palm heart. Low equilibrium prices for export pineapple and palm heart in combination with higher production costs (due to higher biocide prices), result in lower net margins and subsequent lower levels of production. In contrast with export pineapple, less biocide intensive plantain production is favored since net margins of plantain production are less affected by the biocide tax. Moreover, plantain equilibrium prices are relatively stable and high. Similar to the results obtained in the partial simulation, pasture production is hardly altered by the biocide tax, due to the low biocide requirements of pasture production. The biocide tax is efficient in reducing biocide intensive banana production, although less so than in the partial policy simulation.

The shift towards the highly capital and labor intensive plantain production results in a decline in total cultivated area (-4.7%) due to capital and labor constraints, while total labor demand decreases by about 13%, as does agricultural income. Response reactions in terms of labor use and agricultural income are smaller than in the partial simulation, as changes in the production plan are dampened by market-determined product prices. The biocide tax proves to be far less efficient in the aggregate than it is in the partial simulation (a.i. use decreases by 16% instead of 32%), while the increased importance of plantain almost doubles nitrogen depletion.

8.5.5 Improved credit availability

Comparing partial base run with partial simulation

Compared to the partial base run results (Table 8.6), a 20% rise in credit availability results in increased cash crop production (Tables 8.10 and 8.8). Unchanged relative prices and relaxed capital constraints mainly favor plantain production on small and medium size farm types, at the expense of the fallow area, while pineapple and food crop production hardly change. The *hacienda* farm type does not alter its pasture production for livestock fattening, as credit availability is not a constraining factor in view of the sufficient collateral availability (land and cattle). Banana production remains constant as banana mother companies figure as capital suppliers of the farm level operations and the supply is assumed to be unlimited (Section 8.3.4).

Table 8.10. Partial and aggregate policy simulation on the regional level: 20% increase in credit availability

	Partial policy simulation				*Aggregate policy simulation*			
	Land use[1]	*Production*[2]	*Prices*[3]	*Total*	*Land use*[1]	*Production*[2]	*Prices*[3]	*Total*
Production:								
Maize	252	1.8	0.26	-	0	0.0	0.26	-
Beans	0	0.0	0.64	-	0	0.0	0.64	-
Pineapple	16 226	888.0	0.33	-	7848	466.4	0.35	-
Banana	37 509	116.1	8.21	-	34 384	108.1	8.13	-
Plantain	15 858	22.7	5.71	-	13 644	19.6	5.60	-
Palm heart	0	0.0	0.45	-	10 728	106.4	0.17	-
Cassava	161	1.4	0.42	-	0	0.0	0.42	-
Cattle	193 212	20.0	1.77	-	202 586	21.4	1.77	-
Total	263 217	-	-	-	269 189	-	-	-
Income ($ 10^6 y^{-1}):								
Crops and pastures	-	-	-	1416.2	-	-	-	1206.7
Crops	-	-	-	1381.0	-	-	-	1168.8
Labor use (10^3 d mth^{-1})	-	-	-	994.8	-	-	-	865.8
Δ soil N stock (kg ha^{-1} y^{-1}):								
Crops and pastures	-	-	-	-49.7	-	-	-	-46.4
Crops	-	-	-	-48.3	-	-	-	-35.0
a.i. use (kg ha^{-1} y^{-1}):								
Crops and pastures	-	-	-	10.8	-	-	-	9.3
Crops	-	-	-	38.3	-	-	-	34.8

Notes: See Table 8.6.

The cultivated crop area increases slightly (0.9%), mainly due to increased labor-intensive cash crop production (especially plantain). As a result, total labor requirements increase by 2.9%. Higher labor requirements are met by hired labor, made possible by the increased availability of capital. Increased cash crop production at constant output prices leads to a small rise in agricultural income (1.3%). However, the increased importance of nutrient-intensive plantain production results in a nearly 15% increase in nitrogen depletion in crop production.

Comparing partial simulation with aggregate simulation

Comparison of the partial and the aggregate policy simulations again reveals a shift towards more diversified cash crop production. Production of plantain and (especially) pineapple is reduced in favor of palm heart, since prices of the former crops fall as a result of increased production levels in the aggregate policy simulation. On the other hand, food crop production for on-farm consumption disappears as a result of labor and capital constraints. Endogenization of product prices does not affect world market-determined (exogenous) beef prices, and consequently beef production increases given the lower net margins in crop production at lower equilibrium product prices. Similar to the base run situation, banana production diminishes due to a decline in banana prices on the aggregate level of analysis.

Endogenization of product prices shows that the cultivated area is slightly (2.3%) underestimated in the partial policy simulation. This is mainly accounted for by pasture area, which is 5% lower in the partial simulation than it is in the aggregate simulation. A more diversified production pattern of cash crops, including a decreased importance of labor-intensive plantain and pineapple production in favor of labor-extensive palm heart production, leads to a 13% decrease in labor demand. Similarly, both nitrogen depletion and biocide use are overestimated in the partial simulation, since palm heart production is less nitrogen depletive and requires less biocides than pineapple and (especially) plantain production do. Compared to the partial simulation, nitrogen depletion and biocide application levels in crop production are shown to be 27.5% and 9.0% lower on the aggregate level of analysis. Finally, regional agricultural income is also overestimated in the partial simulation, being nearly 15% lower in the aggregate policy simulation.

Comparing aggregate base run with aggregate simulation

Compared to the aggregate base run, a 20% increase in credit availability induces a rise in plantain as well as palm heart production at the expense of pineapple production. Given relative equilibrium prices, plantain and palm heart are the most profitable crops, whose production is increased as the capital constraint is relaxed. Contrary to the partial simulation, pasture production on small and medium size farm households is increased, as net margins from beef production are constant while net margins from crop production depend on equilibrium product prices. Banana production is not affected as capital is not considered a constraining factor in the base run situation.

Pasture area increases by 3.5%, while the crop area remains about unchanged (+0.5%) as resources are shifted from pineapple to plantain and palm heart production. In turn, labor requirements decline by 2.1%, while agricultural income rises by only 0.1% due to increased production levels and correspondingly lower product prices. Agricultural income obtained from crop production decreases by 0.1%. This is mainly because small and medium size farm types maximize their individual utility and full income objectives, using the larger capital availability to hire labor while using their own off-farm labor for the generation of income as well as leisure. Finally, nitrogen depletion remains about constant due to the increased relative importance of palm heart production, which is characterized by relatively low nitrogen depletion levels.

8.6 Conclusions and discussion

This chapter presents a methodology for incorporating equilibrium equations for product markets in regional agricultural land use models, which also include farm characteristics. Representative farm type models are developed for four different farm types (*i.e.*, small farms, medium farms, extensive livestock farms, and banana plantations), each of which is modeled according to its specific objectives, production possibilities and resource restrictions. The competition among farm types is modeled through the endogenization of price formation in product markets. Production possibilities are offered to each farm type model in the form of technical coefficients for crop, pasture and livestock activities. The latter are generated by separate models and include economic, production and sustainability parameters.

In the base run, model simulations performed on the farm level (*i.e.*, with fixed prices) indicate an orientation towards the production of cash crops and livestock, in combination with some basic grain production for farm household consumption. Base run results with endogenous product prices, however, point towards significantly more diversified production patterns, induced by the introduction of product market constraints. Cash crop production becomes more diversified in number as well as in area, while pasture production for livestock fattening becomes relatively more important because of exogenous beef prices determined on the world market. Banana prices decline due to a supply increase, resulting in a termination of banana cultivation on inferior soils. Overall for the base run, the endogenization of product prices leads to more diversified production patterns, with lower prices and resource use. In turn, regional agricultural income decreases, and thus is overestimated when prices are assumed fixed.

Policy simulations include a 20% decrease in transaction costs, a 40% tax on the price of biocides, and a 20% increase in credit availability. Under a fixed price regime, the 20% decrease in transaction costs as well as the 20% rise in credit availability induce larger and less diversified cash crop production, increased resource use and a relatively small rise in agricultural income. Compared to the credit scenario, growth in agricultural income is about four times higher in the transaction costs scenario, while both policies are accompanied by increases in both nitrogen depletion and biocide use. On the other hand, results obtained with the GOAL approach (Chapter 7) suggest that much higher income growth in combination with lower biocide use result from a decrease in transaction costs.

Similar to the results obtained by the REALM approach (Chapter 6), a biocide tax induces a shift towards the cultivation of relatively biocide-extensive crops.

Relaxing the assumption of exogenous product prices indicates that options for specialization in the most profitable products or those requiring fewer biocides are overestimated. Due to price adjustments, production plans remain more diversified, which in turn dampens response reactions. The growth of agricultural income is overestimated for the transaction cost and credit policy scenarios, while income losses from the biocide tax are underestimated. Moreover, the biocide tax proves less efficient in reducing biocide use when prices are determined endogenously.

To summarize, this chapter shows that not taking account of aggregate product demand and supply relationships in regional agricultural policy analysis leads to an overestimation of the degree of specialization in agricultural production. Response reactions are overestimated, while income effects are inflated. Taking account of aggregate product demand and supply relationships in product markets thus leads to improved analysis of the likely effects of agricultural policy measures.

9 Linking information technology and farmer knowledge in a decision support system for improved banana cultivation

JETSE J. STOORVOGEL, ROMANO A. ORLICH,
RONALD VARGAS, and JOHAN BOUMA

Abstract

Environmental and market constraints are increasingly restraining the banana sector in Costa Rica. This chapter reports on the development of a decision support system to enhance a more environment-friendly and economically viable production of banana. The system has been developed in close cooperation with a banana producer and the National Banana Corporation (CORBANA; *CORporación BAnanera NAcional*). The system is based on the principles of precision agriculture, a practice that is being promoted as an alternative form of land management resulting in an economically feasible and environmentally friendly type of agriculture. In precision agriculture spatial and temporal variation in cropping conditions govern farm management. The objective is to obtain maximum quantities of high quality produce without exceeding threshold values of certain environmental indicators. Although precision agriculture is often based on advanced equipment, information technology and deterministic simulation models, in tropical environments and for perennial crops advanced equipment and simulation models are not always available. The decision support system that has been developed is therefore based on information technology in combination with the knowledge of the owner and manager of the farm. Yield maps are being created and linked to other available information for the plantation, including detailed soil maps. Next, areas are identified that have a relatively low production compared to their potential. On the basis of additional analysis, crop management for those sites is modified.

9.1 Introduction

The playing ground for agricultural systems is constantly getting smaller. While economic margins may be declining as a result of, *e.g.*, the globalization of markets, environmental constraints are getting tighter. As a result, there is an urgent need to develop alternative agricultural management systems that comply with the changing economic and environmental conditions. To support the development of alternative management practices, new techniques are being developed. Prototyping, for example, is a recently introduced technique to derive alternatives for integrated and ecological farming systems (Vereijken, 1997). Rapid developments in Information and Communication Technology also open new roads to the definition of new management systems. Precision agriculture is an alternative form of agriculture that has been a direct result of such developments (*e.g.*, global positioning systems, Geographical Information Systems (GIS) and simulation models). Precision agriculture is based on the concept that crop performance and input use efficiency can be increased by matching crop management with the spatial and temporal variation in soil conditions and crop requirements. High-tech applications of precision agriculture, *i.e.*, applications that rely heavily on modern information and communication technology, are not feasible for many production systems as they involve high initial investments and a certain level of know-how. At the same time, some of the tools are only available or applicable for a limited number of crops. Simulation models, for example, are not available for most perennial tropical crops, and yield mapping is currently only operational for grain crops. Nevertheless, precision agriculture can also be applied in a context where only less sophisticated technologies are available. The techniques will have to be adapted to local conditions in which, for example, simulation models are replaced by farmer knowledge as the mechanism by which crop performance is interpreted.

This chapter presents a low-tech approach to precision agriculture that has been developed for a banana plantation in Costa Rica. Banana plantations in Costa Rica have to deal with decreasing economic returns and at the same time are pressured to increase input use efficiency. In precision agriculture, many techniques are used that are not applicable to the very specific management system of banana plantations: harvesting occurs on a weekly basis, machinery is almost absent since the system relies almost entirely on manual labor, and crop growth simulation models are not (yet) available. Nevertheless, there is an increasing call for the development of alternative management practices that decrease the use of external inputs and make the production system more efficient (see Section 2.9). Information and Communication Technology allows us to quantify the spatial and temporal variation in crop performance and link that to soil information. It can be expected that farm managers will be able to achieve higher input use efficiency if they have insight into this spatial and temporal variation.

9.2 The Costa Rican banana sector

9.2.1 History and importance

In Costa Rica, the commercial production of banana started at the end of the 19th century along the railway between San José and Limón from where, in 1880, the first bananas were exported (Pardo, 1984; see also Section 2.5). Since then the area of banana cultivation has grown almost continuously, except for the temporary set backs resulting from diseases, world market conditions and political developments. In the 1930s the Panama disease increasingly affected the crop and production shifted to the less-affected Pacific coast of Costa Rica. With the introduction of the resistant *Gran Cavendish* variety, banana production returned to the Atlantic Zone in 1956. Since then the banana area along the Pacific coast gradually decreased. In the early 1990s the area in the Atlantic Zone of Costa Rica almost doubled to over 50 000 ha. New plantations were established north of the traditional banana belt in an area that was previously considered unsuitable for banana due to drainage problems (see Figure 2.5). These problems were resolved by introducing intensive drainage control.

Since the beginning of the century, input use increased from almost none to the current high use of such external inputs as fertilizers, insecticides, fungicides, etc. The call from both consumers and environmentalists to decrease external inputs is becoming increasingly stronger but at the same time the pressure of pests and diseases requires the use of biocides to control, for example, the black Sigatoka fungus disease and nematodes. To maintain soil fertility, fertilizers need to compensate for nutrient losses through nutrient removal in the harvested products.

Together with coffee, banana has traditionally dominated Costa Rican agricultural exports (Section 2.7). Banana production currently accounts for over a third of total agricultural exports with a total value of over $ 600 million, corresponding to over 100 million boxes of 18.14 kg each or 5.5 10^6 t (Loaiza and Zuñiga, 1999).

9.2.2 Current banana management

In Costa Rica, bananas are cultivated in a continuous production system. Due to climatic differences, production varies throughout the year with a minimum around February after a relatively cold period and a maximum around September. The production of commercial plantations varies between 1500 to 3400 boxes of 18.14 kg ha^{-1} y^{-1} (27.2 t ha^{-1} to 61.7 t ha^{-1}). To maintain this production, it is necessary to make intensive use of external inputs to maintain soil fertility and to control pests and diseases. Fertilizer recommendations for banana in the AZ of Costa Rica vary little and in principle differ only with the volcanic character of the parent material. Total fertilization comprises approximately 500 kg N, 145 kg P and 650 kg K ha^{-1} y^{-1}, corresponding to an application of 100 kg mineral fertilizer per fortnight per ha. Several plantations replace one or more applications of mineral fertilizer with organic fertilizer (*e.g.*, with enriched chicken manure).

To control nematodes, nematicides are being applied routinely at four or six months cycles. Available nematicides are alternated frequently to prevent rapid degradation by soil microflora or resistance build-up by the nematodes (Gowen, 1995). Frequent aerial spraying with a mixture of fungicides (again to avoid resistance build-up) controls fungus diseases (black Sigatoka). On sites with high weed pressure, occasionally herbicides are sprayed, although manual weeding is increasingly substituting chemical weed control. Except for aerial spraying, all management operations are carried out using manual labor. The set-up of banana plantations does not allow for the use of machines to apply fertilizers and/or biocides.

Frequently, laborers check the plantation for bunches that are ready to be harvested. Bunches are transported via a cable system running throughout the plantation towards the packing plant for further processing. Banana plantations are sub-divided in blocks that, for their harvest, correspond to a specific cable. These blocks measure between 3 and 10 ha and are the basic units for farm management. Soto (1994) presents a detailed description of banana management in Costa Rica.

9.2.3 Environmental effects of banana production

In Costa Rica, banana plantations have been criticized for their negative environmental effects (Hallam, 1995). Partially this is based on historical records. Between 1930 and 1950, farmers applied 100-150 kg copper ha^{-1} y^{-1} to combat the fungi infections, a practice that has led to toxic copper concentrations in soils (López and Solis, 1992).

More recently in the 1970s, banana workers experienced serious health problems as a result of improper nematicide management (Ramírez and Ramírez, 1980). In addition, residues of biocides were found in sediments and water (Dinham, 1993). Little is known in relation to the environmental effects of current biocide and fertilizer use. The discussion about the presumed environmental damage is also due to the location of the banana belt upstream from a National Park and a large forest reserve (see also Figure 2.5). However, pollution related to the regular use of biocides and fertilizers is probably limited due to the diluting effect of the perhumid climate with rainfall ranging from 6000 mm y^{-1} upstream of the plantations to 3200 mm y^{-1} in the lower parts of the AZ (Stoorvogel and Eppink, 1995). Nevertheless, pesticides may build up in the fauna. Therefore, high external inputs deserve continuous attention and their use should be minimized as much as possible.

9.2.4 Tools to reduce the environmental effects of banana production

In many cases, environmental effects related to agriculture occur off-farm, leading to so-called externalities. As a result, there often is little or no incentive for farmers to improve their cropping system. Conflicts between environmental quality and agricultural production have grown steadily and concerns related to agro-chemicals are at the core of the sustainability debate. Falconer (1998) lists a number of significant challenges to resolve environmental problems given their invisibility, complexity and potentially geographically undefined ecological effects. Different tools have been developed to gain insight into the relations between the socio-economic environment and land use decisions (Stoorvogel *et al.*, 1998). Some of these tools allow policy makers to get acquainted with the driving forces behind land use and land management changes (*e.g.*, CLUE as described in Chapter 3). Other tools are being developed that allow for the analysis of the trade-offs between agriculture and the environment (*e.g.*, Chapters 5-8; Crissman *et al.*, 1998). These tools allow policy makers to explore policy alternatives and to evaluate their policies in an *ex-ante* manner and check whether they are likely to reduce an unfavorable environmental impact and safeguard food security. Many of these tools incorporate production technologies that are currently known to exist. Other methods and tools are required when it is desired to design whole new technologies, aimed at increased input use efficiency. Such tools should allow on-farm designing, testing and implementing of changes in technology and management that result in less environmental pollution and increased economic returns. Ultimately, the use of such tools should result in *alternative* production technologies that can be effectively used in complying with agricultural policy goals (*e.g.* Chapters 5-8).

9.3 A decision support system for precision agriculture in banana management

This study focuses on the farm level and the development of management alternatives with increased input use efficiency in an attempt to reduce negative externalities and increase productivity of banana plantations in the AZ of Costa Rica. For alternative management systems to be successful, it is necessary to satisfy a number of conditions:

1. To assure a high adoption among banana producers, the initial investment to change to an alternative management system should be relatively low and the system should be economically feasible.
2. The current state of agro-ecological know-how related to the banana crop does not allow for the use of simulation models to analyze the interaction between crop growth and environment. As a result a more participatory approach should be pursued based on expert knowledge.
3. Given the fact that the banana crop does not allow extensive mechanization (except for aerial spraying of fungicides), alternative management systems should be based on the continuing use of manual labor.

Positive experiences with precision agriculture obtained in developed country's agriculture, in the sense of being an environment-friendly alternative that is economically viable (CIBA foundation, 1997), makes precision agriculture an interesting concept to be elaborated for banana. However, the high-tech approaches that are typically being followed for precision agriculture need to be adapted to deal with the above conditions.

9.3.1 Precision agriculture in banana cultivation

Precision agriculture has not been defined in a unique manner and its definition varies from very general ones covering all agriculture making use of information technology (NRC, 1997) to very specific ones dealing with some aspects of precision agriculture only. In general, precision agriculture comprises three major steps:

1. The description of the spatial and temporal variation in soil conditions and crop performance.
2. The translation of (time and location) specific conditions in crop and soil into management recommendations.
3. Farm management that is site and time specific and "optimal" for the prevailing soil and crop conditions.

In most applications of precision agriculture high-tech solutions are chosen (*e.g.*, Robert *et al.*, 1996; CIBA foundation, 1997) for each of the three steps: 1) yield monitoring equipment and sensors are used to register yields and/or soil conditions on-the-go, 2) crop growth simulation models function as a translator for the observations in step 1 into management recommendations, and 3) site-specific management is applied at a certain point in time by using, for example, variable-rate fertilizer spreaders.

Despite the fact that the high tech-approach can not be applied to many tropical crops, farmers in the tropics tend to manage their fields in a site-specific manner. For example, it has been observed that some West African farmers use spatial variability in their fields to reduce risk (Brouwer and Bouma, 1997). Their management is a form of precision agriculture, involving the same three steps. Farmers observe differences in yields and soil conditions and then try to manage them, in this case by a site-specific application of manure. The translation from observation to management is achieved mainly through the farmer's expert knowledge.

An alternative approach to precision agriculture is presented in Figure 9.1. This approach deals with crops about which relatively little is known and enables the identification of problem areas, *i.e.,* areas that present a low productivity given soil and weather conditions. Farmer knowledge is the basis for management of the problem areas, the directives of which are performed by manual labor. In the case of banana plantations, the relatively expensive global positioning system technology necessary for precision agriculture can be replaced by using the cable system for geo-referencing. The cable system divides the plantation into blocks of 2-10 ha (depending on the plantation). If the arched cable supports are numbered, this further divides the area into blocks of 10x100 m (except for the beginning and the end of a cable where larger blocks of approximately 0.3 ha can be found).

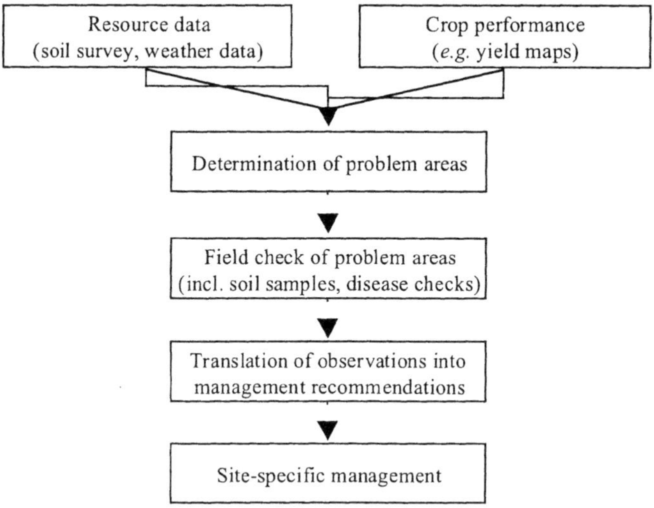

Figure 9.1. A low-tech approach to precision agriculture.

9.3.2 Quantification of variation

Variation in soil resources and crop performance needs to be quantified as a basis for providing variation in crop management. Typically, soil variation is described by a soil survey. Although soil surveys exist for all Costa Rican banana plantations, they are of limited use due to the lack of spatial resolution and of detailed soil descriptions. The original information obtained by augering is in most cases lacking and has been generalized into a soil suitability classification for banana. Therefore, we propose that plantations intending to use precision agriculture carry out a new soil survey with sufficient spatial resolution providing an accurate basis for future management decisions. The soil survey data are stored in digital format using a GIS. Soil differences are quantified by analyzing the different soil layers identified during the survey for their chemical and physical properties. Standard classification procedures are often not appropriate for a specific application because they are based on genetic, soil forming processes that do not necessarily reflect the current potential for the banana crop. By storing the original information, obtained by augering, in a GIS environment, soils can be classified according to different criteria, and the resulting soil data base makes it possible to construct alternative maps. In addition, the data base allows for the evaluation of variation within the soil units, a detail that is overlooked in studies using generalized soil units.

Besides the soil survey, it is essential to monitor crop performance both over time and across space. Yield maps are generated through site-specific yield monitoring. Yield monitoring is increasingly being applied for grain crops where grain-flow monitors in conjunction with global positioning systems are installed on combines, resulting in detailed yield maps (see, *e.g.*, Robert *et al.*, 1996). For other crops, however, yield monitoring is not (yet) being applied because registering systems are lacking. Experiments have, however, revealed that yields in other crops are similarly variable (*e.g.*, Brouwer and Bouma, 1997; Verhagen *et al,* 1995). Banana plantations registering yields per cable found yields to be highly variable, although the cables cover relatively large areas and include a substantial amount of soil variability. However, a methodology for such site-specific yield monitoring was not available and had to be developed.

Depending on weather conditions, specific locations are checked for bunches with bananas with the proper grades approximately once a week. Selected bunches are harvested and transported manually to a grid of cables traversing the plantation. When 20-30 bunches are harvested, they form a "train" that is pulled to the packing plant, where the bananas are processed and packed. To enable yield mapping, the area along the cable is harvested sequentially and the point where the last bunch is harvested is registered. Bunches in a particular train originate from the area between the last bunch of that train and the last bunch of the previous train (or the beginning of the cable, if it is the first train). A weighing device is placed in the cable where the bunches enter the packing plant, registering the following variables for each train:

- location of the last bunch in terms of the cable and support number,
- number of bunches,
- average weight of the bunches,
- minimum and maximum weight, and
- standard deviation of the weight of the bunches.

The yield units, *i.e.,* the spatial units that supply bunches to one specific train, vary in size and location between harvests. To compare and aggregate yields on different days, it is necessary to disaggregate the data into standard "yield registration units" (YRU). The stable YRUs are spatially defined and measure approximately 20 x 100 m. Disaggregation is done by assuming that the yield of a particular train is a good estimate for the central YRU in the yield unit. Subsequently, a surface trend through the YRUs is estimated and corrected for the total measured yield of the cable. This procedure generates yield data per YRU that can easily be aggregated over time, whereas the spatial units are now constant.

9.3.3 Identification of problem areas

The spatial variation includes a limited number of discrete soil units and an almost continuous varying yield throughout the YRUs in the plantation. To facilitate interpretation of yield maps, so-called problem areas are identified, on which production is below the soil potential. As a first step, the average yield per soil unit is calculated for a specific yield map. YRUs include one or more soil types. The yield in a YRU (Y_{YRU} in kg ha^{-1}) is the weighed average of the production of the different soil types in the YRU ($Y_{S,YRU}$ in kg ha^{-1}):

$$Y_{YRU} = \frac{\sum_{S=1}^{m} (A_{S,YRU} \times Y_{S,YRU})}{A_{YRU}}$$

where $A_{S,YRU}$ is the area of soil type S within the YRU (in ha), and A_{YRU} the total area of the YRU (in ha). For each yield map, we have a large number of measured data that can be used to estimate the mean yields for the different soil types ($\overline{Y_S}$). This can be done by solving the optimization problem where the mean yields for the different soil types are estimated so that

$$\sum_{YRU=1}^{n} (E(Y_{YRU}) - Y_{YRU})^2 \text{ is minimal.}$$

$E(Y_{YRU})$ is the expected yield for a YRU on the basis of the soil distribution within a YRU and the average yields for the different soil types:

$$E(Y_{YRU}) = \frac{\sum_{S=1}^{m} (A_{S,YRU} \times \overline{Y_S})}{A_{YRU}}$$

By subtracting the map with the expected yields from the actual yield map, a new map is created that represents the deviation from the expected yields. This map can be classified according to the criteria for problem areas (for example: >20% below the average yield). Likewise, it may be interesting to identify the areas that perform above average, and check the reason for their excellence. This can be done following the same procedure but with different classification criteria (for example: YRUs with a yield of >20 % above their expected yield).

9.3.4 Management alternatives for problem areas

There are many possible reasons for the low productivity of problem areas. Given the fact that we do not have simulation models available to analyze crop performance, we have to rely on expert knowledge. The reasons for the low productivity can not be derived from yield maps. Generally, the different soil types explain only part of the soil variation. As a result, a low productivity in part of a soil unit can be caused by soil variability within that unit. The GIS data base allows us to link the yield maps to the soils data base constructed by recording the individual augerings.

If low productivity cannot be explained by soil variation, a temporal analysis may reveal whether low productivity in the problem area is consistent over time. Farm management might want to focus its attention on areas that have exhibited low productivity for a longer period. If the low productivity is consistent, a field visit becomes necessary. Simple field criteria like, *e.g.*, the vigor of the plants, leaf color or fallen plants with a damaged root system, are good indicators of problems related to, *e.g.*, a specific soil nutrient, nematodes or fungi. As a final step, areas with a consistently low productivity that do not present any visual signs of bad performance can be sampled for more in-depth laboratory analysis to determine possible causes.

9.3.5 A decision support system for banana management

To enable farm management to apply the principles of precision agriculture as described above, a data base management system needs to be developed that registers soil and production data. Given the spatial character of most data, it is logical that a decision support system should be based on a GIS. User-friendliness is a first requisite for the decision support system. The system should enable farm managers to consult the soil map, to generate yield maps and, finally, to identify problem areas. In a second phase, the system should be able to store management operations and enable farmers to retrieve past management decisions (as a possible cause for problem areas). This type of alternative management system can only be implemented on farms if the proper institutional structure is present. Soil surveys, digitalization of information and the interpretation of complex data sets require specific knowledge that is not likely to be present on a banana plantation, nor is specific computer hardware and software likely to be found there. Therefore, a decision support system should also be strongly supported by an institution (*e.g.*, an extension service or, for the specific Costa Rican case, CORBANA). The institution can take care of the soil survey, digitalization of information and backstopping. The farmer, in turn, interprets the information and results on the basis of his specific knowledge of the plantation and its crop.

9.4 Application of precision agriculture at the Rebusca banana plantation

A decision support system for banana management (called BanMan) has been developed and implemented on the Rebusca plantation. The Rebusca plantation (84°01'E, 10°28'N) covers an area of 107 ha and was established in 1991, during the period of expansion of the banana area in the AZ of Costa Rica. A low-tech approach was followed to develop a prototype for banana management. The resulting decision support system allows farm managers to quantify and analyze spatial and temporal variability of the yields on the plantation. Through a consecutive analysis of the results, the system is able to identify the so-called problem areas and analyze temporal trends.

9.4.1 Soil survey

Soil variability on the plantation was described through a detailed soil survey (1:5000) with 454 borings on a 50 m grid (Figure 9.2). In contrast to traditional surveying practices, soil horizons were classified into functional horizons and not into the more traditional genetic horizons. Functional horizons consist of combinations of genetic horizons with similar behavior, *i.e.*, the combined horizons have similar physical and chemical properties (Finke, 1993). In this way, the spatial variability of soil properties is characterized by spatial variation in the thickness of the horizon and the intrinsic variability of the properties of each functional horizon (Finke *et al.*, 1992). The augerings describe the location of the functional horizons in a three dimensional space. Functional horizons were sampled in triplicate for soil chemical and physical analysis. The number of functional horizons is limited and, as a result, the number of required analysis is significantly reduced (compared to traditional surveying techniques).

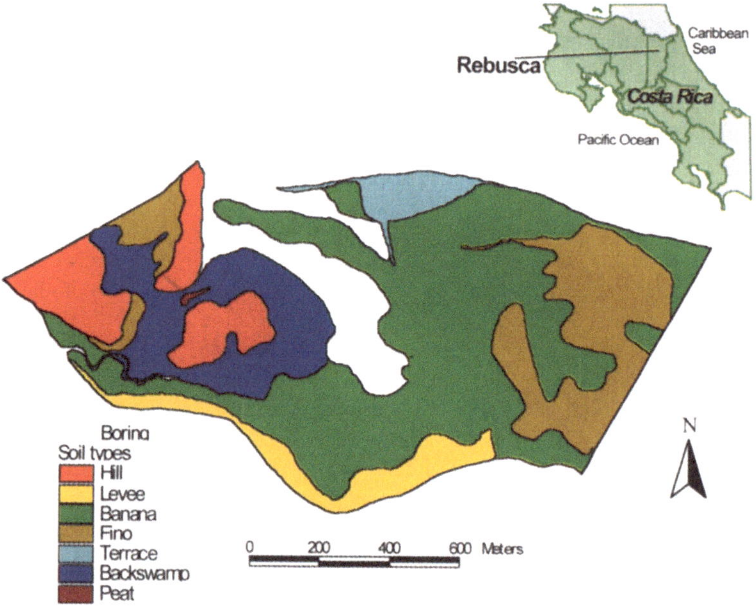

Figure 9.2. 1:5000 Soil map of the Rebusca plantation.

9.4.2 Yield mapping

The farm is subdivided in 450 YRUs for each of which yield is determined. The units vary in size between 0.1 and 0.9 ha. BanMan combines the data of all the trains entering the packing plant and disaggregates the data with respect to the YRUs. Subsequently, BanMan creates a yield map. Yield maps can be created with different temporal resolutions. In most cases farm management works with 4-week periods and yield maps are created for the same period. Spatial resolution is of similar importance. Traditionally yields were registered for the whole farm or per cable. Currently, measurements for yield maps are based on the YRUs. The effect of the measurement scale is demonstrated in the yield maps presented in Figure 9.3. Clearly, the measurements for the whole farm or per cable are too crude for site-specific management. However, the smaller management units are perhaps too detailed from a management point of view.

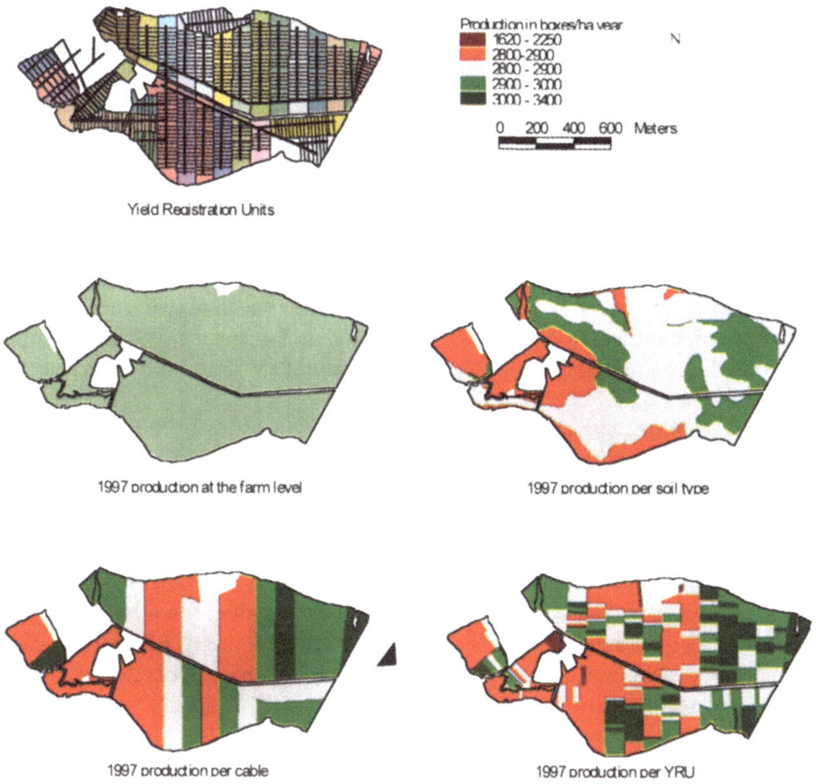

Figure 9.3. 1997 Yield variation on the Rebusca plantation on four scale levels.

9.4.3 Analysis of spatial variation

Yield maps in combination with soil maps enable farm management to perform better analyses of production patterns. A large part of the observed differences is likely to originate from soil variability. Deviations of -30 to +32 % from the mean yield within a single soil unit do occur. All soil variation within soil units, however, does not necessarily result in yield variation. Part of the observed variation is probably due to pests and diseases and part to differences in management, as this is mainly based on manual labor. Deviations to the mean can be classified to detect extreme situations in problem areas (Figure 9.4).

Banana plantations typically apply the same level of fertilization across the entire plantation. It is likely that fertilizer requirements vary with variation in yields, as the amount of nutrients removed by the banana crop varies significantly with differences in yields. Yields in a problem area may be low as a result of nutrient availability and management may decide to increase fertilization for that particular area in order to increase yield. However, if yields are low due to pests or drainage problems, increased fertilization will not lead to increased crop growth, and fertilizer will leach to the environment. As proper simulation models and detailed fertilizer recommendations for the banana crop are lacking, farm managers will have to rely on their expert knowledge to determine optimal fertilizer rates, if they want to improve crop performance in the short run. In other words, by overlaying soil survey data with production maps, one filters for differences in soil type (typically static properties, which can only be changed in the long run). Other differences, *i.e.*, differences within the soil units, are likely to be the result of planting material, diseases and/or management. Additional field observations and chemical analysis of crop and soil may be used to explain the differences. Through the identification of site specific problems, farm management may improve these local limitations, improving the performance of the farm and at the same time reducing costs.

A more accurate approach would be the setup and analysis of specific experiments to determine the input response curves. This will, however, require a substantial investment to be made, and will only yield results in the long run. Without experiments, the data base may be an excellent basis for undertaking model-based studies to assess the emissions of, for example, fertilizer nutrients and biocides to the environment, while providing detailed information on soil characteristics, crop performance and farm management. In a study carried out on the Rebusca plantation, Stoorvogel *et al.* (1999) studied nematicide leaching. Simulation models were used to assess the spatial variation in nematicide leaching and evaluate alternative management practices. Emissions were minimal and restricted to a limited number of so-called "hot spots" with sandy soils (Figure 9.5). In addition, there was a high risk for nematicide leaching only in certain periods of the year, whereas in the rest of the year the risk was almost nil. This kind of analysis forms the basis for an *ex-ante* evaluation of management interventions.

9.4.4 Analysis of temporal variation

Precision agriculture is often regarded as synonymous to site-specific management. However, temporal variation may be similarly or even more important. Instead of working with fixed management calendars for management operations, crop performance and soil conditions can be monitored in time to adapt future management practices (Bouma, 1997). Only if critical, limiting conditions occur (or preferably just before they start to occur) are agro-chemicals applied to resolve these limitations. Although climatic conditions are relatively stable in the AZ of Costa Rica compared to other humid tropical areas, their effect on crop performance is significant. Figure 9.6 presents the variation in production throughout 1994 and 1996. Although crop production fluctuates throughout the year and nutrient demands are highest at times when nutrient leaching is also at its maximum, fertilization takes place uniformly throughout the year. Fine-tuning of fertilization is therefore likely to be an effective strategy.

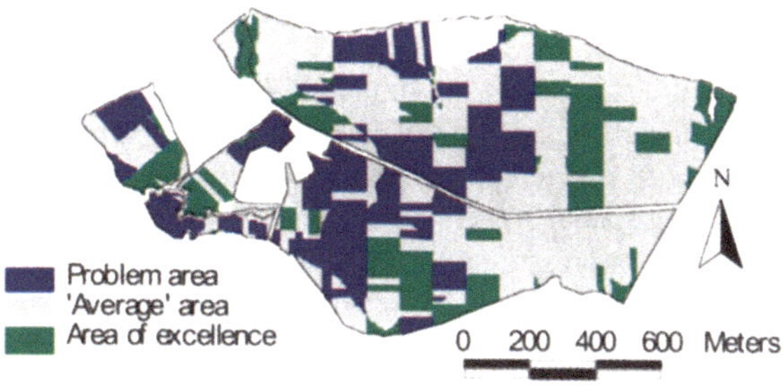

Figure 9.4. Location of the problem areas on the Rebusca plantation (1997 data).

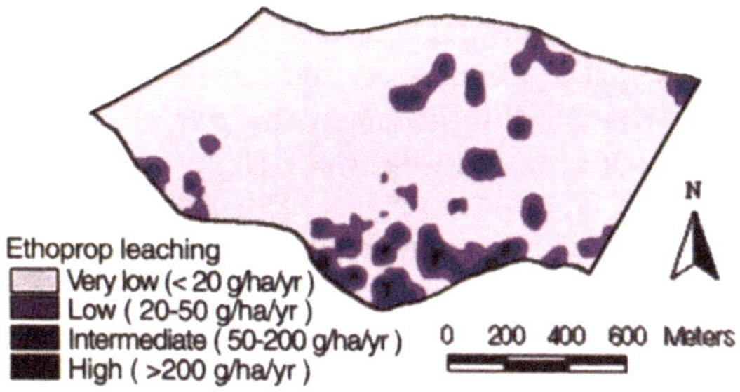

Figure 9.5. Modeled nematicide (Ethoprop) leaching on the Rebusca plantation (Stoorvogel *et al.*, 1999).

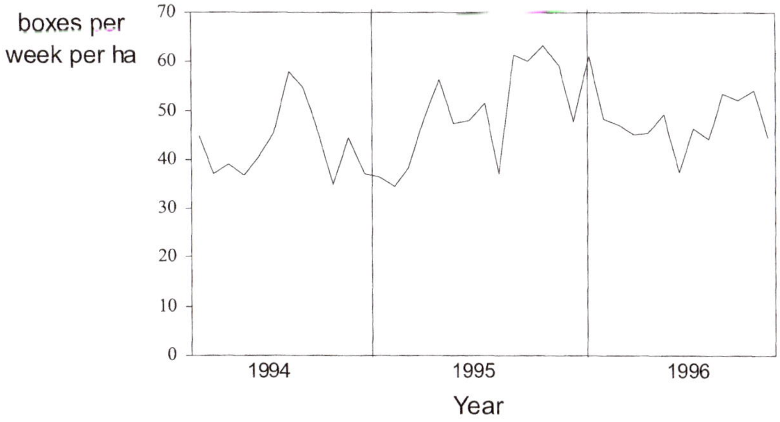

Figure 9.6. Production variability for 1994-1996 on the Rebusca plantation.

9.5 Discussion and conclusion

Precision agriculture is certainly not limited to the high-tech approaches currently being promoted in developed country's agriculture. Low-tech, participatory methods also serve well in practicing site- and temporal-specific forms of agriculture. The main challenge is found in quantifying spatial and temporal variation in crop performance, soil conditions, and pest and disease pressure. Although this forms the basis for a more precise application of agro-chemicals, the future will require a better formal incorporation of this variation in fertilization recommendations. Currently, the knowledge required to manage Costa Rican banana plantations more precisely is insufficient. On-farm experimentation should, therefore, reveal the supply-response function necessary for adequate fertilizer recommendations.

It is important to adequately define the spatial and temporal resolution with which precision agriculture will be applied. Currently, management in most cases is uniform over the plantation, except for drainage, where consideration about the inherent drainage status of the soil must be taken into account. Fertilization and biocide applications do not vary. Yields as well as plant density and some other characteristics are registered for the plantation as a whole or in some cases per cable. Given the fact that cables are positioned in such a way that there is a slight slope towards the packing plant, they typically include high soil variability. This makes the interpretation of data per cable rather awkward. Precision agriculture claims in some cases to manage fields with resolutions of up to one meter. It can be questioned whether this level of detail is necessary, as it will significantly increase the cost of implementing site specific management, whereas management units of, for example, 0.1 ha may adequately record the majority of observed variation.

An independent testing of precision agriculture is virtually unrealizable, as it is not possible to have two identical plantations with the same resources and management. At the same time, part of the system uses knowledge of farmers that again is difficult to test. Nevertheless, it seems that precision agriculture applied in banana production is capable of identifying problem areas in the field, a fact that, depending on the farm manager, can lead to increased input use efficiency.

Acknowledgements

Jetse Stoorvogel's research is funded by the Netherlands Academy of Sciences and Arts and, until 1998, supported by the Research Program on Sustainability in Agriculture (REPOSA, CATIE-MAG-WAU).

10 A toolbox for land use analysis

BAS A.M. BOUMAN, HANS G.P. JANSEN, ROBERT A. SCHIPPER,
JOHAN BOUMA, ARIE KUYVENHOVEN, and MARTIN K. VAN ITTERSUM

Abstract

This chapter provides an overview of the various methodologies for land use analysis presented in this book. It starts with a discussion of objectives, structure, terminology, levels of analysis, and aggregation issues. The methodologies are classified in five groups, (*i*) projective, (*ii*) explorative, (*iii*) predictive, (*iv*) generative and (*v*) design-oriented. Whereas the first three involve land use analysis on (sub-)regional to national scales, the fourth quantifies actual and alternative land use systems, and the fifth provides completely new farm or field management designs. The terminology used in all methodologies closely follows the one originally suggested by the FAO. The land use methodologies cover a series of spatial scales — or levels of analysis — and aggregation issues, involving the land use system, farm, sub-region, region and nation. For all methodologies, the implementation domain is explicitly specified, as well as their mutual complementarity and role in agricultural policy support. Together, the methodologies form a coherent toolbox to support policy design by analyzing (*i*) current land use, (*ii*) likely changes in future land use and their drivers, (*iii*) technical options for future land use, and (*iv*) policies intended to induce land use changes in the (near) future. Sustainability issues addressed are related to the goals of "maintaining the resource base", "protecting the environment" and "optimizing non-renewable resource use efficiency". Whereas the first two concepts are mainly biophysical in nature, the latter focuses more on economics. Specific biophysical sustainability indicators relating to the first two concepts were derived from relevant sustainability issues in the Atlantic Zone of Costa Rica, and implemented in the generative, explorative and predictive methodologies. Finally, reflections are offered on the importance of user involvement in both the development and the application of land use analysis methodologies for successful policy support.

10.1 Introduction

The development of methodologies for sustainable land use analysis requires contributions from various disciplines and may take place on several levels of aggregation. The methodologies presented in this book result from intense collaboration among a number of disciplines including soil science, agronomy, animal husbandry, economics, marketing, and physical geography. Concepts originating in systems analysis and information technology play a central role in each of the methodologies. Tensions between disciplines and aggregation levels may arise when the goals and terminology of different methodologies for land use analysis are insufficiently specified (Rabbinge and Van Ittersum, 1994).[1] Therefore, the scope as well as the terminology of the methodologies used are summarized in the next section. The spatial scale levels and aggregation issues associated with the different methodologies are discussed in Section 10.3. Based on the summary of objectives, scale levels, aggregation issues and characteristics of the main tools involved, complementarity between the different methodologies is discussed in Section 10.4. Sustainability issues, their underlying concepts and specific implementations in the various methodologies, are addressed in Section 10.5. The importance of user involvement in both the development and the application of the methods used in land use analysis is discussed in Section 10.6, where a report on the joint application of the SOLUS methodology involving both REPOSA and the Costa Rican Ministry of Agriculture and Livestock is provided. Finally, a number of conclusions are drawn, including recommendations for successful design and application of land use methodologies and suggestions for further research (Section 10.7).

10.2 Scope and terminology of land use analysis

10.2.1 Scope

The methodologies of land use analysis presented in this book can be differentiated on the basis of their main scope or purpose in the following way (see also Chapter 1):

- *Projective.* The CLUE (Conversion of Land Use and its Effects) modeling framework presented in Chapter 3 projects the likely developments in future land use according to different scenarios. Current distribution of major kinds of land use (also called land covers) are explained by biophysical and socio-economic land use drivers on the basis of statistical regression analysis. Examples of land use drivers are climate, soil, population and degree of urbanization. Likely land use patterns are generated by changing the values of the land use drivers by extrapolating past trends or by projecting certain expectations about their future developments (*e.g.*, population growth). A scenario is a well defined change in (sets of) land use drivers.
- *Explorative.* The SOLUS (Sustainable Options for Land USe) framework presented in Chapters 6 and 7 can be used to explore options for land use with a relatively long time horizon (20-30 years). Technological options for land use are fed into a linear programming model that optimizes for one or more objectives, subject to a number of biophysical and socio-economic constraints. Such constraints typically relate to either resource endowments (and are therefore mostly fixed) or are policy related (*i.e.*, user-defined and therefore flexible). Explorations carried out with this type of model may focus on the quantification of the ultimate biophysical constraints on land use and consider only a limited number of socio-economic factors (Chapter 7). Alternatively, explorations may focus on the options for agricultural development under explicitly simulated socio-economic conditions, such as product demand and labor supply (Chapter 6). The SOLUS methodology thus allows researchers to investigate the aggregate effects of alternative policies on the regional level (including the possibilities of attaining combinations of objectives) as well as to quantify trade-offs between objectives.

[1] An illustrative anecdote is the following: Biophysicists felt offended when economists called some of their model results "perverse". In economics, however, "perverse" is accepted terminology which may be used when results are obtained that are outside the validity domain of the model or its underlying assumptions. For example, a situation in which a higher price of a commodity causes a lower supply of that commodity is typically referred to as a "perverse" supply response by economists (Ghatak and Ingersent, 1984).

- *Predictive.* The policy-oriented UNA-DLV methodology presented in Chapter 8 is suitable for predicting the short-term effects of policy measures on changes in land use decisions, based on the explicit modeling of farmer behavior. This is done by combining programming techniques with econometric farm household models. In the context of the latter, utility functions are defined which are subsequently optimized using linear programming techniques, again subject to both biophysical and socio-economic constraints. In addition to modeling behavior on the farm level (and summing specific farm types within a region), it is possible to model the *aggregate behavior* of producers and consumers on higher spatial scales, such as the region or the nation. An example for the regional level is the REALM implementation of SOLUS (Chapter 6), though its primary aim is *exploration* (see above). As example for the national level, the Spatial Equilibrium Model (SEM) presented in Chapter 4, is a short-term economic optimization model, based on econometric estimation of supply and demand functions for major agricultural commodities as well as transport costs. This model can be used to assess the effects of policy decisions on agricultural production and land use, trade flows, consumption patterns, and welfare.
- The last two methodologies *prototype* or *generate* new land use systems on the farm or field level. Prototyping involves on-farm development of alternative land use systems that meet sustainability requirements while satisfying farmers' socio-economic needs. The generation of new land use systems has the same goal but does not explicitly involve farmer participation or on-farm research. The example of precision agriculture developed in Chapter 9 illustrates the concept of prototyping in a study of a banana plantation in the AZ. This form of analysis involves on-farm development and application of support systems in which the economic and ecological consequences of changes in the production structure of a land use system are analyzed by evaluating the economic and agronomic management decisions on the farm. Typically for precision agriculture, management specifically addresses within-farm spatial and temporal variability. The Technical Coefficient Generators of Chapter 5 quantify inputs and outputs of either actual land use systems or new alternative ones that may not (yet) be in use. Even though the Technical Coefficient Generators were designed to provide the building blocks for the *explorative* and *predictive* land use methodologies, they are also useful as stand-alone tools for *ex ante* analysis of the performance of a land use system on the field level.

10.2.2 Terminology

The terminology used in this book conforms to the definitions and concepts proposed by the FAO (1976) and subsequently elaborated in various publications (FAO, 1983; Driessen and Konijn, 1992; Fresco *et al.*, 1992). In this book, the FAO terminology has been extended with concepts and definitions obtained from the literature dealing with the application of linear programming in land use analysis and production ecology (*e.g.*, Van Ittersum and Rabbinge, 1997). A complete overview of definitions of terminology is given in the section "Concepts and Definitions" in the back of this book.

A basic concept central to the methods of land use analyses presented in this book is the *land use system*, defined as a combination of a single *land unit* with a single *land utilization type*. This combination is uniquely characterized by its inputs and outputs (expressed as technical coefficients) as well as by (possible) land improvement measures, such as irrigation or drainage (after Driessen and Konijn, 1992, and Fresco *et al.*, 1992). *Land use system* in this definition is synonymous with the expression "production activity", which has been defined as the cultivation of a crop or crop rotation in a particular physical environment, completely specified by its inputs and outputs (Van Ittersum and Rabbinge, 1997). A *land utilization type* is a specific kind of land use under well defined biological (*e.g.*, crop, variety, product), socio-economic (*e.g.*, labor use) and technological (*e.g.*, management, sequence of operations, use of inputs) conditions (after Fresco *et al.*, 1992, and Jansen and Schipper, 1995). A *land unit* is defined as a physical area of land that is uniform in its characteristics and qualities (after FAO, 1983). Finally, the expression *major kind of land use* is defined as follows: a widely accepted major subdivision of land use, *e.g.*, annuals, perennials, forest and pasture (after FAO, 1983, and Driessen and Konijn, 1992). A *major kind of land use* refers to a higher spatial integration level than the *land use system*, and as such is less strictly defined. The *major kind of land use* constitutes the unit of analysis used in the CLUE modeling framework.

Land use systems form the building blocks for both the *explorative* land use methodology SOLUS (Chapters 6 and 7) and the *predictive* land use methodology of UNA-DLV (Chapter 8). Land use systems are quantified through the use of Technical Coefficient Generators that calculate input and output coefficients based on a combination of process-based knowledge, expert-knowledge, information from the literature, as well as primary data and field observations. The Technical Coefficient Generators used to generate land use systems (and options for herds and feed supplementation) for the SOLUS and UNA-DLV methodologies are called LUCTOR (Land Use Crop Technical coefficient generatOR) and PASTOR (Pasture and Animal System Technical coefficient generatOR), and are explained in detail in Chapter 5. In both Technical Coefficient Generators, an explicit distinction is made between *actual* and *alternative* land use systems. *Actual* systems represent land use systems as currently practiced by farmers; their technical coefficients are derived mostly from surveys, field observations and measurements in the area under study. *Alternative* systems, on the other hand, are obtained using the so-called target-oriented approach: target production levels (or emission levels) are predefined and the optimal combination of inputs required to realize these target levels is subsequently quantified (De Wit *et al.*, 1988; De Koning *et al.*, 1992; Van Ittersum and Rabbinge, 1997). Input-output relations of alternative systems are defined in such a way that they can be repeated without changing input requirements, *i.e.*, they are sustainable.

Even though technically feasible and sustainable from a biophysical point of view, alternative land use systems generally are not (yet) widely practiced in the area under study. A typical assumption regarding alternative land use systems, based on insights in the ecology of production, is that they use inputs more efficiently than actual systems do, due to (future) efficiency gains in agricultural production (De Wit *et al.*, 1987). Continuing technological progress implies that *i*) alternative land use systems may gradually evolve into actual land use systems, and *ii*) Technical Coefficient Generators need regular updating in order to incorporate new biophysical knowledge and technological advancements. An example of an alternative system that may become an actual system in the near future is the grass-legume combination *Brachiaria brizantha - Arachis pintoi*. This system has successfully been tested on MAG's experimental farm as well as by farmers in the AZ (Ibrahim, 1994; Hernández *et al.*, 1995; Bouman *et al.*, 1999b). Explorations with SOLUS and partial budget analyses have shown that this grass-legume mixture becomes economically attractive on the well-drained soils of the AZ undergoing continued degradation under the current natural pastures (Jansen *et al.*, 1997b; Bouman and Nieuwenhuyse, 1999). An example of technological progress that currently receives wide attention is *precision agriculture* (Finke, 1993; Verhagen *et al.*, 1995), an application of which involving banana production in the AZ is illustrated in Chapter 9. The knowledge obtained in such research can be captured in Technical Coefficient Generators, analyzed on the field level, and subsequently used as building blocks in the *explorative* and *predictive* land use methodologies on higher levels of spatial aggregation.

10.3 Spatial scales and aggregation issues

The spatial scales involved in the methodologies of land use analysis presented in this book include: the land use system, farm, sub-region, region, and nation. Since a land use system is defined as a combination of a homogeneous land unit with a particular land utilization type, a land use system can be conceptualized as a specific field or parcel within a farm used for a specific crop, pasture or forest type. The land use system is the typical spatial level of analysis addressed in the Technical Coefficient Generators of Chapter 5. In the calculation of technical coefficients, however, use is also made of hierarchical levels below that of the land use system. For example in PASTOR, the computation of the feed requirements of cattle grazing on a land unit is based on the detailed modeling of nutrient and energy balances of individual animals (NRC, 1989, 1996). In the prototyping of precision agriculture on banana plantations (Chapter 9), the basic spatial unit of analysis is the land area harvested by a particular cable and receiving a particular management. In this sense, the basic spatial unit corresponds to the definition of a land use system. At the other end of the spatial scale, the Spatial Equilibrium Model in Chapter 4 works with administrative zones to define its basic spatial units (being the six political planning regions in Costa Rica). In this case, specific land properties do not enter the analysis.

The SOLUS and UNA-DLV methodologies address a number of spatial scales, including the land use system, farm, sub-region and region. The basic building blocks in both methodologies are land use systems furnished by Technical Coefficient Generators, which are subsequently aggregated in the linear programming models. The specific aggregation issues at stake in this type of models include the following (Erenstein and Schipper, 1993; Jansen and Stoorvogel, 1998):

1. Regional land use is often considered without detailed information about the behavior of the farm households ultimately responsible for decisions about land use.

2. Individual farmers have resources at their disposal that may differ in relative proportions from the aggregate resources of a region.
3. Variables that are exogenous on one spatial scale may become endogenous on higher levels of aggregation.

The first two aggregation issues, and the economic dimension of the third issue received ample attention in Chapters 6 and 8, and will not be discussed further. However, the third issue has an agronomic parallel to the economic dimension which is worth while mentioning. Prices that are exogenous on the farm level become endogenous on the regional level, if the region accounts for a significant part of the domestic and/or export supply. In (linear) programming models, endogenization of prices can be achieved by maximizing aggregate consumer and producer surplus under the assumption of competitive product markets (as is done in Chapters 4, 6 and 8). Along similar lines, agronomic variables that are exogenous on the field or farm level may become endogenous on the regional level. For example, the pressure of non soil-bound pests and diseases is an exogenous factor on the field level, with crop protection measures applied to achieve desired production goals. However, when optimization with a linear programming model results in a large part of the region being allocated to a certain crop, the population dynamics of pests and diseases that influence the particular crop in question will be affected as well. This would imply that a different level of crop protection is required to maintain the desired production level. Unlike endogeneity in economics, however, most agronomic endogeneity issues can not be solved with programming techniques, because the outcome of the model affects the (fixed) input-output relations of land use systems as calculated in the Technical Coefficient Generators. Even though this issue receives no further attention in this book, it merits consideration in future methodological development.

The spatial levels mentioned above implicitly constitute cartographic *polygons*. Aggregation involves the grouping of units of a lower spatial level into units of higher level and should be performed in such a way that minimizes heterogeneity within these higher level units (Hijmans and Van Ittersum, 1996; Jansen and Stoorvogel, 1998). However, typically little or no information is available on either geographic distribution of, or spatial relationships between the spatial units of analysis. This issue is relevant for most of the methodologies discussed in this book. For example, in the UNA-DLV methodology a region is modeled, so that it consists of a collection of several farm types that are assumed to be homogeneously distributed throughout the area (Chapter 8). Even in the SOLUS methodology, limitations inherent in linear programming techniques restrict a full exploitation of spatial patterns and relationships, despite using geographic aggregation via a GIS (Chapter 6). For example, while the 12 sub-regions distinguished in the Atlantic Zone, as well as the major land units within these sub-regions, are spatially explicitly defined, the spatial distribution of land use systems (selected by the linear programming model during optimization) occupying these combined sub-regions/major land units is not known. An entirely different approach is employed by the CLUE modeling framework (Chapter 3), which uses the *pixel* (or raster) as its basic spatial unit. On the lowest pixel level, sub-pixel information regarding major types of land use and their drivers (*i.e.*, explanatory variables) serve as the basis for modeling. For each pixel, the percentages covered by land units and major types of land use are obtained by map overlaying in GIS. The statistical model used in the CLUE methodology aims at describing the spatial distribution of land use, thus explicitly addressing spatial patterns and relationships. Aggregation is performed by combining pixels of smaller sizes in pixels of larger sizes, and deriving new statistical regressions between major kinds of land use and their drivers. The differences found in the types of land use drivers as well as in the values of their regression coefficients on various levels of aggregation serve as an illustration of yet another kind of aggregation issue, which is different from the three issues mentioned above. In scenario analyses, CLUE uses the results obtained on different spatial aggregation levels to allocate land use in a spatially explicit way. Thus, beside answering the question of *what* might happen under certain scenario assumptions, CLUE can also answer the questions *where* this is most likely to happen.

10.4 Complementarity of methodologies

Table 10.1 illustrates the complementarity among the various methodologies of land use analysis presented in this book by summarizing some key characteristics (see also Table 1.1 in Chapter 1).

Table 10.1. Complementarity of methodologies for land use analysis

Nature	*Name*	*Principal goal*	*Information generated*
Projective	CLUE	Projection of likely future trends in agricultural land use	Likely future land use changes (what, where and when)
Explorative	SOLUS (GOAL-AZ)	Exploration of biophysical options for land use	Possibilities and limitations of agricultural development in relation to policy goals ("policy space")
	SOLUS (REALM)	Exploration of options for land use under combined biophysical and socio-economic constraints	Possibilities and limitations of land use in relation to societal desires ("biophysical space")
Predictive	UNA-DLV	Short-term prediction of policy effects on farmers' land use decisions	Effectiveness of policy measures in inducing adoption of desired land use systems
	SEM	Short-term prediction of policy effects on land use, associated trade flows and welfare	Quantification of policy effects on quantity and distribution of societal welfare
Generation of land use systems	LUCTOR, PASTOR	Provision of building blocks for explorative and predictive methodologies	Quantification of input-output relationships for a large number of land use systems
Design or prototyping	BanMan	On-farm decision support based on principles of precision agriculture	Input-output relations of precision agriculture

Even though the *projective*, *explorative* and *predictive* methodologies for land use analysis, not to mention the methodologies *designing* or *generating* land use systems, all have different objectives, they nevertheless serve the ultimate goal of providing policy support. Each of these approaches uses the practice of scenario development, defined as the analysis of possible future land use given certain expected changes in land use drivers (the *projective* methodology) or certain objectives and policy measures (*explorative* and *predictive* methodologies). Time horizons may vary from short-term (1-5 years) to long-term (20-30 years), depending on the scenario implemented. Together, these methodologies form a coherent "toolbox" that can be used to support policy design by addressing the following three questions:

1. What are the likely changes in future land use if current relationships between land use and their drivers were to continue?
2. What are the options for land use in the future based on feasible or expected technological change?
3. Which effective agricultural policies would induce farmers to adjust their land use in such a way as to satisfy certain policy objectives?

10.4.1 Projective land use analysis

Regarding the first question, information about future land use developments that are likely to take place under *ceteris paribus* conditions is highly relevant to policy makers. The CLUE modeling framework is explicitly designed to generate this type of information. However, since there always remains a certain degree of uncertainty about the (often largely autonomous) development of land use drivers, scenario analyses are performed in which the values of these drivers change in magnitude or direction. An attractive feature of CLUE is that, besides addressing the question of *what* might happen to land use in the future, the methodology also addresses the question of *where* and *when* such land use changes might occur. This gives insight into both the timing and the location of possible "hot-spots" in regional development, and provides policy makers with an approximation about the possible future pathways of land use dynamics.

Information on the likely changes in land use distribution is not only relevant to a number of the sustainability-related issues addressed in this book, but also important in the overall sustainability debate. For example, changes in land use may affect sustainability as a result of subsequent effects on soil erosion, nutrient fluxes, carbon pools, and biodiversity. A clear understanding of the relationships between drivers and land use change, and the scale on which such drivers operate, is a prerequisite for the quantification and mapping of such sustainability. Past applications of CLUE proved the feasibility of linking its results with the modeling of nutrient balances (De Koning *et al.*, 1997), while applications in the field of carbon pools and a dynamic link with nutrient fluxes are currently being developed.

10.4.2 Explorative land use analysis

Projected land use developments based on current relationships between land use and its drivers only allow researchers to look into the future up to a certain extent.

Discontinuities in trends (policy changes or technological progress) can not be taken into account. For example, in the application of CLUE to the Atlantic Zone of Costa Rica, the recent introduction of palm heart and its rapidly increasing role in land use could not be addressed with the same statistical rigor as employed for other land use types because palm heart was not present in 1984, the base year for the statistical models. Extrapolation of past and current relationships between land use patterns and their drivers into the future obviously limits future land use outcomes. New insights in future land use are only possible when the past is not used as the sole measure for the future, but when political and/or social desires are combined with technological developments (Rabbinge and Van Ittersum, 1994; Van Ittersum *et al.*, 1998). The SOLUS methodology facilitates the *exploration* of land use options by allowing new land use systems that are currently not practiced to enter the analysis, and by including explicit optimization of land use for well defined policy objectives. In addition, this methodology allows for the quantification of trade-offs among the various dimensions of biophysical sustainability and socio-economic parameters.

In SOLUS, future land use patterns are constrained by a user-specified combination of *absolute* levels of biophysical and socio-economic restrictions (*e.g.*, quantity of land and labor) and *normative* limitations derived from societal desires. The ultimate biophysical limits to production, under different societal objectives can be explored by omitting most socio-economic constraints (such as the product demand and labor supply) as is done in the GOAL-AZ (General Optimal Allocation of Land use for the AZ) implementation of SOLUS (Chapter 7). In this approach, results obtained from optimizing land use for various objectives, subject to a number of normative constraints derived by considering stakeholder views, make policy debates more transparent and improve general understanding about the possibilities and limitations of agricultural development. Besides exploring the biophysical limits on regional land use, the *combined* effect of biophysical and socio-economic constraints can also be explored with the SOLUS methodology, as is done in the application using REALM (Regional Economic and Agricultural Land use Model; Chapter 6). Socio-economic constraints, such as limited product demand or labor supply, frequently turn out to be more significant than biophysical constraints. However, unlike most biophysical factors, many socio-economic constraints (*e.g.*, transaction costs) can, at least partially, be influenced by agricultural policies. This fact makes it necessary to adopt a complementary use of the two types of explorative land use studies. On the one hand, explorations that relate biophysical conditions to societal desires reveal land use options beyond current and future socio-economic constraints. On the other hand, explorations that include such socio-economic constraints (*e.g.*, market conditions) show how biophysical possibilities are limited by current or projected socio-economic constraints.

10.4.3 Predictive land use analysis

Long-term projective studies provide policy makers with information about likely trends in future land use, while explorative studies show a range of options for land use. The latter are based on technical possibilities and societal desires expressed as goals and normative constraints. In the *short term*, however, policy-makers are interested in policy measures that influence farmers' land use decisions in a certain desired way, *e.g.*, to arrive at a more environmentally friendly land use. Explorative studies that make use of the SOLUS methodology provide information on the technical options and associated long-term trade-offs of different policy goals, and as such inform policy-makers about the size and shape of the policy space in which they can realistically expect to be able to manoeuvre. The effects of policy measures as such can be evaluated with *predictive* land use analysis methodologies that specifically address the behavior of those who ultimately make decisions about land use (*e.g.*, the UNA-DLV methodology; Chapter 8). These decision-makers are often farmers, but they can also be (inter)national corporations, such as the banana companies in the Atlantic Zone of Costa Rica.

Even though predictive land use studies include both biophysical and socio-economic restrictions, investigations into the influence of normative constraints on agricultural development in general (and land use in particular) is usually not an explicit objective of such studies. Rather, these studies aim at identifying policy instruments that are both effective and feasible in order to induce farmers to adopt certain desirable land use systems. Predictive methodologies allow researchers to evaluate a wide spectrum of policy instruments (Colman and Young, 1989; Ellis, 1992; Sadoulet and De Janvry, 1995), including price and market policies, taxes, subsidies, infrastructural improvements, reduced transaction costs, research and extension policies, and financial, trade and exchange rate measures. The UNA-DLV methodology has been made operational on the farm level (through the development and use of a farm typology) as well as on the regional level (through the incorporation of market clearing mechanisms), while the SEM methodology operates on the national as well as on the regional level (by sub-dividing Costa Rica into six planning regions). Besides the SEM and UNA-DLV methodologies, the REALM implementation of the SOLUS methodology can, to some extent, also be used as a predictive methodology on the regional level because it addresses aggregate producer (and consumer) behavior. Unlike the UNA-DLV methodology, however, the REALM implementation of the SOLUS methodology does not incorporate farm types and therefore does not explicitly model farmer behavior. Instead, the sub-regions within the AZ are treated as single farms having the maximization of aggregate economic surplus (consisting of the sum of the producer and consumer surplus in each market) as their objective function. Ignoring farm types has been shown to be a source of aggregation bias (Jansen and Stoorvogel, 1998), stressing the complementarity of explorative and predictive land use methodologies on the regional level.

10.4.4 Generating and designing land use systems

With the exception of the SEM methodology, an important aspect of the explorative and predictive land use methodologies discussed above is the use of *alternative* land use systems. The description of alternative land use systems in quantified input-output relationships, and their subsequent use in optimization models, allow the transition from *projective* modeling to *explorative* and *predictive* modeling. Input-output relationships of land use systems are quantified using Technical Coefficient Generators that combine process-based knowledge, expert knowledge and primary data (obtained from agronomic field experiments and farm surveys). However, technological change is a continuous process in agriculture, and new land use systems keep on being developed. Furthermore, newly designed land use systems typically need testing and further refinement under empirical conditions, as well as adaptation to local conditions (Byerlee and Collinson, 1980; Simmonds, 1986).

An innovative way of designing new land use systems is called "prototyping" (Vereijken, 1997). This concept refers to the design of production technologies through an interactive process between scientists and farmers. Whereas most other methodologies for land use analysis presented in this book serve as tools for policy support and decision making on the level of the sub-region or higher, prototyping explicitly is meant to provide decision support on the field or farm level. In Chapter 9, techniques are described that aim at increasing input use efficiency to serve the dual purpose of increasing farm profit while decreasing environmental contamination. Even though the example of precision agriculture in banana plantations given in Chapter 9 operates on the field level, the information generated can be used for extrapolation to higher levels of aggregation. Also, the paradigm of scientists interactively working with farmers to develop new land use systems could be useful in building and implementing Technical Coefficient Generators. Once the input-output relationships of alternative land use systems are firmly established and validated, they can be incorporated in the Technical Coefficient Generators. SOLUS can subsequently be used to explore the feasibility and the consequences (*e.g.*, in terms of agricultural income or pollution of the environment) of such new land use systems on the regional level. Even though the particular case of precision agriculture for banana plantations in the AZ is not yet sufficiently developed to allow such a regional exploration, new land use systems for pasture (Bouman *et al.*, 1999b) and timber production (Nieuwenhuyse *et al.*, 1999) have been evaluated in this way. Conversely, the results of *explorative* studies can be used to identify attractive land use systems in terms of specific objectives or societal desires. For example, Bouman and Nieuwenhuyse (1999) found that grass-legume mixtures are a promising alternative to natural pastures for maintaining the soil nitrogen stock over time, whereas the use of fertilized, improved grass species proved economically unattractive over a wide range of price conditions. In addition to guiding further agricultural research and prototyping, such results can be used in *predictive* land use models to identify efficient policy instruments for inducing farmers to adopt such systems. Technical Coefficient Generators thus form a crucial bridge between newly designed land use systems and their extrapolation and evaluation on larger spatial scales.

10.5 Sustainability issues

Even though the many debates about the concept of sustainability in the past decade have proven that the term can not be unambiguously defined, it also has become clear that sustainability is intrinsically linked to normative perceptions, risk and acceptance by different (groups of) people in society (*e.g.*, WRR, 1995), making the measurement of the concept as such difficult if not impossible. At best, quantifiable indicators can be defined to monitor different aspects of sustainability, which can subsequently be weighed and discussed by various stakeholders. To facilitate the debate, it is convenient to subdivide the overall concept of sustainability into three major issues, *i.e.*, (*i*) that of maintaining the resource base, (*ii*) that of protecting the environment, and (*iii*) that of efficient use of non-renewable resources. For each issue, different indicators or methods of analysis can be developed. Even though sustainability indicators ideally should be "universal" in the sense that they are well defined in terms of measurement and calculation procedures, the indicators themselves will not always have universal relevance. For example, the salt concentration in groundwater may be a relevant sustainability indicator in irrigated, semi-arid environments, but not so in the humid tropical climate of eastern Costa Rica. In this book, the following sustainability indicators were derived from issues relevant to the northern Atlantic Zone (AZ) of Costa Rica (see Chapters 2, 5-8), and to the above mentioned dimensions of sustainability:

i. Maintaining the resource base: this concept was operationalized through the quantification of changes in soil nutrient stocks for nitrogen (N), phosphorus (P) and potassium (K).

ii. Protecting the environment: this aspect was operationalized through the quantification of nitrogen waste flows that result from leaching, (de)nitrification and volatilization (the latter two also serve as proxies for the emission of N-related trace and greenhouse gases), and through tracking the emission of biocides, expressed both as a quantity (total amount of active ingredients) and as a biocide index (taking into account quantity, toxicity and persistence).

iii. Efficient use of non-renewable resources: this aspect was not translated into an explicit indicator, but is considered in the definition of new, innovative land use systems and implicit in the use of an optimization model as implemented in the explorative and predictive methodologies of land use analysis.

The first two sustainability dimensions are thus operationalized through biophysical indicators only, whereas the third focuses more on economic elements. The presented biophysical sustainability indicators are not exhaustive, but the way they are treated in the presented case studies is illustrative of the ways in which other indicators could be incorporated in land use analyses as well.

10.5.1 Maintaining the resource base

The concept of *maintaining the resource base* is consistent with the first rule on sustainable resource use as postulated by Pearce and Turner (1990): "utilize renewable resources at rates less than or equal to the natural rate at which they regenerate." The soil nutrient stock directly influences land productivity and as such is a key biophysical resource in agricultural production. Land use systems that deplete the soil nutrient stock can be qualified as unsustainable in the sense that their productivity levels decrease over time. In this sense, many *actual* systems in the AZ of Costa Rica are indeed unsustainable (Bouman and Nieuwenhuyse, 1999; Jansen *et al.*, 1995). In exploring the scope of sustainable land use, *alternative* systems are designed that do not deplete soil nutrient stocks and are consequently deemed sustainable, at least from the biophysical point of view of maintaining the soil resource base (Chapter 5).

Making use of an explorative methodology for land use analysis such as SOLUS, and offering both *actual* land use systems and *alternative* systems to the optimization model in a consecutive manner, generates insight in two alternative forms of land use. Maximizing economic surplus using only *actual* systems (i.e., assuming that *alternative* land use systems at present are not (yet) practiced by farmers) quantifies the maximum surplus achievable in the short run (form 1); on the other hand, the exclusive use of *alternative* systems quantifies the level of economic surplus that can be maintained in the long run (form 2). Often, the achievable economic surplus using only *actual*, soil nutrient depleting land use systems (form 1) exceeds the surplus generated through the exclusive use of *alternative* systems (form 2). This result stems from the fact that, in *actual* systems, the soil (temporarily) acts as a free source of nutrients. The use of *alternative* land use systems results in a higher economic surplus only when nutrient recovery rates are sufficiently high and fertilizer costs sufficiently low to compensate for the lost provision of free nutrients from the soil. The relative economic competitiveness of *actual* and *alternative* systems depends on the resource use efficiencies of *alternative* systems, the relative price of fertilizer, and on the level of soil nutrient provision and associated yield of the *actual* systems. In newly reclaimed forest soils of the AZ, soil nutrient stocks and yield levels are relatively high, and *actual* systems are often more competitive than *alternative* ones. However, with increased soil nutrient mining over time, soil nutrient stocks and yield levels of *actual* systems decline, and the relative competitiveness of *alternative* systems increases. Therefore, only a dynamic analysis of soil nutrient depletion in *actual* systems can offer insight into the relative competitiveness of *actual* and *alternative* systems over time.

In a dynamic analysis, linear programming techniques used in explorative and predictive land use analyses offer three ways in which to study temporal aspects of soil nutrient depletion:

1. Multi-period programming. In this technique, the magnitude of soil nutrient depletion in one period is used to reduce productivity levels of land use systems in the next period. When optimizing in terms of economic surplus, multi-period programming models which discount future costs and benefits have a tendency to use sustainable land use options in early periods, and unsustainable options in later periods, thus postponing soil nutrient depletion while minimizing the effect on economic surplus (*e.g.*, Kuiper, 1997; Miranowski, 1984).
2. Quantifying the terminal value of land. In models with a quasi-rent objective function, land resources can be assigned a certain terminal value at the end of the model's time period. Model outcomes will partially depend on the level of this terminal value, which might be derived from the production capacity of the land based on the level of the remaining soil nutrient stock (see Chapter 8).
3. Mimicking the productivity decline of *actual* land use systems, through a recursive sequence of running the linear programming model. This way of modeling the dynamics of soil nutrient depletion involves using a series of inputs in a linear programming model that represent actual land use options in conditions of decreasing productivity, which range from relatively high current levels to low levels associated with nutrient-exhausted soils (Bouman *et al.*, 1999b; compare with the approach followed by Barbier and Bergeron [1999] in a study of the different pathways of

development by Hondurenian hillside farmers). Each model run involves the optimization of economic surplus by selecting between actual and alternative land use systems. Somewhere along the trajectory of declining productivity, alternative land use systems may become more attractive than progressively degrading actual systems.

However, since the linear programming models used in the land use methodologies presented in this book are static, they have a limited scope for studying dynamic processes, such as nutrient depletion. Other methodologies that may be used to better address this issue include dynamic simulation modeling of soil nutrients and soil organic matter (*e.g.*, Parton *et al.*, 1987; Verberne *et al.*, 1990) or optimal control (*e.g.*, Bulte *et al.*, 1999b; McConnell, 1983).

The terminal condition of the land after a period of soil nutrient depletion, and the time span during which this depletion has taken place, are important factors in most people's perception of biophysical sustainability. If the alternative land use systems available to farmers after they have depleted their land of nutrients are unable to provide a satisfactory level of income, such degraded lands may be left fallow or be abandoned altogether. The fate of abandoned land depends on the degree of degradation and the regenerative capacity of the environment (including soil, climate, flora). Old, weathered soils in harsh climates with a small or dispersed bank of seeds and consequently a slow re-growth of (natural) vegetation may hardly recover for agricultural use within a time frame that is interesting to policy makers (recovery may take a few hundred years or more under such conditions; Uhl *et al.*, 1988). On the other hand, where soils are intrinsically rich and weather relatively easily (*e.g.*, the volcanic soils of the Atlantic Zone of Costa Rica), land may regenerate relatively quickly and be covered with natural forest within a couple of decades (see, *e.g.*, Hartshorn [1983] for an account of the regeneration of forest on an abandoned farm in the AZ). Such conditions allow a rotation of land use types within time scales of, say, 20-30 years. However, whether (and to what degree) land degradation is found an acceptable phenomenon by society depends on normative perceptions.

10.5.2 Protecting the environment

The concept of *protecting the environment* is consistent with the second rule on sustainability as put forward by Pearce and Turner (1990): "Keep waste flows to the environment at or below the assimilative capacity of the environment." Waste flows of nutrients and biocides are relatively easily expressed in well-defined indicators (though their exact quantification might still pose operational challenges). However, measuring the assimilative capacity of the environment for nutrients and biocides is notoriously difficult and surrounded by uncertainty. Therefore, this second rule on sustainability is typically operationalized by *normative* values derived from societal desires (WRR, 1995). Where human health is concerned, normative values can be derived from damage levels that are found to be acceptable to the human population at large. On the other hand, the health of an entire ecosystem is more difficult to quantify, and normative values may be derived from discussion among various stakeholders and interest groups. The interactive multiple goal approach of land use analysis has been found to be constructive in the derivation and mutual acceptance of biophysical and socio-economic normative perceptions (Rossing *et al.*, 1998). The development of land use systems that are relatively environmentally friendly plays a large role in reducing the waste flows from agricultural production, and helps to comply with such perceptions, while at the same time maintaining income levels (*e.g.*, the precision agriculture techniques as discussed in Chapter 9).

10.5.3 Optimizing efficiency

An *efficient use of resources* is linked to the third rule on sustainable development formulated by Pearce and Turner (1990): "Optimize the efficiency with which non-renewable resources are used, subject to substitutability between resources and technical progress." In the chapters where alternative land use systems are defined and included in optimization models (Chapters 6, 7 and 8), this rule is explicitly taken into account. This holds particularly true where the objective function is economic in nature (Chapter 6, sum of producer and consumer surplus; Chapter 7, producer surplus; Chapter 8, utility, full income and/or quasi-rent). By maximizing an economic objective function in the process of choosing among alternative land use systems, while simultaneously taking account of limitations in resource availability, resources are used in an optimal, and therefore efficient way.

Substitution of resources is part of this selection process because different land use systems use resources in different proportions. Changing the relative availability of resources and/or their relative prices will result in a different selection of land use systems and use of resources (again "most efficient," but now under the new conditions), implying substitution of resources.

Technological progress is incorporated by using different land use systems involving different technologies as inputs for the optimization model. These technologies can be ranked according to their *technical* efficiency with regard to the use of each resource. In the optimal solution, the most efficient land use systems are selected with regard to the objective function. In models with an economic objective function, inputs used and outputs generated by land use systems are *valued* in terms of prices. Thus, in such models the efficiency concept relates to *economic* efficiency, which by definition combines both *technical* and *allocative* efficiency. Technically more efficient land use systems with an eye to the future can be incorporated into the model. New optimal solutions then again would define a "most efficient" resource use, but this time under changed circumstances.

When the objectives are not (entirely) economic in nature (as is the case in Chapter 7 where the objective function varies between maximization of producer surplus, maximization of employment, minimization of negative soil nutrient stock changes, or minimization of biocide use), a similar reasoning may be applied, but then with the efficiency concept defined in terms of those objectives. However, such objectives do not fall under the Pearce and Turner (1990) definition of sustainable development adopted for the SOLUS framework (Chapter 6), which involves "maximizing the net benefits of economic development, subject to maintaining the services and quality of natural resources over time." On the other hand, in Chapter 7 only *alternative* land use systems are given consideration. Such land use systems are designed in a target-oriented way. Given a certain target yield (which can be sustained in the long run, *i.e.*, without negative changes in soil nutrient stocks), all required inputs (resources) are calculated such that their use is technically efficient, while the requirement of non-depletion over time in the Pearce and Turner definition also is satisfied.

10.5.4 An integrated approach

The manner in which the above three sustainability issues were incorporated in the SOLUS and the UNA-DLV methodologies allows a multi-faceted analysis of sustainability in land use. Biophysical sustainability indicators are objectively formulated and quantified on the level of the land use system in the Technical Coefficient Generators. Alternative land use systems may be generated that fulfil the requirement that the resource base is maintained over time and which (theoretically) maintain their productivity indefinitely. In the linear programming models, sustainability indicators are aggregated to the levels of farm, sub-region or region. Subsequently, normative bounds are placed on the various (summed) indicator values and land use is optimized with respect to a certain goal. This goal may be economic in nature (*e.g.*, maximization of income or employment in agriculture) or environmentally-oriented (*e.g.*, minimization of biocide use). In the latter case, restrictions can be placed on the minimum levels of agricultural production, income or employment to be realized. Ideally, interactive procedures are employed to reach agreement about goals and normative constraints among different stakeholders in the study area. By running the model for different combinations of goals and constraints (scenarios), consequences and trade-offs associated with certain (policy) choices are quantified. In the predictive UNA-DLV methodology, specific policy measures can then be evaluated that effectively lead to the adoption of certain land use systems that are desirable from a sustainability point of view.

10.6 User involvement in land use analysis

The ultimate purpose of land use analysis on the (sub-)regional or national level is to support policy decisions. Policy makers and stakeholders with an interest in the area under study should ideally be active participants in any land use analysis. However, unlike in prototyping, where scientists and farmers together experiment in finding new forms of land use, the methodologies of land use analysis should be in a relatively advanced stage of development before end-users can actively be involved. Issues to be addressed and scenarios to be evaluated should be interactively defined with end-users and stakeholders, but the methodologies to be followed should be clear from the outset. Therefore, when after a period of largely disciplinary research (see Chapter 1) REPOSA in 1990-91 identified a need for methodologies for analyzing land use capable of addressing issues related to socio-economic and biophysical sustainability, the initial focus of the work was on the development of a scientific methodology. Most efforts were geared towards the development of tools that would be able to answer broad policy questions of the type formulated in Section 10.4 and which, according to the researchers involved, were relevant to policy makers in general. Policy makers and stakeholders were not (yet) directly involved because the specific details (or indeed the exact nature) of the resulting methodologies were not yet clear. By the mid 1990s, the methodologies were felt to have sufficiently matured to conduct meetings, workshops and courses meant to familiarize people from different Costa Rican institutions with early results and to disseminate them in general. These institutions included different departments within the Ministry of Agriculture and Livestock of Costa Rica (MAG, *Ministerio de Agricultura y Ganadería*), both on the national and the regional level, the National Production Council (CNP, *Consejo Nacional de Producción*), the Ministry of Planning (MIDEPLAN, *Ministerio de Planificación Nacional y Política Económica*), universities (UNA, *Universidad Nacional Autónoma*; UCR, *Universidad de Costa Rica*), the Tropical Agricultural Research and Higher Education Center (CATIE, *Centro Agronómico Tropical de Investigación y Enseñanza*) and the School of Agriculture in the Humid Tropical Region (EARTH, *Escuela Agrícola de la Región del Trópico Húmedo*), as well as a number of Non Governmental Organizations. In 1994, the UNA-DLV program was initiated in collaboration with the International Centre on Economic Policy for Sustainable Development (CINPE, *Centro Internacional de Política Económica*) to develop predictive land use methodologies in close co-operation with REPOSA. In 1996, new initiatives were started to prototype precision agriculture together with an independent banana farmer in the AZ. Finally, a study was undertaken to identify the development views, objectives and goals of various institutions, stakeholders and policy makers in the AZ in relation to land use policies (Wilhelmus, 1998).

10.6.1 The Aranjuez watershed study

Towards the end of 1997, the Agricultural Research Department (*Dirección Nacional de Investigaciones Agropecuarias*), the Extension Department (*Dirección Nacional de Extensión*) and the Planning Department (SEPSA, *Secretaria Ejecutiva de Planificación del Sector Agropecuario*) of MAG invited REPOSA to jointly execute an explorative land use study in the watershed of the Aranjuez river (*Cuenca del Río Aranjuez*, with a size of about 22 000 ha) located in the Central Pacific region of Costa Rica. The objective was to adopt, adapt and apply the SOLUS methodology to study its usefulness for MAG as an instrument for policy support, analysis and preparation. To execute the study, a team was formed with a nucleus of MAG personnel, REPOSA researchers, and an economist from CINPE trained in the UNA-DLV methodology. This nucleus was complemented with a large number of local MAG field workers and extensionists from the watershed. During the first half of 1998, a number of courses in the different modules of SOLUS (see Figure 6.2 in Chapter 6) were given, followed by the application study itself, including "on-the-job" training for both MAG personnel at the central offices in San José and for local extension workers in the watershed. The joint study was completed in December 1998 and resulted in two technical reports (Hengsdijk, 1999; Sáenz *et al.*, 1999) as well as an independent evaluation of the process of transfering the SOLUS methodology to the different MAG departments (Mera-Orcés, 1999).

The first step in the Aranjuez study consisted in delineating agro-ecological zones within the study area in terms of distinctive soil types, followed by the collection and processing of the technical information of the relevant land use systems (in this case maize, black beans, rice, melons, coffee, sugar cane, mangos, citrus, tomatoes and sweet peppers, along with pastures for cattle breeding, fattening or milk production), and adapting LUCTOR and PASTOR. Furthermore, demographic, farm size and other socio-economic data (in particular prices) were obtained. Apart from using existing studies (*e.g.*, Arroyo *et al.*, 1996 and MAG, 1994), most of the information needed was generated by local extensionists. A linear programming model called MUTCA (*Modelo de Uso de la Tierra en la Cuenca del Río Aranjuez*) was developed to analyze land use in the watershed. Besides sustainability parameters that had already been developed for the AZ, two new indicators relevant to the area under study were defined (*i.e.*, soil erosion and water run off) and incorporated in the Technical Coefficient Generators PASTOR and LUCTOR as well as in MUTCA. Finally, a variety of policy scenarios were designed and evaluated through a number of model runs. The results were discussed extensively among all team members as well as in two workshops. The first of these workshops took place in the study area with a large audience of local MAG extensionists and other stakeholders in the watershed, while the second one was conducted in San José among stakeholders and representatives of various Costa Rican institutions that deal with policy issues related to land use, *e.g.*, the Inter-American Institute for Cooperation on Agriculture (IICA; *Instituto Interamericano de Cooperación para la Agricultura*), the Regional Program for Strengthening Agronomic Research on Grains in Central America (PRIAG; *Programa Regional de Reforzamiento a la Investigación Agronómica sobre los Granos en Centroamérica*), the National Banana Corporation (CORBANA; *Corporación Bananera Nacional*), and the UNA.

10.6.2 Lessons learned

The main conclusions that can be derived from the Aranjuez watershed study are twofold, the first relating to the use and transfer of knowledge in the form of models and software, and the second to the use of the SOLUS methodology on the sub-regional level. Regarding knowledge transfer, the intense discussions that took place between members of the MAG-UNA-REPOSA project team as well as between team members and other MAG personnel proved highly valuable. These discussions related to the development of technical coefficients, defining appropriate sustainability indicators, setting up the optimization model, defining scenario runs, and analyzing the model outcomes. For example, it was clearly demonstrated that labor is an important bottleneck for any attempt to increase agricultural production in the Aranjuez watershed. Another salient conclusion was that the current price of so-called "organic" coffee (*i.e.*, coffee that is produced without inorganic fertilizer and with low biocide use) is far too low to become more attractive than coffee produced in the traditional way. However, before arriving at such conclusions, the participating MAG officials had to go through a phase of technology transfer. MAG officials collaborating in the Aranjuez study clearly were not satisfied with discussing scenarios and results of models the functioning of which they did not fully understand. However, the functional details of the SOLUS methodology are rather complex and indeed require substantial prior knowledge and training. This holds particularly true for the linear programming and optimization module of SOLUS. For example, the MUTCA model was developed mainly by the CINPE economist who, as a former member of the UNA-DLV team, had had more than two years of intensive experience in the use of linear programming in land use analysis. On the other hand, the Technical Coefficient Generators PASTOR and LUCTOR were (somewhat unexpectedly) found to raise considerable interest as independent tools for evaluating alternative land use systems on the farm level. Both PASTOR and LUCTOR stimulated the elicitation of expert knowledge among local MAG extensionists, forcing them to quantify their knowledge to help generate technical coefficients for the linear programming model. Some of them even extended PASTOR and LUCTOR with their own suggestions and ideas. Generally for the SOLUS methodology, however, a substantial amount of time and on-the-job training were essential to generate a sufficient level of understanding among the different partners in the project. For MAG personnel belonging to the central San José office, involvement in other projects and administrative duties limited the available time which could be spent on the Aranjuez study.

The second lesson learned concerns a methodological issue. Decisions regarding land use take place on the farm rather than on a (sub-)regional level, such as the Aranjuez watershed, and authorities can only formulate certain land use guidelines or restrictions. On the farm, income maximization usually is an important objective (possibly in addition to other objectives like, for example, food security and risk avoidance). Furthermore, a (sub-)region is not isolated from the rest of the country. Products from the Aranjuez watershed are marketed to a large extent outside the area, while at the same time labor exchange with other areas takes place. Even though output markets are largely located outside the watershed, they obviously influence production opportunities within the area. For example, the production of certain crops such as tomatoes, sweet peppers and mangos can be increased considerably from their current levels as long as sufficient labor is available. As the demand for these products is limited, sufficiently large quantities of these products from the Aranjuez watershed could affect national prices. However, modeling markets for these products is outside the scope and purpose of the model for the Aranjuez watershed. A similar reasoning applies to labor. In the watershed, the labor market is generally tight. Even though labor could theoretically be attracted from outside the area (*e.g.*, from other areas in Costa Rica as well as from Nicaragua), for various reasons that are outside the scope of the Aranjuez watershed study this does not happen in reality (except during the coffee and sugar cane harvests). Therefore, it may be argued that on the sub-regional level (such as the Aranjuez watershed) land use studies should focus on either (*i*) models of different farm (household) types (*e.g.*, Chapter 8), (*ii*) models that clearly explore the biophysical limitations on production without taking product or labor market considerations into account (*e.g.*, Chapter 7), or (*iii*) models of larger regions that include markets for major commodities as well as labor (*e.g.*, Chapter 6).

Despite the above caveats, most team members involved in the Aranjuez watershed study were quite satisfied with the experience gained with the SOLUS methodology. Currently, a project proposal is being formulated by the Office of the Minister of the MAG (in collaboration with Costa Rican and Dutch researchers) with the objective to gain more experience with SOLUS as a tool for land use analysis and policy evaluation.

10.7 Conclusions

At the end of a book that deals with a wide range different methodologies for land use analysis, it is difficult to draw general conclusions. For details on the functioning and application domain of the methodologies that receive attention in this book, the reader is referred to Chapters 3 to 9, while Chapters 1 and 10 are of a more comparative and reflective character. Nonetheless, the following concluding remarks are considered worth making:

1. In order to be able to address land use issues and related policies in a comprehensive way useful to policy makers, land use analysis should have a multidisciplinary and multilevel character, involving both biophysical and socio-economic elements. It also should consider at least two, preferably adjacent, levels out of the five spatial levels distinguished in this book (land use system, farm, sub-region, region and nation).
2. The "toolbox" for land use methodologies presented in this book contains methods that can be used to answer specific, user-defined questions related to land use. As such, each of the methodologies has a stand-alone value. However, overall policy or regional development issues can be more effectively addressed when the methodologies are implemented in a sequential or, even better, iterative manner. For example, one may start with a *projective* analysis of past trends (*e.g.*, CLUE), then realize biophysically oriented (*e.g.*, SOLUS/GOAL) and bio-economically oriented (*e.g.*, SOLUS/REALM) *explorative* studies, followed by *predictive* policy oriented studies of the UNA-DLV type, taking farm household behavior explicitly into account. In these explorative and predictive methodologies, part of the model coefficients can be provided by Technical Coefficient Generators (*e.g.*, LUCTOR or PASTOR). *Designing* or *prototyping* methodologies, of which BanMan is an example, can play an important role to carry the synergy between economic aspects and biophysical sustainability further. Explorations and predictions on higher levels of aggregation can help in targeting the design of new land use systems on the field/farm level. During or after the explorative studies one may also undertake a SEM type of study to analyze the regional welfare and trade effects of development policies formulated on the national level. Results of such a SEM-type of study may again be used to formulate policy scenarios which can be further analyzed in explorative and predictive studies.
3. Research into land use analysis does not end with this book. Many methodological questions surrounding spatial and temporal aggregation remain unresolved. Furthermore, if the tools at hand are to play an important role in evaluating policies for sustainable agricultural production and development, these tools should be made more transparent for policy makers and more user-friendly for analysts. Finding the right balance between scientific rigor and practical usability constitutes a major challenge for future research.

References

Agne, S., 1996. Economic analysis of crop protection policy in Costa Rica. Pesticide policy project publication series no. 4, Institute of Horticultural Economics, Hannover University, Hannover, Germany.

Alexandratos, N. (Ed.), 1995. World agriculture towards 2010: a FAO study. FAO, Rome, Italy. John Wiley and Sons, Chichester. UK.

Alvarado, G.E., 1989. Los volcanes de Costa Rica. UNED, San José, Costa Rica.

Alfaro, R., 1993. Análisis de inventario en una comunidad campesina de la Zona Atlántica de Costa Rica: el caso de Agrimaga. Report No. 43, Phase 2, Atlantic Zone Program (CATIE/MAG/WAU), Turrialba, Costa Rica.

Arroyo, L., M. Ugalde, R. Méndez and V. Salazar, 1996. Diagnóstico para la evaluación de tierras en la cuenca del Río Aranjuez. Departamento de Suelos y Evaluación de Tierras, Dirección de Investigaciones Agropecuarias, Ministerio de Agricultura y Ganadería (MAG). San José, Costa Rica. 56 pp.

Askari, H. and J.T. Cummings, 1976. Agricultural supply response: a survey of the econometric evidence. New York, Praeger.

Bade, J., H. Hengsdijk, G. Kruseman, R. Ruben and P.C. Roebeling, 1997. Farm household modelling in a regional setting: the case of Cercle Koutiala, Mali. DLV-report No. 6. AB-DLO/WAU, Wageningen, The Netherlands. 39 pp.

Bakker, E.J., H. Hengsdijk and K. Sissoko, 1998. Sustainable land use in the Sudano-Sahelian zone of Mali: exploring economically viable options using multiple-goal linear programming. Netherlands Journal of Agricultural Science 46: 109-122.

Barbier, B. and G. Bergeron, 1999. Natural resource management in the hillsides of Honduras: bioeconomic modeling at the micro-watershed level. EPTD Discussion Paper No. 32. International Food Policy Research Institute, Washington.

Belder, M., 1994. Land use and land use dynamics in the Atlantic Zone of Costa Rica. Field Report 84, Phase 2, REPOSA, CATIE/MAG/WAU, Turrialba, Costa Rica. 47 pp.

Bell, C., P. Hazell and R. Slade, 1982. Project evaluation in regional perspective: a study of an irrigation project in Northwest Malaysia. The Johns Hopkins University Press, Baltimore, USA. 326 pp.

Bessembinder, J., 1997. Uncertainty and temporal aspects in long-term explorations of sustainable land use; with reference to the Northern Atlantic Zone of Costa Rica. Ph.D. thesis, Wageningen Agricultural University, Wageningen, The Netherlands. 166 pp., plus appendices.

Bosworth, D., P. Dawkins and T. Stromback, 1996. The economics of the labor market. Longman, Harlow, UK.

Bouma, J. 1997. The role of quantitative approaches in soil science when interacting with stakeholders. Geoderma 78: 1-12.

Bouman, B.A.M., H. van Keulen, H.H. van Laar and R. Rabbinge, 1996. The 'School of de Wit' crop growth simulation models: pedigree and historical overview. Agricultural Systems 52: 171-198.

Bouman, B.A.M., A. Nieuwenhuyse and H. Hengsdijk, 1998a. PASTOR: a technical coefficient generator for pasture and livestock systems in the humid tropics, version 2.0. Quantitative Approaches in Systems Analysis No. 18. AB-DLO/C.T. de Wit Graduate School for Production Ecology. Wageningen, The Netherlands. 59 pp., plus appendices.

Bouman, B.A.M., A. Nieuwenhuyse, R.A. Schipper, H. Hengsdijk and H.G.P. Jansen, 1998b. An integrated methodology for sustainable land use exploration using GIS. Pages 230-237 in: Proceedings of the first international conference on geospatial information in agriculture and forestry, Lake Buena Vista, Florida. 1-3 June 1998. Volume II. ERIM International, Ann Harbor, USA.

Bouman, B.A.M., R.A. Schipper, A. Nieuwenhuyse, H. Hengsdijk and H.G.P. Jansen, 1998c. Quantifying economic and biophysical sustainability trade-offs in land use exploration at the regional level: a case study for the Northern Atlantic Zone of Costa Rica. Ecological Modelling 114: 95-109.

Bouman, B.A.M. and A. Nieuwenhuyse, 1999. Exploring options for sustainable beef cattle ranching in the humid tropics; a case study for the Atlantic Zone of Costa Rica. Agricultural Systems 59(2): 145-161.

Bouman, B.A.M., H.G.P. Jansen, R. A. Schipper, A. Nieuwenhuyse, H. Hengsdijk and J. Bouma, 1999a. A methodology for integrated biophysical and economic land use analysis at different scales. Agriculture, Ecosystems and Environment 75(1-2): 55-73.

Bouman, B.A.M., A. Nieuwenhuyse and M. Ibrahim, 1999b. Pasture degradation and restoration by legumes in humid tropical Costa Rica. Tropical Grasslands 33(2): 98-110.

Bouman, B.A.M., R.A.J. Plant and A. Nieuwenhuyse, 1999c. Quantifying economic and biophysical sustainability trade-offs in tropical pastures. Ecological Modelling 120(1): 35-50.

Brooke, A., D. Kendrick and A. Meerhaus, 1992. GAMS release 2.25: a user's guide. The Scientific Press, South San Francisco, USA. 289 pp.

Brouwer, J. and J. Bouma, 1997. Soil and crop growth variability in the Sahel: highlights of research (1990-95) at ICRISAT Sahelian Center. Information Bulletin no. 49. ICRISAT, Niamey, Niger and WAU, Wageningen, The Netherlands. 42 pp.

Bulte, E.H., M. Joenje and H.P.G. Jansen, 1999a. Tropical forest functions and the optimal forest stock in developing countries: is there too much or too little forest in the Atlantic zone of Costa Rica? Canadian Journal of Forest Research. (in press)

Bulte, E.H., B.A.M. Bouman, R.A.J. Plant, A. Nieuwenhuyse and H.G.P. Jansen, 1999b. The economics of soil nutrient stocks and cattle ranching in the tropics: optimal pasture degradation in humid Costa Rica. European Review of Agricultural Economics. (in press)

Buurman, P., K. de Boer and Th. Pape, 1997. Laser diffraction grain-size characteristics of Andisols in perhumid Costa Rica: the aggregate size of allophane. Geoderma 87: 71-91.

Byerlee, D, and M. Collinson, 1980. Planning new technologies appropriate to farmers: concepts and procedures. CYMMYT, Mexico.

Cadisch, G., R.M. Schunke and K.E. Giller, 1994. Nitrogen cycling in a pure grass pasture and a grass-legume mixture on a red latosol in Brazil. Tropical Grasslands 28: 43-52.

Cartín, S. and I. Piszk, 1980. La producción de granos básicos en Costa Rica. Revista de Ciencias Sociales, Instituciones Estatales y Fuerzas Sociales. Periódo de diversificación económica, 19-20: 25-35. University of Costa Rica, San José.

Castillo, L., E. de la Cruz and C. Ruepert, 1997. Ecotoxicology and pesticides in tropical aquatic ecosystems of Central America. Environmental Toxicology and Chemistry 16: 41-51.

Castro, C., G. Barrantes and F. Sáenz, 1996. Información socioeconómica sobre la Zona Atlántica de Costa Rica. CINPE, Universidad Nacional Autónoma, Heredia, Costa Rica. 43 pp.

Castro, V., 1985. Estudio climático de Costa Rica para la zonificación agropecuaria y forestal. Vol. 1 - manual de uso. SEPSA - MIDEPLAN, San José, Costa Rica.

CATIE (Centro Agronómico Tropical de Investigación y Enseñanza), 1989. Sistemas silvopastoriles para el trópico húmedo bajo. Informe primera fase. MAG-IDA-CATIE/CIID, CATIE, Turrialba, Costa Rica. 50 pp.

CCE, 1990. Desarollo rural integrado. Concepto o finalidad? Una experiencia en Costa Rica. Comisión de la Comunidad Europea, el proyecto CEE NA/82-12. Agrofototechnica, Firenze, Italy. 100 pp.

Céspedes, V.H., 1998. Crecimiento y condiciones sociales. Pages 187-250 in: Jiménez, R. (Ed.) Estabilidad y desarrollo económico en Costa Rica: las reformas pendientes. Academia de Centroamerica, San José, Costa Rica.

Chandler, W. and R.D. Norton, 1977. Multi-level programming and development policy. World Bank Staff Working Paper 25, World Bank, Washington, USA.

Chandler, W., J. Fortuny-Amat and B. Mc Carl, 1981. The potential of multi-level programming in agricultural economics. American Journal of Agricultural Economics 63: 521-531.

Chavarría, M.I. and R. Valerio, 1993. Guia preliminar de parametros silviculturales para apoyar los proyectos de reforestación en Costa Rica. Dirreción General Forestal, MIRENEM, San José, Costa Rica. 202 pp.

CIAT, 1995. Informe biannual 1994-1995. Programa de forajes tropicales. Documento de Trabajo 153. Centro International de Agricultural Tropical, Cali, Colombia.

CIBA foundation, 1997. Precision agriculture: spatial and temporal variability of environmental quality. CIBA Foundation Symposium 210. John Wiley and Sons. New York, Toronto.

CNP [Consejo Nacional de Producción] and SEPSA [Secretaría Ejecutiva de Planificación Sectoral Agropecuaria], 1990. Encuesta ganadería nacional 1988. San Jose, Costa Rica.

Colman, D. and T. Young, 1989. Principles of agricultural economics: markets and prices in less developed countries. Cambridge University Press, Cambridge, UK. 323 pp.

Coopemontecillos. Revista informativa de comité de educación y bienestar social. Alajuela, Costa Rica. Various issues.

CORBANA, 1997. Costa Rica Informe annual de estadísticas de exportación de banano 1997. CORBANA, San José, Costa Rica.

Crissman, C.C., J.M. Antle, and S.M. Capalbo, 1998. Economic, environmental, and health tradeoffs in agriculture: pesticides and the sustainability of Andean potato production. Kluwer Academic Publishers, Dordrecht in cooperation with the International Potato Center, Lima.

Crotty, R., 1980. Cattle, economics and development. Farnham Royal: Commonwealth Agricultural Bureau, England. 253 pp.

Deaton, A. and A. Case, 1988. Analysis of household expenditures. LSMS Working Paper no. 28, The World Bank, Washington, USA.

De Koning, G.H.J., H. van Keulen, R. Rabbinge and H. Janssen, 1995. Determination of input and output coefficients of cropping systems in the European Community. Agricultural Systems 48: 485-502.

De Koning, G.H.J., P.J. van de Kop and L.O. Fresco, 1997. Estimates of sub-national nutrient balances as sustainability indicators for agro-ecosystems in Equador. Agriculture, Ecosystems and Environment 65: 127-139.

De Koning, G.H.J., A. Veldkamp and L.O. Fresco, 1998a. Land use in Equador: a statistical analysis at different aggregation levels. Agriculture, Ecosystems and Environment 70: 231-247.

De Koning, G.H.J., A. Veldkamp, P.H. Verburg, K. Kok and A.R. Bergsma, 1998b. CLUE: a tool for spatially explicit and scale sensitive exploration of land use changes. Pages 59-68 in: Stoorvogel, J.J., J. Bouma and W.T. Bowen (Eds) Information technology as a tool to assess land use options in space and time. Quantitative Approaches in Systems Analysis no. 16, AB-DLO/C.T. de Wit Graduate School for Production Ecology. Wageningen, The Netherlands.

Delvaux, B, 1995. Soils. Pages 230-257 in: Gowen, S. (Ed.). Bananas and plantains. Chapman & Hall, London, UK.

Den Daas, J.L. 1993. Producing pineapple in the Atlantic Zone of Costa Rica: agronomic and marketing aspects. Field Report 51, Phase 2, REPOSA (CATIE/MAG/WAU), Turrialba, Costa Rica.

Detwiler, R.P. and C.A.S. Hall, 1988. Tropical forests and the global carbon cycle. Science 239: 42-47.

De Vries, P., 1990. State-smallholder relations in a tropical colonization area: an institutional perspective on the Atlantic Zone. Pages 8-19 in: Box, L. (Ed.). From common ignorance to shared knowledge: knowledge networks in the Atlantic Zone of Costa Rica. Wageningen Studies in Sociology No. 28, Wageningen Agricultural University, The Netherlands.

De Wit, C.T., 1979. The efficient use of land, labour and energy in agriculture. Agricultural Systems 5: 279-287.

De Wit, C.T., 1994. Resource use analysis in agriculture: a struggle for interdisciplinarity. Pages 41-55 in Fresco, L.O., L. Stroosnijder, J. Bouma and H. van Keulen (Eds) The future of the land. Mobilizing and integrating knowledge for land use options. John Wiley and Sons, Chichester.

De Wit, C.T., H. Huisman and R. Rabbinge, 1987. Agriculture and its environment: are there other ways? Agricultural Systems 23: 211-236.

De Wit, C.T., H. Van Keulen, N.G. Seligman and I. Spharim, 1988. Application of interactive multiple goal programming techniques for analysis and planning of regional agricultural development. Agricultural Systems 26: 211-230.

DGEC [Dirección General de Estadística y Censos], 1987a. Censo agropecuario 1984. Ministerio de Economía, Industria y Comercio, San José, Costa Rica.

DGEC [Dirección General de Estadística y Censos], 1987b. Censo de población 1984, tome 1 y 2. Ministerio de Economía, Industria y Comercio, San José, Costa Rica.

DGEC [Dirección General de Estadística y Censos], 1988. Metodología. Encuesta Nacional de Ingresos y Gastos de los Hogares, Informe No. 2. Ministerio de Economía, Industria y Comercio, San José, Costa Rica.

DGEC [Dirección General de Estadística y Censos], 1990. Avance de resultados. Encuesta Nacional de Ingresos y Gastos de los Hogares, Informe No.1. Ministerio de Economía, Industria y Comercio, San Jose, Costa Rica.

DGEC [Dirección General de Estadística y Censos], 1997a. Costa Rica: calculo de población por provincia, canton y distrito al 1[a] de julio de 1996. Ministerio de Economía, Industria y Comercio, San José, Costa Rica. 17 pp.

DGEC [Dirección General de Estadística y Censos], 1997b. Encuesta de hogares de proposito multiples modulo de empleo julio de 1996. Ministerio de Economía, Industria y Comercio, San José, Costa Rica. 71 pp.

Diltz, D., 1980. Methodology for qualitative economic land suitability evaluation for agriculture. AG: DP/SIL/73/002; Technical Report 9, UNDP/FAO, Freetown, Sierra Leone. 145 pp.

Dinham, B., 1993. The pesticide hazard: a global health and environmental audit. Zed, London.

Driese, K.L. and W.A. Reiners, 1999. Thematic mapper of the Atlantic Zone of Costa Rica 1996. Unpublished data of Wyoming University, USA.

Driessen, P.M. and Konijn, N.T., 1992. Land use systems analysis. Department of Soil Science and Geology, Wageningen Agricultural University, Wageningen, The Netherlands. 230 pp.

Duloy, J.H. and R.D. Norton, 1983. The CHAC demand structures. Pages 58-82 in: Norton, R.D and L. Solís (Eds) The book of CHAC: programming studies for Mexican agriculture. Johns Hopkins University Press, Baltimore, USA.

Duloy, J.H. and R.D. Norton, 1973. CHAC: a programming model of Mexican agriculture. Pages 291-337 in: Goreux, L.M. and A.S. Manne (Eds). Multi-level planning: case studies in Mexico. North-Holland, Amsterdam, The Netherlands.

Dumanski, J., W.W. Pettapiece and R.J. McGregor, 1998. Relevance of scale dependent approaches for integrating biophysical and socio-economic information and development of agroecological indicators. Nutrient Cycling in Agroecosystems 50: 13-22.

Easterling, W.E., 1997. Why regional studies are needed in the development of full-scale integrated assessment modelling of global change processes. Global Environmental Change 7: 337-356.

Easterly, W., N. Loayza and P. Montiel, 1997. Has Latin America's post-reform growth been disappointing? Journal of International Economics 43: 287-311.

Echandi, R., 1998. Deterioro de la infraestructura del sector transportes, implicaciones y soluciones. Pages 359-392 in: Jiménez, R. (Ed.) Estabilidad y desarrollo económico en Costa Rica: las reformas pendientes. Academia de Centroamerica, San José, Costa Rica.

Ellis, F., 1992. Agricultural policies in developing countries. Cambridge University Press, Cambridge, UK. 357 pp.

Erenstein, O.C.A. and R.A. Schipper, 1993. Linear programming and land use planning: a case study of Matara district, Sri Lanka. Wageningen Economic Studies 28, Pudoc, Wageningen, The Netherlands. 95 pp.

Estado de la Nación, 1997. Estado de la nación en desarollo humano sostenible. San José, Costa Rica.

Faber, A., 1997. Study on investigations on pesticides and the search for alternatives in Costa Rica. Field Report 128, Phase 2, REPOSA (CATIE/MAG/WAU), Turrialba, Costa Rica. 23 pp., plus appendices.

Falconer, K.E., 1998. Managing diffuse environmental contamination from agricultural pesticides: an economic perspective on issues and policy options, with particular reference to Europe. Agriculture, Ecosystems and Environment 69: 37-54.

FAO [Food and Agriculture Organization of the United Nations], 1976. A framework for land evaluation. FAO Soils Bulletin 32, FAO, Rome, Italy. 72 pp.

FAO [Food and Agriculture Organization of the United Nations], 1983. Guidelines: land evaluation for rainfed agriculture. FAO Soils Bulletin 52, FAO, Rome, Italy. 237 pp.

FAO [Food and Agriculture Organization of the United Nations], 1993. Guidelines for land use planning. FAO Development Series 1, FAO, Rome, Italy. 96 pp.

FAO [Food and Agriculture Organization of the United Nations], 1998. The FAOSTAT databases. http://apps.fao.org/lim500/agri_db.pl or via the FAO homepage: http://www.fao.org/

Fields, G.S., 1988. Employment and economic growth in Costa Rica. World Development 16: 1493-1509.

Finke, P.A., 1993. Field scale variability of soil structure and its impact on crop growth and nitrate leaching in the analysis of fertilizing scenarios. Geoderma 60: 89-107.

Finke, P.A., J. Bouma and A. Stein, 1992. Measuring field variability of disturbed soils for simulation purposes. Soil Science Society of America Journal 56: 187-192.

Fisher, M.J., I.M. Rao, M.A. Ayarza, C.E. Lascano, J.I. Sanz, R.J. Thomas and R.R. Vera, 1994. Carbon storage by introduced deep-rooted grasses in the South American savannas. Nature 371: 236-238.

Fresco, L.O. and S.B. Kroonenberg, 1992. Time and spatial scales in ecological sustainability. Land Use Policy 9: 155-168.

Fresco, L.O., H. Huizing, H. van Keulen, H. Luning and R.A. Schipper, 1992. Land evaluation and farming systems analysis for land use planning, FAO-working document, Rome, Italy/ITC, Enschede, The Netherlands/Wageningen Agricultural University, Wageningen, The Netherlands. 209 pp.

Frimpong-Ansah, J.H., 1992. The vampire state in Africa: the political economy of decline in Ghana. Africa World Press, Trenton, N.J. and James Currey, London.

Geurts, J.A.M.M., H.G.P. Jansen and A. van Tilburg, 1997. Domestic demand for food in Costa Rica: a double hurdle analysis. Serie Técnica. Informe Técnico No. 286, CATIE, Turrialba, Costa Rica. 100 pp.

Ghatak, S. and K. Ingersent, 1984. Agriculture and economic development. Wheatsheaf Books, Sussex. 380 pp.

Gittinger, J.P., 1982. Economic analysis of agricultural projects. John Hopkins University Press, Baltimore, USA. 505 pp.

González M. H., 1994. Desarrollo agropecuario y políticas macroeconómicas en la decada del 80 en Costa Rica. Pages 53-73 in: Mora Alfaro, J., O. Oviedo Sánchez and L.F. Fernández Alvarado (Eds) El impacto de las políticas macroeconómicas en el agro costarricense. Facultad de Ciencias Sociales, Universidad Nacional, Editorial Universidad Nacional, Heredia, Costa Rica.

González, R., E. Bustamente, P. Shannon and C. Ruiz, 1997. El control bióIogico en el manejo integrado de *Mycosphaerella fijiensis*. Actas de la III Semana Científica, CATIE, Turrialba, Costa Rica. pp. 302-305.

Gowen, S. (Ed.), 1995. Bananas and plantains. Chapman & Hall, London.

Guardia, J., A. di Mare, T. Vargas, V.H. Céspedes, J. Corrales and M. Baldares, 1987. La política de precios en Costa Rica. COUNCEL-CIAPA-AID, Trejos publishers, San José, Costa Rica.

Habekotté, B., 1994. TCG-CROP, a model for the calculation of production and environmental variables of different crops, developed for the project 'Introduction integrated arable farming' (in Dutch). Simulation Reports CABO-TT No. 35. AB-DLO, Wageningen, The Netherlands. 40 pp., plus appendices.

Hallam, D. 1995. The world banana economy. Pages 509-533 in Gowen, S. (Ed.) Bananas and plantains. Chapman & Hall, London, UK.

Hansen-Kuhn, K., 1993. Sapping the economy. Structural adjustment policies in Costa Rica. The Ecologist 23: 179-185.

Harsthorn, G.S., 1983. Plants. Pages 118-157 in: D.H. Janzen (Ed.) Costa Rican natural history. The University of Chicago Press, Chicago, USA.

Hausmann, R., 1998. Avances parciales y estanamientos en el proceso de reforma estructural costarricense. Pages 3-17 in: Jiménez, R. (Ed.) Estabilidad y desarrollo económico en Costa Rica: las reformas pendientes. Academia de Centroamerica, San José, Costa Rica.

Haynes, R.J. and P.H. Williams, 1993. Nutrient cycling and soil fertility in the grazed pasture ecosystem. Advances in Agronomy 49:119-199.

Hazell, P.B. and R.D. Norton. 1986. Mathematical programming for economic analysis in agriculture. Collier MacMillan Publishers, London, UK. 400 pp.

Heady, E.O. and G.F. Vocke (Eds), 1992. Economic models of agricultural land conservation and environmental improvement. Iowa State University Press, Ames, USA.

Heilig, G. 1996. World population prospects. Analyzing the UN population projections. IIASA, Working paper WP-96-146. IIASA, Laxenburg, Austria.

Henderson J.M. and R.E. Quandt, 1980. Microeconomic theory: a mathematical approach. McGraw-Hill Book Company, 420 pp.

Hengsdijk, H., W. Quak, E.J. Bakker and J.J.M.H. Ketelaars, 1996. A technical coefficient generator for land use activities in the Koutiala region of south Mali. DLV report no. 6, AB-DLO/WAU, Wageningen, The Netherlands. 96 pp., plus appendices.

Hengsdijk, H, 1999. LUCTOR-Aranjuez: sistema de experto para cuantificar sistemas de cultivo en la cuenca Aranjuez. Report No. 140, REPOSA (CATIE/MAG/WAU), Turrialba, Costa Rica. 42 pp., plus appendices.

Hengsdijk, H., A. Nieuwenhuyse and B.A.M. Bouman, 1998a. LUCTOR: land use crop technical coefficient generator. A model to quantify cropping systems in the northern Atlantic Zone of Costa Rica. Version 2.0. Quantitative Approaches in Systems Analysis No. 17. AB-DLO/C.T. de Wit Graduate School for Production Ecology. Wageningen, The Netherlands. 65 pp., plus appendices.

Hengsdijk, H., M.K. van Ittersum and W.A.H. Rossing, 1998b. Quantitative analysis of farming systems for policy formulation: development of new tools. Agricultural Systems 58: 381-394.

Hengsdijk, H., B.A.M. Bouman, A. Nieuwenhuyse and H.G.P. Jansen, 1999. Quantification of land use systems using technical coefficient generators: a case study for the Northern Atlantic Zone of Costa

Rica. Agricultural Systems 61(2): 109-121.

Henneberry, S. R., 1986. A review of agricultural supply responses for international policy models. Agricultural Policy Analysis Project, US Agency for International Development and Oklahoma State University, Stillwater, Oklahoma, USA.

Hernández, M., Argel, P.J., Ibrahim, M.A. and L. 't Mannetje, 1995. Pasture production, diet selection and liveweight gains of cattle grazing *Brachiaria brizantha* with or without *Arachis pintoi* at two stocking rates in the Atlantic Zone of Costa Rica. Tropical Grasslands 29: 134-141.

Herrera, W., 1985. Clima de Costa Rica. UNED, San José, Costa Rica.

Hijmans, R.J. and M.K. van Ittersum, 1996. Aggregation of spatial units in linear programming models to explore land use options. Netherlands Journal of Agricultural Science 44: 145-162.

Hoekstra, S.J., 1996. Field research on transportation costs, farm-gate and farmer market prices in the Atlantic Zone of Costa Rica. Field Report 110, REPOSA (CATIE/MAG/WAU), Turrialba, Costa Rica. 89 pp.

Holling, C.S., 1992. Cross-scale morphology, geometry, and dynamics of ecosystems. Ecological Monographs 62: 447-502.

Houghton, R.A., 1991. Tropical deforestation and atmospheric carbon dioxide. Climatic Change 19: 19-118.

Ibrahim, M.A, 1994. Compatibility, persistence and productivity of grass-legume mixtures for sustainable animal production in the Atlantic Zone of Costa Rica. Ph.D. thesis, Wageningen Agricultural University, Wageningen, The Netherlands. 129 pp.

Jansen, D.M. and R.A. Schipper, 1995. A static descriptive approach to quantify land use systems. Netherlands Journal of Agricultural Science 43: 31-46.

Jansen, D.M., J.J. Stoorvogel and R.A. Schipper, 1995. Using sustainability indicators in agricultural land use analysis: an example from Costa Rica. Netherlands Journal of Agricultural Science 43: 61-82.

Jansen, D.M., J.J. Stoorvogel and H.G.P. Jansen, 1997a. A quantitative tool for regional land-use analysis. Pages 399-412 in Teng, P.S., M.J. Kropff, H.F.M. ten Berge, J.B. Dent, F.P. Lansigan and H.H. van Laar (Eds) Applications of systems approaches at the farm and regional levels, Volume 1. Kluwer Academic Publishers, Dordrecht, The Netherlands.

Jansen, H.G.P, M.A. Ibrahim., A. Nieuwenhuyse, L. 't Mannetje, M. Joenje and S. Abarca, 1997b. The economics of improved pasture and sylvipastoral technologies in the Atlantic Zone of Costa Rica. Tropical Grasslands 32: 34-44.

Jansen, H.G.P. and A. van Tilburg (with J. Belt and S. Hoekstra), 1996. Agricultural marketing in the Atlantic Zone of Costa Rica: a production, consumption and trade study of agricultural commodities produced by small and medium-scale farmers. CATIE Serie Técnica, Informe Técnico No. 271, Turrialba, Costa Rica. 120 pp.

Jansen, H.G.P. and J.J. Stoorvogel, 1998. Quantification of aggregation bias in regional agricultural land use models: application to Guácimo county, Costa Rica. Agricultural Systems 58: 417-439.

Jansen, H.G.P., E. Uytewaal and J J. Stoorvogel, 1998. Health externalities and pesticide use in the Atlantic Zone of Costa Rica: an economic evaluation. Ambio (submitted).

Jarvis, L.S., 1986. Livestock development in Latin America. The World Bank, Washington, USA. 214 pp.

Jarvis, P.G., 1993. Prospects for bottom-up models. Pages 115-126 in: Ehleringer J.R. and C.B. Field (Eds) Scaling physiological processes. Leaf to globe. Academic Press, San Diego, USA.

Jones, J.R., 1990. Colonization and environment - land settlement projects in Central America. United Nations University Press, Tokyo, Japan. 155 pp.

Kaimowitz, D., 1996. Livestock and deforestation in Central America in the 1980s and 1990s: a policy perspective. Center for International Forestry Research, Jakarta, Indonesia. 88 pp.

Keller, M., E. Veldkamp, A.M. Weitz and W.A. Reiners, 1993. Effect of pasture age on soil trace-gas emissions from a deforested area of Costa Rica. Nature 365: 244-246.

Kesel, R.H. and D.R. Lowe, 1987. Geomorphology and sedimentology of the Toro Amarillo fan in a humid tropical environment. Geografiska Annaler 69A: 85-99.

Krishnaiah, J., and S. Krishnamoorthy, 1988. Inter-regional allocation of major food grains in Andhra Pradesh: an application of spatial equilibrium model. Indian Journal of Agricultural Economics 43: 35-43.

Kropff, M.J., P.S. Teng and R. Rabbinge, 1995. The challenge of linking pest and crop models. Agricultural Systems 49: 413-434.

Kruseman, G., H. Hengsdijk, R. Ruben, P. Roebeling and J. Bade, 1997. Farm household modelling system for the analysis of sustainable land use and food security: theoretical and mathematical description. DLV Report No.7. AB-DLO/WAU, Wageningen, The Netherlands. 56 pp.

Kruseman, G., R. Ruben and H. Hengsdijk, 1994. Agrarian structure and land use in the Atlantic Zone of Costa Rica. DLV Report No. 3. AB-DLO/WAU, Wageningen, The Netherlands.106 pp.

Kruseman, G., R. Ruben, H. Hengsdijk and M.K. van Ittersum, 1995. Farm household modelling for estimating the effectiveness of price instruments in land use policy. Netherlands Journal of Agricultural Science 43: 111-123.

Kruseman, G., R. Ruben, H. Hengsdijk, J. Bade and H. Djouara, 1996. Policy instruments to address market and institutional constraints for sustainable land use at farm household level. Pages 299-308 in Bierschenk, T., P.-Y. Le Meur and M. von Oppen (Eds) Institutions and technologies for rural development in West Africa. Margraf Verlag Weikersheim, Germany.

Kuiper, M., 1996. The Neguev revisited: a study of the agricultural changes between 1987 and 1996 in the Neguev settlement, Costa Rica. Report 112, REPOSA (CATIE/MAG/WAU), Turrialba, Costa Rica. 64 pp.

Kuiper, M., 1997. The impact of land degradation on agricultural productivity: a multi-period economic land use model. A case study of the Neguev settlement, Costa Rica. Report 124, REPOSA. CATIE/MAG/WAU, Turrialba, Costa Rica.

Kutcher, G.P., 1972. Agricultural planning at the regional level: a programming model of Mexico's Pacific Northwest. Ph.D. thesis, University of Maryland.

Kutcher, G.P., 1983. A regional agricultural programming model for Mexico's Pacific Northwest. Pages 317-351 in Norton R.D. and L. Solis (Eds) The book of CHAC: programming studies for Mexican agriculture, The John Hopkins University Press, Baltimore, USA.

Kutcher, G.P. and P.L. Scandizzo, 1981. The agricultural economy of Northeast Brazil. Johns Hopkins University Press, Baltimore, USA. 271 pp.

Kuyvenhoven, A., R. Ruben and G. Kruseman, 1995. Options for sustainable agricultural systems and policy instruments to reach them. Pages 187-212 in: Bouma, J., A. Kuyvenhoven, B.A.M. Bouman, J.C. Luyten and H.G. Zandstra (Eds) Eco-regional approaches for sustainable land use and food production. Kluwer Academic Publishers, Dordrecht, The Netherlands.

La Gaceta, 1998. Decreto N°26639-Ministerio de Agricultura y Ganadería. Año CXX, número 23: 1-4.

Lambin, E.F., 1994. Modelling deforestation processes: a review. TREES Series B, research report 1. European Commission, EUR 15744 EN, Luxembourg.

Latin America Monitor, June 1998. Business Monitor International, London, UK.

Levin, S.A., 1993. Concepts of scale at the local level. Pages 7-19 in: Ehleringer J.R. and C.B. Field (Eds) Scaling physiological processes. Leaf to globe. Academic Press, San Diego, USA.

Lizano, E., 1994. The impact of policy reform on food consumption in Costa Rica. Ph.D. thesis, Tufts University, U.S.A.

Loaiza C.E. and M.M. Zuñiga, 1997. Costa Rica. Informe anual de estadísticas de exportación de banano. Corporación Bananera Nacional, S.A., Dirección de política Bananera y Estadística, San José, Costa Rica.

López, A. and P. Solís, 1992. Acumulación de cobre en suelos dedicados al cultivo de banano en el Pacífico Sur de Costa Rica. I. Estado actual de las plantaciones. CORBANA 16: 9-19.

Maddala, G.S., 1992. Introduction to econometrics. Second edition. MacMillan Publishing Company, New York, U.S.A.

MAG, 1994. Estudio de zonificación agropecuaria en la región Pacífico Central. Convenio MAG-MIDEPLAN, Realizado por el Centro Científico Tropical. 2 Volumes. Ministerio de Agricultura y Ganadería, San José, Costa Rica.

MAG-FAO, 1997. Planificación participativa para la agricultura conservacionista: opciones técnicas para cumplir con los principios de la agricultura conservacionista. Informe técnico No. 8, Ministry of Agriculture and Livestock - Food and Agriculture Organization of the United Nations, San José, Costa Rica.

Mamingi, N., 1997. The impact of prices and macroeconomic policies on agricultural supply: a synthesis of available results. Agricultural Economics 16: 17-34.

Mantel, S., 1993. Soil physical properties of a fertile, poorly drained soil and their adequacy for use in quantified land evaluation. Report No. 24, Phase 2, Atlantic Zone Program (CATIE-WAU-MAG), Turrialba, Costa Rica.

Martin, L.J., 1981. Quadratic single and multi-commodity models of spatial equilibrium: a simplified exposition. Canadian Journal of Agricultural Economics 29: 21-48.

Martin, L.J. and A.C. Zwart, 1975. A spatial and temporal model of the North American sector for the evaluation of policy alternatives. American Journal of Agricultural Economics 55: 55-67.

May, P.H. and O.S. Bonilla, 1997. The environmental effects of agricultural trade liberalization in Latin America: an interpretation. Ecological Economics 22: 5-18.

McConnell, K.E., 1983. An economic model of soil conservation. American Journal of Agricultural Economics 64: 83-89.

Melson, W.G., J. Barquero, R. Sáenz and E. Fernández, 1985. Erupciones explosivas de importancia en volcanes de Costa Rica. Boletín de Vulcanología 15: 15-20, Heredia, Costa Rica.

Mera-Orcés, V., 1999. Evaluación de la transferencia de la metodología SOLUS (de REPOSA) al Ministerio de Agricultura y Ganadería (MAG), Costa Rica. Report No. 138, REPOSA (CATIE/MAG/WAU), Turrialba, Costa Rica. 53 pp., plus appendices.

Mesalles, L., 1998. Condiciones macroeconómicas de la economía costarricense. Pages 81-106 in: Jiménez, R. (Ed.) Estabilidad y desarrollo económico en Costa Rica: las reformas pendientes. Academia de Centroamerica, San José, Costa Rica.

MIDEPLAN, 1991. Plan de desarrollo region Huetar Atlántica. MIDEPLAN, San José, Costa Rica.

Miller, C.P. and T.G.H. Stockwell, 1991. Sustaining productive pastures in the tropics 4. Augmenting native pasture with legumes. Tropical Grasslands 25: 98-103.

Ministry of Trade and Commerce, 1998. Unpublished information.

Miranowski, J.A., 1984. Impacts of productivity loss on crop production and management in a dynamic economic model. American Journal of Agricultural Economics 65: 61-71.

Moll, H.A.J. and R.A. Schipper, 1994. Introduction to regional agricultural planning: notes for the 1994 course (D100-207) on regional agricultural development: analysis and planning. Department of Development Economics, Wageningen Agricultural University, Wageningen.

Montenegro, J. and Abarca, S., 1998. La ganadería en Costa Rica: tendencias y proyecciones, 1984-2005. Ministerio de Agricultura y Ganaderia, San José, Costa Rica. 63 pp.

Mora Alfaro, J., O. Oviedo Sánchez and L.F. Fernández Alvarado. 1994. El impacto de las políticas macroeconómicas en el agro costarricense. Facultad de Ciencias Sociales, Universidad Nacional. Editorial Universidad Nacional, Heredia, Costa Rica.

Myers, R.J.K. and G.B. Robbins, 1991. Sustaining productive pastures in the tropics 5. Maintaining productive sown grass pastures. Tropical Grasslands 25: 104-110.

Nerlove, M., 1958. Distributed lags and estimation of long-run supply and demand elasticities. Journal of Farm Economics 40: 301-10.

Nieuwenhuyse, A., 1996. Landscape formation and soil genesis in volcanic parent materials in humid tropical lowlands of Costa Rica. Ph.D. thesis, Wageningen Agricultural University, The Netherlands. 131 pp.

Nieuwenhuyse, A., 1996. Soils, geology and soil-related sustainability aspects of the perhumid tropical Limon basin, Costa Rica. Serie Técnica. Informe Técnico 272. CATIE, Turrialba, Costa Rica. 85 pp.

Nieuwenhuyse, A. and S.B. Kroonenberg, 1994. Volcanic origin of Holocene beach ridges along the Caribbean coast of Costa Rica. Marine Geology 120: 13-26.

Nieuwenhuyse, A., H. Hengsdijk, B.A.M. Bouman, R.A. Schipper and H.P.G. Jansen, 1999. Can forestry be a competitive land use option? Model simulations from humid tropical Costa Rica. Forest Ecology and Management. 4993: 1-18.

Norton, R.D., 1995. Review of the agricultural models for Costa Rica being developed by Wageningen Agricultural University. Consultant's report. Norton Advisory Services, Inc., Panama.

NRC [National Research Council], 1989. Nutrient requirements of dairy cattle. Sixth revised edition update 1989. National Academy Press, Washington D.C., USA. 157 pp.

NRC [National Research Council], 1996. Nutrient requirements of beef cattle. National Academy Press, Washington D.C., USA. 242 pp.

NRC [National Research Council], 1997. Precision agriculture in the 21th century. National Academy Press. Washington, D.C., USA.

Nuhn, H., 1978. Regionalisierung und Entwicklungsplanung in Costa Rica: ein Beitrag zur angewandten Geographie und Regionalplanung unter Einsatz von EDF. Beitraege zur geographischen Regionalforschung in Lateinamerika ; Bd. 2. Universitaet Hamburg.

O'Neill, R.V., D.L. DeAngelis, J.B. Waide and T.F.H. Allen, 1986. A hierarchical concept of ecosystems. Monographs in population biology 23. Princeton University Press, Princeton, New Yersey, UK. 253 pp.

OFIPLAN (Oficina de Planificación Nacional y Políticas Económicas), 1982. Evolución socioeconómica de Costa Rica 1950-1980. San José, Costa Rica.

Opschoor, J.B. and L. Reijnders, 1991. Towards sustainable development indicators. Pages 7-27 in: Kuik, O. and H. Verbruggen (Eds) In search of indicators of sustainable development. Kluwer, Dordrecht, The Netherlands.

Opschoor, J.B., 1992. Sustainable development, the economic process and economic analysis. Pages 25-52 in: Opschoor, J.B. (Ed.) Environment, economy and sustainable development. Wolters-Noordhoff, Groningen, The Netherlands.

Pardo T. J., 1984. El cultivo del banano. Cultivos mayores No. 7. UNED, San José, Costa Rica.

Parton, W.J., D.S. Schimel, C.V. Cole and D.S. Ojima, 1987. Analysis of factors controlling soil organic matter levels in great plain grasslands. Soil Science Society of America Journal 51: 1173-1179.

Pearce, D.W. and R.K. Turner, 1990. Economics of natural resources and the environment. Harvester Wheatsheaf, New York, USA. 378 pp.

Penning de Vries, F.W.T., H. van Keulen and R. Rabbinge, 1995. Natural resources and limits of food production in 2040. Pages 65-87 in: Bouma, J., A. Kuyvenhoven, B.A.M. Bouman, J. Luyten and H. Zandstra (Eds) Eco-regional approaches for sustainable land use and food production. Kluwer Academic Publishers, Dordrecht, The Netherlands.

Phimister, E., 1993. Savings and investments in farm households: analysis using life cycle. Aveburry, Great Britain. 187 pp.

Pieri, R.G., K.D. Meilke and T.G. MacAuley, 1977. North American-Japanese pork trade: an application of quadratic programming. Canadian Journal of Agricultural Economics 25: 61-79.

Plant, R.A.J. and B.A.M. Bouman, 1999. Modeling nitrogen oxide emissions from current and alternative pastures in Costa Rica. Journal of Environmental Quality 28(3): 866-872.

Pomareda, C., 1995. Instituciones y políticas para contribuir al manejo integrado de los recursos naturales. Paper presented at the regional consultative meeting of ICRAF, Lima, Peru, 13-17 November.

Pomareda, C., 1996. Institutional aspects of sustainable development. Pages 41-54 in: Castro, E. and G. Kruseman (Eds) Policies for sustainable land use in Costa Rica. Editorial Guayacán, CINTERPEDS-UNA / WAU, Heredia, Costa Rica.

Pomerada, C., 1998. Las políticas públicas y el sector agropecuario. Ministerio de Planificación Nacional y Política Económica, San José, Costa Rica.

Portig, W.H., 1976. The climate of Central America. Pages 405-478 in: W. Schwerdtfeger (Ed.). Climates of North and South America. World survey of climatology Vol.12. Elsevier, Amsterdam, The Netherlands.

Quesada Mateo, C., 1990. Estrategía de conservación para el desarrollo sostenible de Costa Rica. ECODES/Ministry of Natural Resources, Energy and Mining, San José, Costa Rica.

Quirós, R., F. Barrantes, H. Clemens and M. Ugalde, 1997. Estudio de oferta y demanda de crédito rural en Costa Rica, a partir de estudios de casos en tres regiones del país. Mesa Nacional Campesina, Centro de Estudios para el Desarrollo Rural-Universidad Libre de Amsterdam, Instituto para el Desarrollo y la Acción Social. San José, Costa Rica. 65 pp.

Rabbinge, R., 1993. The ecological background of food production. Pages 2-29 in: proceedings of the Ciba foundation symposium 177 on crop protection and sustainable agriculture. Wiley, Chichester, UK.

Rabbinge, R., C.A. Van Diepen, J. Dijsselbloem, G.J.H. De Koning, H.C. Van Latesteijn, E. Woltjer and J. Van Zijl, 1994. Ground for choices: a scenario study on the perspectives for rural areas in the European Community. Pages 95-122 in: Fresco, L.O., L. Stroosnijder, J. Bouma and H. Van Keulen (Eds). The future of the land. Mobilising and integrating knowledge for land use options. Wiley, Chichester, UK.

Rabbinge, R. and M.K. van Ittersum, 1994. Tensions between aggregation levels. Pages 31-40 in Fresco, L.O., L. Stroosnijder, J. Bouma and H. Van Keulen (Eds). The future of the land. Mobilising and integrating knowledge for land use options. Wiley, Chichester, UK.

Rabbinge, R. and H. van Latesteijn, 1992. Long term options for land use in the European Community. Agricultural Systems 40: 195-210.

Ramírez S., A. and T. Madonado U., 1988. Desarollo socioeconómico y el ambiente natural de Costa Rica: situación actual y perspectivas. Editorial Heliconia, San Jose, Costa Rica. 160 pp.

Ramírez, A.L. and C.M. Ramírez, 1980. Esterilidad masculina causada por la exposición laboral al nematicida 1,2-dibromo-3-cloropropano. Actos Medicos Costaricense 23 (3): 219-222.

Rao, J.M., 1989. Agricultural supply response: a survey. Agricultural Economics 3: 1-22.

Reagan, M.K., 1987. Turrialba volcano, Costa Rica: magmatism at the southeast terminus of the Central American arc. Ph.D. thesis, University of California, Santa Cruz.

Reynolds, J.F., D.W. Hilbert and P.R. Kemp, 1993. Scaling ecophysiology from the plant to the ecosystem: a conceptual framework. Pages 127-140 in: Ehleringer J.R. and C.B. Field (Eds) Scaling physiological processes. Leaf to globe. Academic Press, San Diego, USA.

Robert, P.C., R.H. Rust, and W.E. Larson, 1996. Proceedings of the third international conference on precision agriculture, June 23-26, 1996, Minneapolis, Minnesota. ASA-CSSA-SSSA, Madison, WI.

Roebeling, P.C., F. Sáenz, E. Castro and G. Barrantes, 2000. Agrarian policy responsiveness of small farmers in Costa Rica. Pages 76-102 in: Pelupessy, W. and R. Ruben (Eds) Agrarian policies in Central America.

MacMillan Press Ltd, Great Britain, pp. 209.

Roebeling, P.C., R. Ruben and F. Sáenz, 1998. Políticas agrarias para la intensificación sostenible del sector ganadero: una aplicación en la Zona Atlántica de Costa Rica. Pages 156-174 in: Castro, E. and R.Ruben (Eds) Políticas agrarias para el uso sostenible de la tierra y la seguridad alimentaria en Costa Rica. UNA-CINPE / WAU-DLV, Costa Rica.

Roebeling, P.C., H.G.P. Jansen, A. van Tilburg and R.A. Schipper, 1999. Spatial equilibrium modeling for inter-regional trade flow estimation and agricultural policy analysis in Costa Rica. Report No 142, REPOSA, CATIE/MAG/WAU, Turrialba, Costa Rica. 56 pp.

Röling, N., 1994. Platforms for decision-making about ecosystems. Pages 385-393 in Fresco, L.O., L. Stroosnijder, J. Bouma and H. Van Keulen (Eds). The future of the land. Mobilising and integrating knowledge for land use options. Wiley, Chichester, UK.

Romero, C., 1993. A research on Andalusian farmers' objectives: methodological aspects and policy implications. Paper presented to the VII-EAAE congress, Stresa, Italy.

Rossing, W.A.H., J.M. Meynard and M.K. Van Ittersum, 1997. Model-based explorations to support development of sustainable farming systems: case studies from France and the Netherlands. European Journal of Agronomy 7: 271-283.

Rosswall, T., R.G. Woodmansee and P.G. Risser (Eds), 1998. Scales and global change. Spatial and temporal variability in biospheric and geospheric processes. SCOPE 35, John Wiley & Sons Ltd., Chichester, UK. 355 pp.

Ruben, R., G. Kruseman and H. Hengsdijk, 1994. Farm household modelling for estimating the effectiveness of price instruments on sustainable land use in the Atlantic Zone of Costa Rica. DLV Report No. 4. AB-DLO/WAU, Wageningen, The Netherlands. 44 pp., plus appendices.

Ruiz-Silvera, C., E. Bustamente, R. González, M. González and A. Gamboa, 1997. Sustratos o enmiendas foliares como elemento de control biológico de la *Sigatoka negra*. Actas de la III semana científica, CATIE, Turrialba, Costa Rica. pp. 143-146.

Sader, S.A. and A.T. Joyce, 1988. Deforestation rates and trends in Costa Rica, 1940 to 1983. Biotropica 20: 11-19.

Sadoulet, S., and A. De Janvry, 1995. Quantitative development policy analysis. The Johns Hopkins University Press, Baltimore, USA. 397 pp.

Sáenz, F., R. Ruben, P. Roebeling, E. Castro and G. Barrantes, 1998. Comportamiento de la inversión en pequeñas fincas ante cambios en la política agraria: un estudio de caso en la Zona Atlántica de Costa Rica. Pages 138-155 in: Castro E. and R. Ruben (Eds) Políticas agrarias para el uso sostenible de la tierra y la seguridad alimentaria en Costa Rica. UNA-CINPE/WAU-DLV. CINPE, Heredia, Costa Rica.

Sáenz, F., R.A. Schipper, H. Hengsdijk, R. Azofeifa and H.G.P. Jansen, 1999. Análisis del efecto de políticas agrícolas sobre el uso de la tierra en Costa Rica: un estudio de caso en la cuenca del río Aranjuez. Report No. 139, REPOSA (CATIE/MAG/WAU), Turrialba, Costa Rica. 27 pp., plus appendices.

Salazar, M.A.G., 1977. Efecto de la frecuencia de corte y de cuatro niveles de fertilización nitrogenado sobre el rendimiento y valor nutritivo del pasto estrella (*Cnynodon nlemfuensis*). San Pedro de Montes de Oca. Costa Rica. 66 pp.

Samuelson, P.A., 1952. Spatial price equilibrium and linear programming. American Economic Review 41: 283-300.

Schipper, R.A., 1993. Una caracterización de fincas en Neguev, Río Jiménez y Cocori, Zona Atlántica, Costa Rica: informe en base a una encuestra del marco de estudios de base. Field Report 49, REPOSA (CATIE/MAG/WAU), CATIE, Turrialba, Costa Rica. 34 pp., plus appendices.

Schipper, R.A., 1996. Farming in a fragile future. Economics of land use with applications in the Atlantic Zone of

Costa Rica. Ph.D. thesis, Wageningen Agricultural University, Wageningen, The Netherlands. 282 pp.

Schipper, R.A., H.G.P. Jansen, J.J. Stoorvogel and D.M. Jansen, 1995. Evaluating policies for sustainable land use: a sub-regional model with farm types for Costa Rica. Pages 377-395 in: Bouma, J., A. Kuyvenhoven, B.A.M. Bouman, J. Luyten and H. Zandstra (Eds) Eco-regional approaches for sustainable land use and food production. Kluwer Academic Publishers, Dordrecht, The Netherlands.

Schipper, R.A., H.G.P. Jansen, B.A.M. Bouman, H. Hengsdijk, A. Nieuwenhuyse and F. Sáenz, 1998. Evaluation of development policies using integrated bio-economic land use models: applications to Costa Rica. Field Report 136 — Program Paper 16, REPOSA (CATIE/MAG/WAU), CATIE, Turrialba, Costa Rica. 80 pp.

Schoorl, J.M., A. Veldkamp and L.O. Fresco., 1997. The conversion of land use and its effects (CLUE-CR), a regression based model applied to Costa Rica (Pascal version 1.2). Quantitative Approaches in Systems Analysis No. 8. AB-DLO/C.T. de Wit Graduate School for Production Ecology. Wageningen, The Netherlands. 51 pp., plus appendices.

Sellers, P.J., M.D. Heiser, F.G. Hall, S.B. Verma, R.L. Desjardins, P.M. Schuepp and J.I. MacPherson, 1997. The impact of using area-averaged land surface proportions — topography, vegetation condition, soil wetness — in calculations of intermediate scale (approximately 10 km^2) surface-atmosphere heat and moisture fluxes. Journal of Hydrology 190: 269-301.

SEPSA [Secretaría Ejecutiva de Planificación Sectoral Agropecuaria], 1989. Programa nacional sectoral de raíces y tuberculos. San José, Costa Rica.

SEPSA [Secretaría Ejecutiva de Planificación Sectoral Agropecuaria], 1992. Información básica del sector agropecuario 6. San José, Costa Rica. 58 pp.

SEPSA [Secretaría Ejecutiva de Planificación Sectoral Agropecuaria], 1995. Sostenibilidad, estrategía de desarrollo del sector agropecuario. Ministry of Agriculture and Livestock (MAG), San José, Costa Rica.

SEPSA [Secretaría Ejecutiva de Planificación Sectoral Agropecuaria], 1997. Políticas del sector agropecuario (revisión y ajuste). Ministry of Agriculture and Livestock (MAG), San José, Costa Rica.

Seyfried, H., A. Astorga, H. Amann, C. Calvo, W. Kolb, H. Schmidt, and J. Winsemann, 1991. Anatomy of an evolving island arc: tectonic and eustatic control in the south Central American fore-arc area. Special Publications of the International Association of Sedimentologists 12: 217-240.

Sheehy, S.J. and R.H. McAlexander, 1965. Selection of representative bench mark farms for supply estimation. Journal of Farm Economics 47: 681-695.

Simonds, N.W., 1986. A short review of farming systems research in the tropics. Experimental Agriculture 22: 1-13.

Singh, I., L. Squire and J. Strauss, 1986. Agricultural household models: extensions, applications and policy. Johns Hopkins Press for the World Bank, Baltimore, U.S.A.

Sklar, F.H. and R. Constanza, 1991. The development of dynamic spatial models for landscape ecology: a review and prognosis. Pages 239-288 in: Turner M.G. and R.H. Gardner (Eds) Quantitative methods in landscape ecology, Ecological Studies 82. Springer Verlag, Berlin, Germany.

Soil Survey Staff, 1992. Keys to soil taxonomy, 5th ed. SMSS technical monograph No. 19. Pocahontas Press, Inc. Blacksburg, Virginia.

Soto, M., 1995. Banano cultivo y comercialización. Litografía e Imprenta LIL S.A., San José, Costa Rica.

Spaans, E.J.A., G.A.M. Baltissen, J. Bouma, R. Miedema, A.L.E. Lansu, D. Schoonderbeek and W.G. Wielemaker, 1989. Changes in physical properties of young and old surface volcanic soils in Costa Rica after clearing of tropical rain forest. Hydrological Proc. 3: 383-392.

Spharim, I., R. Spharim and C.T. De Wit, 1992. Modelling agricultural development strategy. Pages 159-192 in: Alberda, Th., H. Van Keulen, N.G. Seligman and C.T. De Wit (Eds) Food from dry lands. An integrated approach to planning of agricultural development. Kluwer Academic Publishers, Dordrecht.

Sprechmann, P. (Ed.), 1984. Manual de geología de Costa Rica 1: estrategrafía. Editorial Univ. de Costa Rica, San José, Costa Rica.

Spronk, J. and F.R. Veeneklaas, 1983. A feasibility study of economic and environmental scenarios by means of interactive multiple goal programming. Regional Science and Urban Economics 13: 141-160.

Stomph, T.J., L.O. Fresco and H. van Keulen, 1994. Land use system evaluation: concepts and methodology. Agricultural Systems 44: 243-255.

Stoorvogel, J.J., 1993. Optimizing land use distribution to minimize nutrient depletion: a case study for the Atlantic Zone of Costa Rica. Geoderma 60: 277-292.

Stoorvogel, J.J. 1995. Geographical information systems as a tool to explore land characteristics and land use with reference to Costa Rica. Ph.D. thesis, Wageningen Agricultural University, Wageningen, The Netherlands. 151 pp.

Stoorvogel, J.J. and G.P. Eppink, 1995. Atlas de la Zona Atlántica Norte de Costa Rica. Programa Zona Atlántica (CATIE-UAW-MAG), Guápiles, Costa Rica. 84 pp.

Stoorvogel, J.J., R.A. Schipper and D.M. Jansen, 1995. USTED: a methodology for a quantitative analysis of land use scenarios. Netherlands Journal of Agricultural Science 43: 5-18.

Stoorvogel, J.J., J. Bouma and W. Bowen (Eds), 1998. Information technology as a tool to assess land use options in space and time. Proceedings of an international workshop Lima, September 28 - October 4, 1997. Quantitative Approaches in Systems Analysis No. 16. AB-DLO/C.T. de Wit Graduate School for Production Ecology, Wageningen; International Potato Center, Lima; International Consortium for Agricultural Systems Application.

Stoorvogel, J.J., L. Kooistra and J. Bouma, 1999. Spatial and temporal variation in nematocide leaching, management implications for a Costa Rican banana plantation. In: Corwin, D.L., T.R. Ellsworth and K. Loague (Eds) Application of GIS, remote sensing, geostatistics and solute transport modeling to the assessment of nonpoint source pollutants in the vadose zone. Joint AGU-Chapman conference and SSSA Outreach conference, Riverside California. USA. October 19-24, 1997 (in press).

Stroosnijder, L., S. Efdé, T. van Rheenen and L. Agustina, 1995. QFSA: a new method for farm level planning. Pages 299-331 in: Fresco, L.O., L. Stroosnijder, J. Bouma and H. Van Keulen (Eds) The future of the land. Mobilising and integrating knowledge for land use options. Wiley, Chichester, UK.

Strycker, J.D., 1990. Trade, exchange rate, and agricultural pricing policies in Ghana. Washington D.C.: World Bank.

Takayama, Y. and G.G. Judge, 1964. Spatial equilibrium and quadratic programming. Journal of Farm Economics 46(1): 67-93.

Takayama, Y. and G.G. Judge, 1971. Spatial and temporal price and allocation models. North Holland, Amsterdam.

Teitzel, J.K., M.A. Gilbert and R.T. Cowan, 1991. Sustaining productive pastures in the tropics 6. Nitrogen fertilized grass pastures. Tropical Grasslands 25: 111-118.

Thomas, R.J., C.E. Lascano, J.I. Sanz, M.A. Ara, J.M. Spain, R.R. Vera and M.J. Fisher, 1992. The role of pastures in production systems, Pages pp 121-144 in: pastures for the tropical lowlands, CIAT's contribution. Publication 211, Centro Internacional de Agricultura Tropical (CIAT), Cali, Colombia.

Timmer, P., 1981. Is there curvature in the Slutsky matrix? Review of Economics and Statistics 63: 395-402.

Turner II, B.L., D.L. Skole, S. Sanderson, G. Fischer, L.O. Fresco and R. Leemans, 1995. Land-use and land-cover change. Science/Research Plan. IGBP report No. 35 and HDP report No. 7, Stockholm (Norway) and Geneva (Switzerland). 132 pp.

Uhl. C., Buschbacher, R. and E.A.S. Serrão, 1988. Abandoned pastures in Eastern Amazonia. I. Patterns of plant succession. Journal of Ecology 76: 663-681.

UN [United Nations], 1997. World population prospects 1950 — 2050. The 1996 revision. United Nations Population Division, New York, USA.

Upton, M., 1989. Livestock productivity assessment and herd growth models. Agricultural Systems 29: 149-164.

Upton, M., 1993. Livestock productivity assessment and modelling. Agricultural Systems 43: 459-472.

US Census Bureau, 1998. International Data Base. http://www.census.gov/ftp/pub/ipc/www/idanew.html

USDA [United States Department of Agriculture], 1998. USDA agricultural baseline projections to 2007. USDA, World Agricultural Outlook Board, Staff report WAOB-98-1. Washington, USA. 128 pp.

Van Dam, D., E. Veldkamp and N. van Breemen, 1997. Soil organic carbon dynamics: variability with depth in forested and deforested soils under pasture in Costa Rica. Biogeochemistry 39: 343-375.

Van der Grinten, P., M.T. Baayen, L. Villalobos, R.H. Dwinger and L. 't Mannetje, 1992. Utilisation of kikuyu grass (*Pennisetum clandestinum*) pastures and dairy production in a high altitude region of Costa Rica. Tropical Grasslands 26: 255-262.

Van der Pol, F., 1993. Soil mining. An unseen contributor to farm income in Southern Mali. Amsterdam: Royal Tropical Institute Bulletin No. 325, 47 pp.

Van der Valk, A.L., 1999. Elasticities i Claro que sí !: an analysis of the domestic Costa Rican food demand. M.Sc. thesis. Department of Marketing and Marketing Research, Wageningen Agricultural University, Wageningen, The Netherlands.

Van der Ven, G.W.J., 1992. Grasmod, a grassland management model to calculate nitrogen losses from grassland. CABO-DLO report 158. CABO-DLO, Wageningen, The Netherlands. 108 pp.

Van der Werf, H.M.G., 1996. Assessing the impact of pesticides on the environment. Agriculture, Ecosystems and Environment 60: 81-96.

Van Hijfte, P.A., 1989. La ganadería de carne en el Norte de la Zona Atlántica de Costa Rica. Report No. 31, Atlantic Zone Program (CATIE/MAG/WAU), Turrialba, Costa Rica. 52 pp., plus appendices.

Van Ittersum, M.K., R. Rabbinge and H.C. van Latesteijn, 1998. Exploratory land use studies and their role in strategic policy making. Agricultural Systems 58: 309-330.

Van Ittersum, M.K. and R. Rabbinge, 1997. Concepts in production ecology for analysis and quantification of agricultural input-output combinations. Field Crops Research 52: 197-208.

Van Keulen, H. and J. de Wolf, 1986. Modelling of agricultural production: weather, soils and crops. Pudoc, Wageningen, The Netherlands. 464 pp.

Van Leeuwen, A.C.J. and A.M. Hofstede, 1995. Forests, trees and farming in the Atlantic Zone of Costa Rica: an evaluation of the current and future integration of trees and forests in farming systems in the Atlantic Zone of Costa Rica. Serie Técnica. Informe Técnico no. 257. Centro Agronómico Tropical de Investigación y Enseñanza, Turrialba, Costa Rica. 55 pp., plus appendices.

Van Loon, M., 1997. Characterization of beef cattle farming in the Atlantic Zone of Costa Rica. Field report 127, REPOSA (CATIE/MAG/WAU), CATIE, Turrialba, Costa Rica. 46 pp.

Van Seeters, R.J.M., 1993. A study on geomorphology, mineralogy and geochemistry of the Toro Amarillo-Tortuguero and the Chirripó-Matina river systems in the Atlantic Zone of Costa Rica. Report No. 47, REPOSA (CATIE/MAG/WAU), CATIE, Turrialba, Costa Rica.

Vargas, G.E., 1984. Tabla de composición de alimentos para animales de Costa Rica. Universidad de Costa Rica, San José, Costa Rica. 111 pp.

Vargas, T., 1998. Reforma fiscal: políticas y instituciones. Pages 127-164 in: Jiménez, R. (Ed.) Estabilidad y desarrollo económico en Costa Rica: las reformas pendientes. Academia de Centroamerica, San José, Costa Rica.

Veeneklaas, F.R., H. van Keulen, S. Cisse, P. Gosseye and N. van Duivenboode, 1994. Competing for limited resources: options for land use in the Fifth Region of Mali. Pages 227-247 in: Fresco, L.O., L. Stroosnijder, J. Bouma and H. Van Keulen (Eds) The future of the land. Mobilising and integrating knowledge for land use options. Wiley, Chichester, UK.

Veldkamp, A. and L.O. Fresco, 1996a. CLUE: a conceptual model to study the conversion of land use and its effects. Ecological Modelling 85: 253-270.

Veldkamp, A. and L.O. Fresco, 1996b. CLUE-CR: an integrated multi-scale model to simulate land use change scenarios in Costa Rica. Ecological Modelling 91: 231-248.

Veldkamp, A. and L.O. Fresco, 1997. Exploring land use scenarios, an alternative approach based on actual land use. Agricultural Systems 55: 1-17.

Veldkamp, E., 1993. Soil organic carbon dynamics in pastures established after deforestation in the humid tropics of Costa Rica. Ph.D. thesis, Wageningen Agricultural University, Wageningen, The Netherlands. 117 pp.

Veldkamp, E., M. Keller and M. Nuñez, 1998. Effects of pasture management on N_2O and NO emissions from soils in the humid tropics of Costa Rica. Global Biogeochemistry Cycles, 12: 71-79.

Veldkamp, E., A.M. Weitz, I.G. Staritsky and E.J. Huising, 1992. Deforestation trends in the Atlantic Zone of Costa Rica: a case study. Land Degradation and Rehabilitation 3: 71-84.

Verberne, E.L.J., J. Hassink, P. de Willigen, J.J.R. Groot and J.A. Veen, 1990. Modeling organic matter dynamics in different soils. Netherlands Journal of Agricultural Science 38: 221-238.

Verburg, P.H. and A. Veldkamp, 1997. Modelling the spatial pattern of land use change in China. Pages 69-83 in: proceedings of a workshop Wageningen-China May 1997. Report 84 AB-DLO, Wageningen, The Netherlands.

Verburg, P.H., A. Veldkamp and J. Bouma, 1999a. Land use change under conditions of high population pressure: the case of Java. Global Environmental change 9(4): 303-312.

Verburg, P.H., A. Veldkamp and L.O. Fresco, 1999b. Simulation of changes in the spatial pattern of land use in China. Applied Geography 19: 211-233.

Verburg, P.H., G.H.J. de Koning, K. Kok, A. Veldkamp and J. Bouma, 1999c. A spatially explicit allocation procedure for modelling the pattern of land use change based upon actual land use. Ecological Modelling 116: 45-61.

Vereijken, P., 1997. A methodological way of prototyping integrated and ecological arable farming systems (I/EAFS) in interaction with pilot farms. European Journal of Agronomy 7: 235-250.

Verhagen, A., H.W.G. Booltink and J. Bouma, 1995. Site-specific management: balancing production and environmental requirements at farm level. Agricultural Systems 49: 369-384.

Vicente-Chandler, J., F. Abruna, R. Caro-Costas, J. Figarella, S. Silva and R.W. Pearson, 1974. Intensive grassland management in the humid tropics of Puerto Rico. Bulletin 233. Agricultural Experiment Station Rio Piedras, Puerto Rico.

Villareal, M., 1992. Experiencias con el pasto Ratana (*Ischaemum indicum*): producción de foraje, valor nutritivo y alternativas de manejo. Seminar: el pasto Ratana en Costa Rica, alternativa o problemática en nuestra ganadería? Asistencia Técnica Dos Pinos, San Carlos, Costa Rica.

Von Düszeln, J., 1990. Pesticide quality control in Costa Rica, Central America. Food Laboratory News 21: 26-29.

Wesseling, C., 1997. Health effects from pesticide use in Costa Rica: an epidemiologic approach. Ph.D. thesis, Karolinska Institute, Stockholm, Sweden.

Wesseling, C., L. Castillo and C.G. Elinder, 1993. Pesticide poisonings in Costa Rica. Scandinavian Journal for Work and Environmental Health 19: 227-235.

Wetering, R. and J.B. Opschoor, 1994. Towards environmental performance indicators based on the notion of environmental space. Publication RMNO 96, Advisory Council for Research on Nature and Environment (RMNO), Rijswijk, The Netherlands.

Weyl, R., 1980. Geology of Central America. 2nd, revised ed. Beitrage zur regionalen Geologie der Erde 15. Borntraeger, Berlin and Stuttgart.

Wielemaker, W.G. and A.W. Vogel, 1993. Un sistema de información de suelos y tierras para la Zona Atlántica de Costa Rica. Report no. 22, REPOSA (CATIE/MAG/WAU), Guápiles, Costa Rica. 82 pp.

Wilhelmus, C.M., 1998. Supporting consensus on land use policy: the merits and limitations of computer modelling. A case study in Costa Rica. Report 131, REPOSA (CATIE/MAG/WAU), CATIE, Turrialba, Costa Rica. 45 pp.

Willey, R., 1971. An introduction to American archaeology, vol.2: South America, pp. 338-343. Prentice-Hall, London.

Williams, J. and C.J. Chartres, 1991. Sustaining productive pastures in the tropics 1. Managing the soil resource. Tropical Grasslands 25: 73-84.

World Bank, 1997. World development report 1997: the state in a changing world. Oxford University Press, Oxford, UK.

WRR [Netherlands Scientific Council for Government Policy], 1992. Ground for choices, four perspectives for the rural areas in the European Community. Reports to the Government No. 42. Sdu Uitgeverij, Den Haag, The Netherlands. 144 pp.

WRR [Netherlands Scientific Council for Government Policy], 1995. Sustained risks: a lasting phenomenon. Reports to the Government No. 44. Sdu Uitgeverij, Den Haag, The Netherlands. 205 pp.

Zarate, C., 1978. Comportamiento del viento en Costa Rica. Instituto Meteorológico Nacional, San José, Costa Rica.

Zuñiga, N., 1996. Estimación de la tasa de interés promedio. El Financiero, 22-28 January 1996.

Abbreviations

Abbreviation	Explanation
ACP	African, Caribbean and Pacific countries of the Lomé convention
AU	Animal Units; 1 AU = 400 kg live weight
AZ	Atlantic Zone
BanMan	Banana Management decision support system
BIOA	Total amount of active ingredients in biocides
BIOI	Biocide Index
¢	*Colón*; Currency of Costa Rica
CAT	Export tax credit certificate (*Certificado de Abono Tributario*)
CATIE	Tropical Agricultural Research and Higher Education Center (*Centro Agronómico Tropical de Investigación y Enseñanza*)
c.i.f.	cost-insurance-freight
CENADA	National Foodstuffs Distribution Center (*Centro Nacional de Distribución de Alimentos*)
CINPE	International Centre on Economic Policy for Sustainable Development (*Centro Internacional de Política Económica*)
CLUE	Conversion of Land Use and its Effects
CNP	National Production Council (*Consejo Nacional de Producción*)
CORBANA	National Banana Corporation (*Corporación Bananera Nacional*)
CP	Crude Protein
CPI	Consumer Price Index
DAI	Import tax (*Derecho Arancelario de Importaciones*)
$	US $; US Dollar
DLV	Sustainable Land Use and Food Security (in Dutch: *Duurzaam Landgebruik en Voedselzekerheid*)
EARTH	School of Agriculture in the Humid Tropical Region (*Escuela Agrícola de la Región del Trópico Húmedo*)
EU	European Union
FAO	Food and Agriculture Organization of the United Nations
f.o.b.	free-on-board
GDP	Gross Domestic Product
GIS	Geographic Information System
GNP	Gross National Product
GOAL-AZ	General Optimal Allocation of Land use for the northern Atlantic Zone
ICAFE	Costa Rican Coffee Institute (*Instituto Costarricense del Café*)
IDA	Institute for Agricultural Development (*Instituto de Desarrollo Agropecuario*)
IICA	Inter-American Institute for Collaboration in Agriculture (Instituto Interamericano de Cooperación para la Agricultura)
K	Potassium

LEFSA	Land Evaluation and Farming Systems Analysis
LAICA	Consortium of Sugar Cane Processors (*Liga Agrícola Industrial de la Caña de Azúcar*)
LUCTOR	Land Use Crop Technical coefficient generatOR
MAG	Ministry of Agriculture and Livestock of Costa Rica (*Ministerio de Agricultura y Ganadería*)
ME	Metabolizable Energy
MIDEPLAN	Ministry of Planning (*Ministerio DE PLANificación Nacional y Política Económica*)
MGLP	Multiple Goal Linear Programming
MUTCA	Land Use Model for the Aranjuez River Watershed (*Modelo de Uso de la Tierra en la Cuenca del Río Aranjuez*)
N	Nitrogen
P	Phosphorus
PA	Precision Agriculture
PASTOR	Pasture and Animal System Technical coefficient generatOR
PRIAG	Regional Program for Strengthening Agronomic Research on Grains in Central America (*Programa Regional de reforzamiento a la Investigación Agronómica sobre los Granos en Centroamérica*)
REALM	Regional Economic and Agricultural Land use Model
REPOSA	Research Program on Sustainability in Agriculture
SEM	Spatial Equilibrium Model
SFP	Soil Fertile Poorly drained
SFW	Soil Fertile Well drained
SIW	Soil Infertile Well drained
SOLUS	Sustainable Options for Land USe
UCR	University of Costa Rica (*Universidad de Costa Rica*)
UNA	National University (*Universidad Nacional Autónoma*)
WAU	Wageningen Agricultural University
WHO	World Health Organization of the United Nations
YRU	Yield Registration Unit

Concepts and definitions employed in land use analysis

The terminology related to land use analysis employed in this book conforms to the definitions and concepts proposed by the FAO (1976), and subsequently elaborated by Driessen and Konijn (1992), FAO (1983, 1993) and Fresco *et al.* (1992). The FAO terminology has been extended by definitions used in most of the literature dealing with the application of linear programming techniques in land use analysis and production ecology (*e.g.*, Van Ittersum and Rabbinge, 1997). Finally, some new definitions and concepts have been introduced to clarify the methodologies of land use analysis presented in this book (*e.g.*, the product definition). Below, concepts and definitions are given that are relevant to the main body of this book; specific concepts used in individual chapters are explained in those chapters.

Biocides	All agrochemicals used to protect crops against pests and diseases (together called pesticides) and to combat weeds (called herbicides).
Constraint (in linear programming)	Restriction under which the objective function of an optimization problem is maximized (or minimized). In optimization problems in land use analysis, three types of constraints are distinguished: *resource* constraints (*e.g.*, available land and labor), *normative* constraints (*e.g.*, upper limits on sustainability indicators or lower limits on production levels), and *balance* constraints (*e.g.*, the number of animals purchased by cattle fattening systems should be equal to the number of animals produced by cattle breeding systems).
GIS	Geographic Information System. A computer system for storage, analysis and retrieval of information, in which all data are spatially referenced by their geographic coordinates (FAO, 1993).
Herbicides	Agrochemicals to combat weeds.
Inputs	Material inputs (*e.g.*, seed, fertilizer, biocides, fuel) and other inputs (*e.g.*, labor, services) applied to the use of land (after FAO, 1983).
Land	An area of the earth's surface, the characteristics of which embrace all reasonably stable, or predictably cyclic, attributes

	of the biosphere above and below this area including those of the atmosphere, the soil and underlying geology, the hydrology, the plant and animal populations, and the results of past and present human activity, to the extent that these attributes exert a significant influence on present and future uses of land by man (FAO, 1983; Fresco *et al.*, 1992).
Land characteristic	Measurable property of land, used to distinguish land units, *e.g.*, slope, soil depth, mean annual rainfall (after Fresco *et al.*, 1992).
Land cover	Synonymous with "major kind of land use".
Land quality	Complex cluster of land characteristics that covers a basic requirement of land use, and influences land suitability more or less independently of other land characteristics or aggregations of land characteristics (Driessen and Konijn, 1992).
Land unit	Physical area of land that is uniform (in its characteristics and qualities) (after FAO, 1983).
Land use system	A combination of one land unit with one land utilization type, uniquely characterized by its inputs and outputs, *i.e.* its technical coefficients, and possibly land improvement systems such as irrigation or drainage (after Driessen and Konijn, 1992, and FAO, 1993). In this definition, the "plant and animal population" are implicitly no longer properties of land (units) as they are in its original definition (FAO, 1983; Fresco *et al.*, 1992); instead, plants and animals are included in "land utilization type." Pasture land use systems are a sub-system of livestock systems, others being herds and possible feed supplements (Bouman *et al.*, 1998a; Fresco *et al.*, 1992). The term land use system is synonymous with the term "production activity" as defined in production ecology (see Production activity).
Land utilization type	A specific kind of rural land use under well described biological (*e.g.*, crop, variety, product), socio-economic (*e.g.*, labor use) and technological (*e.g.*, management, sequence operations, use of inputs) conditions (after Fresco *et al.*, 1992, and Jansen and Schipper, 1995). See also the "product" definition below. Here, land utilization type is synonymous with land use type.
Linear programming	Mathematical technique to solve optimization problems using linear objective functions and constraints.

Livestock system — A system comprising herds, possibly stocked on pasture with or without feed supplements, transforming plant biomass into animal products.

Major kind of land use — A major subdivision of rural land use, *e.g.*, annuals, perennials, forest and pasture (after FAO, 1983, and Driessen and Konijn, 1992).

Objective function — The function $Z = f(X)$ in the optimization problem "maximize (or minimize) Z", subject to constraints on X. In applications in land use analysis, X represents a vector of land use systems with corresponding technical coefficients, and Z represents the variable to be optimized, *e.g.*, economic surplus or use of biocides.

Outputs — Products and other benefits (either positive, *e.g.*, biodiversity and carbon sequestration, or negative, *e.g.*, waste losses of biocides and nutrients) of land use systems. Other benefits may be internal to the land use system, such as soil nutrient depletion, or external, such as emission of greenhouse gases.

Pesticides — All agrochemicals protecting crops against pests and diseases, such as insecticides, nematicides and fungicides.

Pixel — Spatial grid cell, as used in Geographic Information Systems (GIS).

Precision agriculture — A management strategy that uses information technology to bring data from multiple sources to bear on decisions associated with crop production (NRC, 1997).

Product — A material output of a land utilization type for a specific market and with specific price formation, as determined by its type and quality. One land utilization type may yield various products. For instance, a certain pineapple land utilization type may yield the product "first class pineapples" for export and the product "second class pineapples" for the local market. However, when a land utilization type is geared towards the production of a product of a specific type or quality, *e.g.*, by the use of a specific variety and specific management technology, different land utilization types may be distinguished. For instance, a pineapple land utilization type producing "first class pineapples" is distinct from a pineapple land

	utilization type producing "second class pineapples" when different technologies, inputs or varieties are used (thus resulting in different technical coefficients of the land use system)
Production activity	The cultivation of a crop or crop rotation in a particular physical environment, completely specified by its inputs and outputs (Van Ittersum and Rabbinge, 1997).
Production technique	A well described sequence of operations and corresponding inputs used in a land utilization system. Synonymous with production technology.
Production technology	A well described sequence of operations and corresponding inputs used in a land utilization system. Synonymous with production technique.
Stakeholder (in land use)	Individuals, communities or governments that have a traditional, current or future right to co-decide about the use of land (FAO, 1993).
Technical coefficients	Inputs and outputs of land use systems. In linear programming models, technical coefficients are the elements of the matrix (tableau).

Introduction to the CDROM

The CDROM accompanying this book contains many of the tools, models and data presented in the various chapters. Some tools are completely included, with manuals and introductory exercises, allowing users to work interactively with them, either using the data provided or trying to use data of their own. Other tools are included as demonstration versions only, mainly because of copyright or other legal restrictions. The list below gives an idea of the contents of the CDROM, while a more complete introduction is given on the CDROM itself. Even though much care has been taken by all authors to make the software easy to handle and free of errors, it should be remembered that the software originated in an academic environment and was not designed for commercial purposes. All software runs on IBM compatible PCs, either under DOS (or the Windows DOS box) or under Windows95. Minimum system requirements are software dependent. Some models require commercial software (*e.g.*, Access, Excel, GAMS or GIS-software), whereas others can be run independently.

Directory	*Content*	*Software required*	*Chapter reference*
NAZGIS	GIS data maps of the northern Atlantic Zone[1]	GIS-software/None	2
CLUE	Demonstration version of CLUE	None	3
SEMCORI	SEM optimization model	GAMS	4
	Complete set of data files	Excel/ Lotus123	
PASTOR	PASTOR Technical Coefficient Generator	None	5
	PASTOR user manual	Word	
	PASTOR introductory exercises	None	
	PASTOR linear programming exercises	GAMS	
LUCTOR	LUCTOR Technical Coefficient Generator	Excel	5
	LUCTOR user manual	Word	
	LUCTOR introductory exercises	Excel	
	LUCTOR linear programming exercises	Excel	
REALM	REALM linear programming model	GAMS	6
	REALM model description	Word	
	Complete set of ASCII data files	-	
GOALAZ	GOAL-AZ linear programming model	GAMS	7
	Complete set of ASCII data files	-	
FHHM	Farm linear programming model	GAMS	8
	Complete set of data files	Excel/ Lotus123	
BANMAN	Demonstration of BanMan for precision agriculture in banana plantations	None	9
PUBLICATION	Complete list of publications of REPOSA	Access	-

[1] Courtesy J. Stoorvogel

Authors' affiliations

Name	Affiliation
RODRIGO ALFARO	Escuela de Ciencias Agrarias, Universidad Nacional Autónoma (UNA), Heredia, Costa Rica.
JANETTE BESSEMBINDER	Laboratory of Theoretical Production Ecology, Department of Crop Science, Wageningen Agricultural University, P.O. Box 430, 6700 AK Wageningen, The Netherlands; and Agronomy Group, Department of Crop Science, Wageningen Agricultural University, P.O. Box 37, 6700 AA Wageningen, The Netherlands. Current address: Casilla 443, Potosí, Bolivia
JOHAN BOUMA (chairman of the Costa Rica Steering Committee)	Laboratory of Soil Science and Geology, Department of Environmental Sciences, Wageningen Agricultural University, P.O. Box 37, 6700 AA Wageningen, The Netherlands.
BAS A.M. BOUMAN	Research Program on Sustainability in Agriculture (REPOSA - CATIE/MAG/WAU), Costa Rica; and DLO-Research Centre for Agrobiology and Soil Fertility, P.O. Box 14, 6700 AA Wageningen, The Netherlands. Current addres: International Rice Research Institute (IRRI), P.O. Box 3127, Makati Central Post Office (MCPO), 1271 Makati City, Philippines.
EDMUNDO CASTRO	Centro Internacional de Política Económica (CINPE), Universidad Nacional Autónoma (UNA), Apartado 555-3000, Heredia, Costa Rica.
HUIB HENGSDIJK	Research Program on Sustainability in Agriculture (REPOSA - CATIE/MAG/WAU), Costa Rica.

Current address: Laboratory of Theoretical Production Ecology, Department of Crop Science, Wageningen Agricultural University, P.O. Box 430, 6700 AK Wageningen, The Netherlands.

HANS G.P. JANSEN (REPOSA Program Coordinator)

Research Program on Sustainability in Agriculture (REPOSA - CATIE/MAG/WAU), Costa Rica; and Development Economics Group, Department of Economics and Management, Wageningen Agricultural University, P.O. Box 8130, 6700 EW Wageningen, The Netherlands.

Current address: Agricultural Economics Research Institute (LEI-Wageningen UR), P.O. Box 29703, 2502 LS The Hague, The Netherlands

KASPER KOK

Agronomy Group, Department of Crop Science, Wageningen Agricultural University, P.O. Box 37, 6700 AA Wageningen, The Netherlands.

ARIE KUYVENHOVEN

Development Economics Group, Department of Economics and Management, Wageningen Agricultural University, P.O. Box 8130, 6700 EW Wageningen, The Netherlands.

ANDRÉ NIEUWENHUYSE

Research Program on Sustainability in Agriculture (REPOSA - CATIE/MAG/WAU), Costa Rica.

Current address: proyecto ZONISIG, Casilla 14533, La Paz, Bolivia

ROMANO A. ORLICH

Finca La Rebusca, Inversiones Orlich, Apartado 4921, 1000 San José, Costa Rica

PETER C. ROEBELING

Centro Internacional de Política Económica (CINPE), Universidad Nacional Autónoma (UNA), Apartado 555-3000, Heredia, Costa Rica; and Research Program on Sustainability in Agriculture (REPOSA - CATIE/MAG/WAU), Costa Rica.

Current address: Development Economics Group,
Department of Economics and Management,
Wageningen Agricultural University, P.O. Box 8130,
6700 EW Wageningen, The Netherlands.

RUERD RUBEN

Development Economics Group, Department of
Economics and Management, Wageningen Agricultural
University, P.O. Box 8130, 6700 EW Wageningen,
The Netherlands.

FERNANDO SÁENZ

Centro Internacional de Política Económica (CINPE),
Universidad Nacional Autónoma (UNA), Apartado
555-3000, Heredia, Costa Rica.

ROBERT A.SCHIPPER

Development Economics Group, Department of
Economics and Management, Wageningen Agricultural
University, P.O. Box 8130, 6700 EW Wageningen,
The Netherlands.

JETSE J. STOORVOGEL

Laboratory of Soil Science and Geology,
Department of Environmental Sciences,
Wageningen Agricultural University, P.O. Box 37,
6700 AA Wageningen, The Netherlands.

MARTIN K. VAN ITTERSUM

Laboratory of Theoretical Production Ecology,
Department of Crop Science,Wageningen Agricultural
University, P.O. Box 430, 6700 AK Wageningen,
The Netherlands.

RONALD VARGAS

Corporación Bananera Nacional (CORBANA),
Dirección de Investigaciones, Apartado 390-7210,
Guápiles, Costa Rica

TOM VELDKAMP

Laboratory of Soil Science and Geology,
Department of Environmental Sciences,
Wageningen Agricultural University, P.O. Box 37,
6700 AA Wageningen, The Netherlands.

Index

Climate 9, 13, 21, 29, 33, 97, 148, 202, 214, 224, 226

Systems Approaches for Sustainable Agricultural Development

1. Th. Alberda, H. van Keulen, N.G. Seligman and C.T. de Wit (eds.): *Food from Dry Lands.* An Integrated Approach to Planning of Agricultural Development. 1992
ISBN 0-7923-1877-3
2. F.W.T. Penning de Vries, P.S. Teng and K. Metselaar (eds.): *Systems Approaches for Agricultural Development.* Proceedings of the International Symposium (Bangkok, Thailand, December 1991). 1993 ISBN 0-7923-1880-3; Pb 0-7923-1881-1
3. P. Goldsworthy and F.W.T. Penning de Vries (eds.): *Opportunities, Use, and Transfer of Systems Research Methods in Agriculture to Developing Countries.* Proceedings of a International Workshop (The Hague, November 1993). 1994
ISBN 0-7923-3205-9
4. J. Bouma, A. Kuyvenhoven, B.A.M. Bouman, J.C. Luyten and H.G. Zandstra (eds.): *Eco-regional Approaches for Sustainable Land Use and Food Production.* Proceedings of a Symposium (The Hague, December 1994). 1995 ISBN 0-7923-3608-9
5. P.S. Teng, M.J. Kroppf, H.F.M. ten Berge, J.B. Dent, F.P. Lansigan and H.H. van Laar (eds.): *Applications of Systems Approaches at the Farm and Regional Levels.* 1996
ISBN 0-7923-4285-2
6. M.J. Kroppf, P.S. Teng, P.K. Aggarwal, J. Bouma, B.A.M. Bouman, J.W. Jones and H.H. van Laar (eds.): *Applications of Systems Approaches at the Field Level.* 1996
ISBN set volume 5 & 6: 0-7923-4287-9; ISBN 0-7923-4286-0
7. G.Y. Tsuji, G. Hoogenboom and P.K. Thornton (eds.): *Understanding Options for Agricultural Production.* 1998 ISBN 0-7923-4833-8
8. B.A.M. Bouman, H.G.P. Jansen, R.A. Schipper, H. Hengsdijk and A. Nieuwenhuyse (eds.): *Tools for Land Use Analysis on Different Scales.* With Case Studies for Costa Rica. 2000 ISBN 0-7923-6479-1; Pb ISBN 0-7923-6480-5

KLUWER ACADEMIC PUBLISHERS – DORDRECHT / BOSTON / LONDON

GPSR Compliance

The European Union's (EU) General Product Safety Regulation (GPSR) is a set of rules that requires consumer products to be safe and our obligations to ensure this.

If you have any concerns about our products, you can contact us on ProductSafety@springernature.com

In case Publisher is established outside the EU, the EU authorized representative is:

Springer Nature Customer Service Center GmbH
Europaplatz 3
69115 Heidelberg, Germany

Batch number: 10755332

Printed by Printforce, the Netherlands